THE SCIENTIFIC METHOD

THE SCIENTIFIC METHOD

An Evolution of Thinking from Darwin to Dewey

HENRY M. COWLES

Harvard University Press
Cambridge, Massachusetts London, England 2020

Printed in the United States of America

Second printing

Library of Congress Cataloging-in-Publication Data

Names: Cowles, Henry M., 1985– author.
Title: The scientific method : an evolution of thinking from Darwin to Dewey / Henry M. Cowles.
Description: Cambridge, Massachusetts : Harvard University Press, 2020. | Includes index.
Identifiers: LCCN 2019041820 | ISBN 9780674976191 (cloth)
Subjects: LCSH: Science—Methodology—History. | Science—Philosophy—History. | Evolution.
Classification: LCC Q174.8 .C69 2020 | DDC 507.2/1—dc23
LC record available at https://lccn.loc.gov/2019041820

For Marge, who always asked

You must bring out of each word its practical cash-value, set it at work within the stream of your experience. It appears less as a solution, then, than as a program for more work, and more particularly as an indication of the ways in which existing realities may be changed. *Theories thus become instruments, not answers to enigmas, in which we can rest.* We don't lie back upon them, we move forward, and, on occasion, make nature over again by their aid.

—William James

CONTENTS

THE SCIENTIFIC METHOD

1

AGE OF METHODS

The scientific method does not exist. But "the scientific method" does. This is a distinction with a difference. Scientists will tell you that there is no single method that characterizes all that they do, much less a simple set of steps that binds everything called "science" together. Scientific labor is complex and diverse, brutally difficult and impossible to encapsulate. If you think you have found a unifying principle, no doubt it leaves out some important aspect of scientific thinking or excludes a branch of what we now call the sciences. In the unlikely event that it does not, then the principle is probably *overly* inclusive, capturing too many practices to mean much at all. And it is not just scientists who doubt whether such a method exists. Historians are skeptical of it as well—for good reason. One glance back at the history of science reveals even more diversity than exists today, making a single set of steps uniting all the sciences that much harder to imagine. Scientists and historians do not always agree, but they do on this: there is no such thing as the scientific method, and there never was.[1]

And yet, "the scientific method" is alive and well. The idea of a set of steps that justifies science's authority has persisted in the face of constant denials of its existence. Why? Because "the scientific method" is a myth—and myths are powerful things. How we talk about science, how we account for its origins and argue for its results, instills mythical authority in some claims and invalidates others. The myth of "the scientific method" matters, even if (or perhaps, *because*) the reality it attests is ambiguous at best. Between the doubtful existence of the scientific method and the unquestionable power of "the scientific method," a history remains to be told. Doing so means exploring how these two phenomena interact, how the way we *talk* about thinking has shaped the quiet, even tacit process of thinking itself. As the historian of science Steven Shapin has argued: "A practice without an attendant myth is likely to be weak, hard to justify, hard even to make visible as a distinct kind of activity." If Shapin is right that we are now "dubious of claims that there is anything like 'a scientific method'—a coherent, universal, and efficacious set of procedures for making scientific knowledge," we must recognize the power that inheres in the *myth* of such a method and its complex relationship to how science is actually done.[2]

In the United States, where many schoolchildren are taught that "the scientific method" is the source of science's unique power, this myth can be traced back to 1910. That year, in a little book called *How We Think*, the philosopher and psychologist John Dewey tried to sum up two decades of research on how children learn in the classroom. Halfway through the book, Dewey offered a brief list of the steps that made up what he called "a complete act of thought":

> Upon examination, each instance reveals, more or less clearly, five logically distinct steps: *(i)* a felt difficulty; *(ii)* its location and definition; *(iii)* suggestion of possible solution; *(iv)* development by reasoning of the bearings of the suggestion; *(v)* further observation and experiment leading to its acceptance or rejection; that is, the conclusion of belief or disbelief.

If you learned "the scientific method" as a set of steps in a textbook or on a poster, you are familiar with the basic elements of Dewey's list. And that is no accident. As the historian of education John Rudolph has shown, Dewey's five steps, which were supposed to represent so much more than science, were quickly narrowed into something else: "the scientific method," a list of rules that set science apart from ordinary, everyday thinking.[3]

To understand how that happened, how Dewey's expansive account of "how we think" was made into the narrower myth of "the scientific method," we must treat his publication not as the beginning of the story, but as its end. Rather than follow his steps forward, into the textbooks and classrooms of the twentieth century, we must track them in the other direction, into the nineteenth-century debates out of which they emerged. When we do, what we find is not an unbroken genealogy of yearning for the one true method, but rather a series of methodological moments, each populated by its own protagonists and animated by new, often diverging concerns. Peeling back the layers of this story means uncovering a natural history of mind and method, one that takes us from Dewey's work in the classroom backward to field observations in anthropology, from laboratory experiments on animals to theories of evolution at a grand scale. In the century leading up to *How We Think,* science came to seem like something natural and adaptive, an organic capacity of evolving minds. Making sense of "the scientific method" means coming to grips not only with what one historian has called "methods discourse," but also with how methodological ideas operated when methods were not the explicit topic of discussion. Methodology could be both obvious and hidden, flitting between public claims about the natural world and private self-reflection on the thinking process.[4]

The reduction of science to method, and its transformation from natural to artificial, was not simply a tragic misunderstanding. Rather, the seeds for method's denaturalization, for the calcification of "the scientific method" into a manual for technical practice, were there almost from the start. At its root, this is an anxious story. Concerns about science's authority lay behind everything from the flowering of methodological discourse in the 1830s to the solidity bestowed upon Dewey's steps after 1910. Charles Darwin's search for a method to organize his specimens, his followers' use of that theory as an analogy for mental life, and Dewey's translation of their evolutionary psychology into a program for experimental pedagogy were riddled with anxieties about the limits of human cognition. These worries, in turn, helped blur the boundaries between the objects of debate and the means with which they were debated. Evolution and experiment, in these shifting contexts, came to seem like versions of the same fundamental process. From Darwin to Dewey, uncertainty about the status of science led practitioners to blend method and theory together, seeing their own minds in the world they were studying—and vice versa. Over a generation or two, nature was reimagined as a grand experiment conducted with the tool of evolution, while scientific inquiry came to seem like nothing more than an extension of nature's method.[5]

This image of science as a natural, organic process can seem a long way off from how we talk about "the scientific method" today. A set of steps rooted in asking questions and proposing answers, Dewey's list has been recast less as a natural aptitude and more as a means of mastering nature. Hypothesis and test, experiment and solution—the language of science today is often one of stability and control. The uncertain, unstable terrain of Darwin's day somehow became, over the century that followed, something solid, if not stultified. By the turn of the twentieth century, what had begun as an anxious attempt to justify the use of bold hypotheses became a default account of science's obvious authority. More than a myth about how science works, "the scientific method" became an explanation of what makes it special. Whether or not we believe in such a method as an academic matter today, it is still taught as a handy approximation of what sets science apart, a uniquely human tool for unlocking nature's secrets. The history of this myth and its power, of "the scientific method," is at bottom a history of unintended consequences, of a naturalistic vision enabling the denaturalization of its ultimate object.

What made that history possible is the subject of this book. The story will take us from Darwin's methodological morass in the 1830s to Dewey's accidental inauguration of "the scientific method" in the early twentieth century. At its heart, this is a story about how evolution and experiment were woven together so tightly that they came to define one another. During these decades, experiment—and by extension, science—came to seem *natural.* Today, when learning to do science is sometimes framed as unlearning what comes naturally, the idea of "the scientific method" as *how we think* is surprising. But a century ago it was obvious. Evolution and experiment were so intertwined that, when Dewey tried to explain their relationship, he faltered. "One must add the rashness of the prophet to the stubbornness of the partisan," he wrote, "to venture a systematic exposition of the influence upon philosophy of the Darwinian method." Dewey's ensuing effort to explain experiment in evolutionary terms capped off a century of such attempts, each as rash and stubborn as his own.[6]

SCIENCE EVOLVES

Like many books, this one began at the end. The history of Dewey's steps quickly became one among many paths through the methodological moments of the nineteenth century. Dewey's *How We Think* was the last in this series of moments, distilling an experimental essence out of evolutionary studies of learning that had themselves translated experimental protocol into the terms of natural selection. Each moment indexed a broader pattern, one in which ideas from earlier periods were remade in the service of new aims, turning hard-won results into hypotheses for testing in an ever-widening array of disciplines and debates. This give-and-take, a genealogy of ways to adapt ideas to new contexts, was itself the sort of pattern observed and theorized by the figures in question. Those involved did not present themselves in a unified way, of course. Some laid claim to tradition while others tried to shed the legacies of their ideas; at times, both modes of self-presentation were pursued at once. The psychologist William James, a character in the history that follows, was a case in point. Constantly expanding his network in pursuit of ideas and inspiration, he also insisted that his own work was merely descended from those who came before him. His philosophy of pragmatism was, in his own words, "a new name for some old ways of thinking." Neither humility nor bravado can be taken at face value, of course, but the attention James and others gave to matters of inheritance helps to outline the ragged edges of the history of mind and method that follows.[7]

Working backward, from Dewey's steps to Darwin's search, a strange sort of teleology sets in. It is not the familiar teleology, in which progressive assumptions about the past shape the narrative toward a predetermined endpoint. A teleological history in that sense would argue that each methodological moment had the same goal, that each led inexorably to what followed. Yet something like the opposite is true. In this history, episodes point in different directions. If you rewind from Dewey to Darwin and run the tape forward, you see all the ways this history might have been otherwise. Darwin's theory of natural selection was the result of conscious aims and unconscious slippages; the same was true for the ideas of his followers, who used his theory in ways Darwin did not—and with results that surprised them, too. Each generation, each author, arrived at answers they did not expect, often to questions they were not even asking. Without a plan, what seemed like progress and pitfalls occurred in equal measure. Beginning by looking backward, from Dewey to Darwin, it might seem tidy, each generation picking up something from the preceding one. By flipping the tape and running it forward, we realize how contingent, how messy and confusing the history of "the scientific method" really was. What follows is not *the* evolution of thinking; it is *an* evolution, one path among many over an entangled bank.

Before any of this began, before "science" was a tool for thinking, the term was closer to something like knowledge. That is, the term signaled more content than process, an accumulation of facts and theories you could almost point to. This stable vision soon fell apart, as science went from meaning the products of mental work to the mental work itself. This is how science came to seem less like knowledge, a category that was (and is) of little use to psychologists, and more like thinking—an embodied, organic, measurable activity. Because the sciences of mind zoomed in on such activities in the nineteenth century, and because they often did so in the language of evolutionary theory, scientific thinking came to seem more and more *adaptive*. That is, it took on the qualities of natural selection as that account was applied to studying minds. As theories of mind changed, so did methodology. This intertwining of evolution and experiment, the back-and-forth between accounts of mental adaptation and scientific method, played out on cultural and cognitive levels at once.[8]

In the process, a version of what Thomas Kuhn called "the essential tension" emerged. As Kuhn saw it, scientific thinking always struck a balance between innovation and tradition, a tension that enabled the gradual progress science was supposed to embody. Far from the grand dichotomy of "normal" and "revolutionary" science for which he is best remembered, Kuhn's "essential tension" was something more immediate. It was a tension every scientist could, even *should,* grapple with in a day's work: "To do his job the scientist must undertake a complex set of intellectual and manipulative commitments. Yet his claim to fame, if he has the talent and good luck to gain one, may finally rest upon his ability to abandon this net of commitments in favor of another of his own invention. Very often the successful scientist must simultaneously display the characteristics of the traditionalist and of the iconoclast." Although he noted that the tension was greatest at the level of "the professional group rather than the individual," Kuhn insisted that both radical and conservative elements could be found in every scientist. The task was to achieve a careful cognitive balance. And that required effort.[9]

Kuhn did not invent the essential tension; he was naming something recognized, off and on, since the nineteenth century. And whether or not his model fits every period or branch of science, it is an apt description of the specific methodological developments traced in this book. From Darwin to Dewey, the idea of equilibrium was fundamental to visions of both nature and science. Experiment and evolution meant progress through conflict, a struggle out of which the best ideas (or animals) ultimately emerged. Significantly, it was a *felt* tension. Caught between ambitions to do something new and worries about going too far, men of science in those years made this tension definitional for scientific thinking. Doing so helped make science seem like something natural, like a human extension of an evolutionary process that was itself structured as a give-and-take. The idea that science evolved might seem to tarnish its claims to unique insight, but something like the opposite was true. And while traces of the natural history of scientific thinking are hard to find in "the scientific method" as it is discussed today, they are there. Beneath its abstract, even artificial veneer, the authority that method has enjoyed was partly built on the idea that science once lived and breathed—or seemed to.[10]

Why is the idea of a natural history of science hard to imagine today? Because, consciously or not, we think of science as something both narrower and more expansive than the things that have such histories. Our narrow sense of science mostly took hold after the events narrated in this book. Many of its causes—patterns in postwar science education and funding, concomitant changes in the public image of science as "open"—are too far-removed to explore in much detail. But at least two of the reasons for science's reduction to a single, unified method stretch back to the developments recounted in this book. The first was paradoxical. The impulse to locate science in a natural context, to see it as "nature's method," actually helped denaturalize it—because of the specific shape that impulse took. By framing science as evolutionary, seeing it as akin to natural selection, its naturalizers turned mental functions into adaptive behaviors. This transformation was gradual, and in some respects even accidental, but its impact was profound. If science was one more evolved (and evolving) tool in the competition over scarce resources, then the way was paved for seeing it as a method that was simple and shared, like any other mental process with an evolutionary history.[11]

Seeing experiment as evolutionary was the first step toward an artificial, algorithmic scientific method, but the second step was detaching that method from the mind. This had to do less with the rise of evolutionary psychology in the late nineteenth century and more with the increasing emphasis on instrumental value in the same period. Framing not just the goal but also the nature of science in terms of problem solving helped narrow the meaning of science. It is not that science had never been instrumental—or that it was narrower everywhere. Debates about "useful knowledge" are ancient, after all, and anxieties about applying theoretical knowledge are pervasive in the history of science. But in these Anglo-American debates over the evolution of experiment, there took hold a new embodied, even biological, approach to both science and utility. The capacity for "experimental reasoning" was extended across lines of race, gender, culture, age, and even species in the decades around 1900, often to the consternation of those attempting to constrain it. As psychology's subject pool expanded, it became difficult to contain the modes of reasoning on which that expansion was predicated: everywhere psychologists looked, they saw their own methods. But expansion entailed a contraction as well, a thinning of science into an instrument one could find or wield anywhere. The expansion of its meaning helped make science seem smaller, more portable, even artificial.[12]

We have ideas, solve problems, and concoct theories all the time. Some might say it is what makes us human. But explaining how we think, or why it matters, often means ignoring all those mundane realities in favor of idealized accounts. The stories we tell about science stray far from what it is like to think scientifically. In the real world, we make mistakes and get bogged down; it is only in hindsight that thinking seems clean and rational. Looking back, we tell stories about how we solve problems even as elegance evades us in the moment. Method describes both phases: how we work and the stories we tell about doing it. To the figures examined below, scientific thinking was complex and imprecise, full of errors and shortcuts. They aimed to capture science as it really was—which, strange as it seems, meant studying it scientifically. By folding it on itself, science came to seem less like a body of facts and more like an embodied *relationship* to those facts. Accomplishing that transformation started with acknowledging our deep fallibility. This is where methodology began: as a proposal to study science with its own tools, a psychology of science that was honest to what really happened in the minds of those pursuing it. Making good on that proposal seemed to require a peculiar kind of education.[13]

A LIBERAL EDUCATION

Method was everywhere in the nineteenth century—or so it seemed. "We live," Charles Peirce told an audience at Johns Hopkins University in 1882, "in the very age of methods." Part philosopher and part scientist, Peirce saw method as a way to fuse those identities. Indeed, he saw it everywhere. "Modern methods have created modern science," he continued, "and this century, and especially the last twenty-five years, have done more to create new methods than any former equal period." Peirce had reason to believe in the age of methods. He was pitching the idea as part of a new course, on a subject he knew many in his audience found desperately dry: logic. So what he set out to do was convince them that, far from formalism, logic was the only way to adapt to a world being upended by modern science. Offering a new way of thinking, a "method of methods," Peirce promised to help his audience not just survive but thrive in this new scientific era: "You cannot play billiards by analytical mechanics nor keep shop by political economy. But when new paths have to be struck out, a spinal cord is not enough; a brain is needed, and that brain an organ of mind, and that mind perfected by a liberal education. And a liberal education—so far as its relation to the understanding goes—means *logic*. That is indispensable to it, and no other one thing is." Peirce's pitch ended on a high note: "It must be confessed that we students of the science of modern methods are as yet but a voice crying in the wilderness, and saying prepare ye the way for this lord of the sciences which is to come." Adapting to this regime meant molding your mind to it, using logic to turn oneself into a scientific thinker in step with a new scientific world.[14]

And Peirce was right. Understanding science, getting inside it and mastering a wide range of potential areas of application, meant thinking in a new way—specifically, thinking *about thinking* in a new way. But he was wrong about the shape such thinking would take. When the lord of the sciences arrived, it was not Peirce's beloved, expansive logic. Rather, science's public image is ruled by something much narrower: a single, shared set of steps for thinking correctly—in other words, by "the scientific method." Though some see an algorithmic, multistep approach to science as deeply, even definitionally logical, its narrowness would have struck Peirce as flat wrong, more a symptom of a crisis of confidence than a courageous step forward. His "method of methods" was meant to be both more natural and more human than the set of steps eventually installed in its place. In the shift from methods to method, from scientific studies of science to an idealized account of how it worked, an alternative history of science rose and fell. In the age of methods, science was human—and then, suddenly, it was not.[15]

That history was made by a network of philosophers, psychologists, biologists, and others in Great Britain and the United States. Tracing the journals in which they published, the students they mentored, and the projects they pursued, a cohort coalesces around ideas of evolution and experiment as applied to, and as products of, the mind. The edges of that cohort are blurry, to be sure, extending over multiple generations and across national boundaries. But its lines were not arbitrary. The group was bound together by a conceptual scheme and a sense of how that scheme had emerged. What began as an almost political project of scientific renewal in Great Britain in the 1830s became, from the 1860s on, a concerted effort to turn British empiricism into a new kind of American science. From shame about Britain's international scientific standing to pride in a legacy of pragmatic open-mindedness, a new sense of science took hold in both countries. The cementing of English as science's *lingua franca* and a nativist yearning for a special relationship between Britain and the United States only tightened the network. By century's end there had taken shape an Anglophone science that was naturalistic and practical, evolutionary and experimental—a product of conscious and unconscious boundary-work at once.[16]

Of course, methodology was not everything for everyone. These same decades saw several developments that altered how science was understood but that were not, on the surface, methodological. For one, science was becoming professionalized. What began as a pastime for gentlemen with the wealth to pursue it was, by century's end, something to do for a living—at least for some. Professionalization, especially in the United States, intersected with questions of method, to be sure. This was how a name for practitioners—"scientist"—gained traction after lying dormant for years, helping pave the way for new vocational possibilities in the sciences as well. The rise of the scientific journal, to take just one example, is part of the transformation in how science was perceived in this period. Here, too, changes intersected with the story of method. By radically increasing the *amount* of science being published, these journals contributed to science's shift from content to method—ironically, it seems, as a way of dealing with just how much content there now was. Like the rise of "scientists," this shift in scientific publishing heightened the sense that something was needed to hold science together. The increase in quantity made possible a qualitative shift in how science was perceived.[17]

The virtues associated with science were changing, too. Perhaps the most prominent of these changes was the rise of what has been called "mechanical objectivity," meaning more or less the same thing we mean by "objectivity" today. According to this new norm, scientific workers had to suppress their identities in order to present nature as it "really was." Today we think of this as a core scientific ideal, but that was never an inevitability. As Lorraine Daston and Peter Galison show in their book *Objectivity,* the idea that scientists' identities were irrelevant, even harmful, was very much a nineteenth-century notion. Both before and since, features of those same identities—such as one's ability to fill in the gaps that nature left—deeply mattered for producing accurate scientific accounts. As practices of observation and publication shifted, so did the features of practitioners and the virtues they were supposed to uphold. The rise of such norms has been accounted for in various ways, but historians agree: many of the ideals we link to science today arose amid or alongside debates over appropriate modes of thinking (and teaching) in the nineteenth century. As evolutionary and experimental vocabularies became intertwined, science's virtues were "discovered" in an adaptive procedure increasingly called method.[18]

Americans and Britons were not alone in thinking about thinking, of course. Nor was Peirce's "age of methods"—the era examined in this book—the only one in which science came in for special attention. Methods were changing, and being scrutinized, in a wide variety of contexts, as they had been for some time. Discussions of a unique scientific method stretch at least as far back as Aristotle, whose logical works were later collected under the title *Organon,* or "instrument." These texts, and especially the *Posterior Analytics,* were so foundational that they were often explicitly referenced in later methodological titles, including

Francis Bacon's *Novum Organum* in the seventeenth century and William Whewell's *Novum Organum Renovatum* in the nineteenth. The early modern period, and especially what is often called "the Scientific Revolution," was a high-water mark for articulations of the experimental approach to thinking. As we will see, disputes over the use of hypotheses in this period—especially in the work of Bacon and Isaac Newton—became myth-making touchstones in the centuries that followed. Debates in later methodological moments were conducted in vocabularies from earlier periods, even as their significance and definitions were consciously being remade.[19]

In France and German-speaking central Europe, a mixture of anxiety and hope about science—*la science, Wissenschaft*—differed from Britain or the United States in the nineteenth century. The legacies of Enlightenment and Romantic ideals, of the revolutions of 1789 and 1848, gave particular shape to a unity of "the sciences" in each context. In France, for example, Auguste Comte's hierarchy of *les sciences* embodied the Enlightenment taxonomic fascination that had powered the vast *Encyclopédie* in the previous century. While Comte's impact was profound, his *Cours de Philosophie Positive* had as many enemies as friends. Some British figures, such as John Stuart Mill, were deeply shaped by encounters with Comte—but it was Mill's *rejection* of Comte, rather than his allegiance, that left the longest impression. Later in the century, something similar occurred with the French philosopher Henri Bergson. His theory of "creative evolution" resembled the work of James and others, but differences matter as much as their similarities. What motivated their evolutionary accounts of mental life and the uses to which they put them tell both convergent and divergent stories.[20]

The same goes for German science, in which the organization of thinking was stamped with cultural specificity even as it intersected with research occurring elsewhere. "Science" is a notoriously poor translation of the German *Wissenschaft,* which encompasses something broader than is conveyed by the English term. Even in the nineteenth century something was lost in translation, a problem that was compounded in more specific terms like *Naturwissenschaft* (roughly, natural science) or *Naturforscher* (even more roughly, scientist). Historians have shown how issues of translation like these were also always issues of political and cultural authority, where the meaning and scope of science involved intense national, regional, and even local stakes. Disputes over the lines between learned and popular discourse, for example, or how scientific collections were presented to the public, were crucial to definitional debates. Thus, methodological disputes over whether the human sciences were reducible to the natural sciences, or the *Methodenstreit* in late nineteenth-century economics, were part of broader shifts *and* specific to their time.[21]

Every history has its own method, one rooted in the discipline's conflicted relationship with science and its shifting standards of evidence and argument, of proof and pluralism. This particular history is motivated by a consideration of where it terminates: a five-step procedure, distilled from a century of natural history and psychology, that was cemented as "the scientific method" and is still taught to children today. By returning to a time when science seemed natural to some, comparable to problem solving in children and nonhuman animals, this book recovers a new history of scientific authority and its application to human affairs. Examining a network of scientific authors, disparate yet specific, reveals how reflexive scientific thinking helped make the mind and its processes seem natural. In the decades between Darwin's translation of an experimental idiom into an evolutionary theory and Dewey's use of that theory to buttress new approaches to experimentalism, we find a kind of chaotic coherence, a patterned assembly of reference and self-reflection.[22]

SITUATED LOOPING

The age of methods coincided with significant changes in the study of human nature. Fields like psychology and anthropology were coalescing around scientific efforts to describe the mind and society, bringing their chosen methods under new scrutiny. This was in part because the claims such fields were making—about how individual humans think and how groups of humans work together—could be reflexively applied to themselves. Science was becoming a way of thinking, a practice in which individuals and groups both took part. Psychological and anthropological studies of embodied cognition were turned backward on themselves almost immediately, with experimental techniques and field observations subjected to experimental and observational analysis of their own. The human sciences were not just oriented "out there," to life beyond the laboratory. They could be, and were, also targeted internally, remaking the nature and the stakes of the scientific thinking being pursued. The combined force of these developments—investment in the scientific study of humans and turning that study on itself—helped make science into a scientific object. The instrument that so many now trusted to explain the natural world was subject to its own scrutiny, each advance shifting the goalposts of scientific self-awareness. Appeals to science were thus appeals to a moving target: any claim about how science worked was self-referential and, potentially, self-reinforcing.[23]

Making sense of this recursive history means reckoning with scientists' dual role as both subjects and objects of scientific research. Psychologists became not only a new class of experts on scientific thinking in these years, but also models of that same thinking—and, by extension, of the human mind as a whole. This makes the story of nature's method a special example of what the philosopher Ian Hacking calls "the looping effect of human kinds." The idea is subtle but profound. Labeling human behavior changes the very thing you are describing, because humans react to labels. As a result, a feedback loop is generated:

> To create new ways of classifying people is also to change how we can think of ourselves, to change our sense of self-worth, even how we remember our own past. This in turn generates a looping effect, because people of the kind behave differently and so are different. That is to say the kind changes, and so there is new causal knowledge to be gained and perhaps, old causal knowledge to be jettisoned.

The human sciences are special because their objects respond to the practice of science. If you study the mind with psychology, you change the minds you study. Labeling them scientific, or pathological, creates an immediate response as people react to those labels. The looping initiated by labeling in the human sciences is impossible to stop once it is set in motion.[24]

The concept of looping effects has generally been applied to labels like "schizophrenic," but it works just as well to explain what happens when scientific behavior is described in new ways. All such labels alter what they claim to mark. In the case of scientific attention to method, it just happens more rapidly: using a method to study something, and rethinking that method as you do, changes the conditions and meaning of such studies as they happen. What your method is, how it works, when and where you can use it—all are altered by thinking about science in scientific terms. Looping effects occur at the nexus of mind and matter, where how you talk about what you do and the actual doing of it intersect. This makes them central to methodology in its fullest sense. After all, if method is anything, it is a vocabulary for describing how hidden activities of the scientific mind intersect with observable behavior. Its terms are how scientists tell one another what they have done, how their ideas got tested by experience. Method is where mind and matter meet. When it comes to how we think experimentally, method is the name of the game; and when such thinking is itself subjected to experimental attention, what we get is a special version of the looping effect of human kinds.[25]

Intriguingly, this talk of "kinds" would have made a lot of sense to the actors involved. Both the idea of kinds as general phenomena and their application to science were introduced by one of the era's central figures—indeed, by the very man who coined the term "scientist." His name was William Whewell, and his famous terminological proposal marked a foundational moment for the age of methods that followed. Although the term "scientist" did not take off for almost a century, the idea of such a name was symptomatic of ongoing changes in how science was imagined as a single, unified entity. If Whewell is recalled today, it is for the term he proposed. But *how* he proposed it is more meaningful. Deciding how science worked and what made it special would require, Whewell argued, a "science of science," or the reflexive study of scientific thinking. This call gave a double meaning to human science: it meant "a science of humans," understanding humans in scientific terms, but it also meant seeing science as itself a human activity. The idea of a science of science was not Whewell's (or the nineteenth century's) alone, but the constellation of practices at play—the rise of psychology and the return of the hypothesis—produced a looping effect with specific, profound consequences.[26]

Looping is not the only way to describe what happened, however, nor does it capture what is arguably the era's most important result. The philosopher Donna Haraway reminds us that knowing is situated in both time and place, governed by unspoken assumptions. Looping effects can prop up existing structures and reify ideas about reliability even as they change the phenomena under study. To see how, we need to look beyond the scientist as a kind of agent of stasis. Accidents happen; objects talk back. One way to flesh out the dynamic equilibrium these figures idealized and set into motion is by addressing what Haraway calls "situated knowledges." Doing so can help us push back on claims to universality while acknowledging the work such claims can do. "Situated knowledges," Haraway argues, "require that the object of knowledge be pictured as an actor and agent, not as a screen or a ground or a resource, never finally as slave to the master that closes off the dialectic in his unique agency and his authorship of 'objective' knowledge." Situatedness is "paradigmatically clear in critical approaches to the social and human sciences, where the agency of people studied itself transforms the entire project of producing social theory." Thus, Haraway and Hacking agree: there is something special about the human sciences, something centered on their power to remake their ostensible object—human nature—even as they describe it. It is a vibrating, often violent power.[27]

As Haraway recognizes, the same goes for the *history* of those sciences. In other words, this book is no exception to the situatedness she identifies. The scientific ideal is far from lost on historians today, with their language of evidence and archives and attention to cause and effect. But this is doubly true in this book, the subject of which—the messy, all-too-human nature of scientific thinking—was also the focus of the figures being analyzed. Just as historians want to see the past "as it really happened"—to pick apart assumptions underlying stated positions—so these scholars wanted to abandon idealized portraits of science in favor of more naturalistic accounts. We are still looping, in other words, as historical commitments overlap with subjects' own work on science-as-cognition. The fact that so much history is practiced in the vernacular of this period, with its emphasis on cultural factors and practices, turns any account of the age of methods into a way to resituate the scientific method as it is invoked and debated today.

And resituating matters. One way of combining Haraway and Hacking is to draw on the terms of the past while analyzing it. Situated looping would be more than what cultural theorists call "immanent critique" or what anthropologists call "emic analysis." It would take up not only actors' categories, but also actors' concepts. In the case of the age of methods, this means telling a history that mirrors the give-and-take tensions identified by those involved. Robert Richards has done just that in his masterful book *Darwin and the Emergence of Evolutionary Theories of Mind and Behavior,* proposing at the end a "natural selection model of science" drawn from the ideas he traces. For Richards, science (and history) are evolutionary: they progress in a fitful way as ideas adapt to circumstances. Whether or not we agree with Richards that natural selection is an *exportable* analytic, applicable to other contexts, it is clearly an *endogenous* one, forged in the image of its own materials. Endogenous analysis is a kind of situated looping, an approach that attends to local concerns without aspiring to universal generalization. An endogenous history of the age of methods stays with the trouble of the essential tension, situating it in the development of a dynamic equilibrium that stretched from Darwin to Dewey.[28]

THE POLITICS OF METHOD

The transformation of science from product to process, from thought to thinking, was one of the crowning achievements of the nineteenth century. Indeed, this was noted a century ago. The philosopher Alfred

North Whitehead counted this shift in science's meaning as nothing less than a turning point in Western history: "The greatest invention of the nineteenth century was the invention of the method of invention. A new method entered into life. In order to understand our epoch, we can neglect all the details of change, such as railways, telegraphs, radios, spinning machines, synthetic dyes. We must concentrate on the method in itself; that is the real novelty, which has broken up the foundations of the old civilization." Whitehead's work was part of a broader movement that has been labeled "process philosophy," a view of the world resonant with both Hacking and Haraway as well as figures in the age of methods. Peirce was an inspiration for Whitehead, as was Peirce's friend James, who saw more clearly than anyone else the promise of the mind as an instrument. *"Theories thus become instruments,"* he once wrote, *"not answers to enigmas, in which we can rest."* By the time of Whitehead's reflection, science was both product and process, a question and a way of answering it.[29]

Like any instrument, this approach to science as a way of thinking had a politics of its own, one that stretched back from James and Dewey's progressive optimism, through what has come to be called social Darwinism, to the reformist attitudes of Darwin and those who inspired his early work. In its broadest sense, it was a politics of gradual reform that these figures and many others thought they recognized in the struggle among animals as well as between ideas in their own minds. Casting scientific thinking and its chosen objects in the same terms was part of what bound these overlapping generations into a shared community of discourse. Whether they saw it as a counterweight to other cultural forces or as a delicate balancing act of its own, these thinkers framed science as a stabilizing force for the individuals who pursued it. Their projects, taken together or separately, reproduced an image of stable progress, an image that was achieved and defined by the mechanism they increasingly saw in the natural world as well. But this equilibrium did not come easily—it was *dynamic* for a reason. It required work. Just like James's instrumental theories, stability meant labor, not rest. In the age of methods, balance—social, political, even scientific—was a verb.[30]

Now, identifying an alternative way of thinking about thinking—an earlier moment in which it seemed possibly to portray science as simple and universal, natural and powerful—is not the same as celebrating that view of science or taking it at face value. The idea of an evolution of thinking strikes some as progressive, even promising; others find claims to universality and progress to be off-base, even dangerous. That is how politics works: agreements about definitions often mask deep ethical or normative divisions. The politics of science as an evolutionary process is no exception. From Darwin to Dewey, looping generated intense debate and divergent programs. Education, war, retrenchment—scientific method and its discontents laid bare many of the contradictions inherent to reform liberalism. It can be easy, looking back, to let discussions of openness paper over just how far science was, and is, from being truly open. Making sense of appeals to universality and persistent failures to live up to them requires both cognitive and political work. The narrow foundations of scientific method's sweeping claims, its political history combined with its antipolitical rhetoric, is a paradox that needs unpacking. One way of proceeding is by turning to the past.

Fascinatingly, this is just what many of the figures in the age of methods did. History of science, an emerging genre in the 1830s, was one place where appeals to balance were made early and often. Authors of these new histories cast science as a savior for a world locked in political struggle. Science, on their view, did not just reduce tensions by counteracting bias (a role to which it is often reduced today). To some, thinking scientifically meant leaning into its political nature, taking those biases on board rather than denying they existed. This was one way to embody equipoise: to incorporate competing impulses rather than seek out a third way. In this sense, science was not just a source of stability—it was a *symbol* of stability itself, representing the whole process rather than one side of it. This is how the politics of method became the politics of equilibrium, how the celebration of gradual change inflected science past and present. Part of the broader appearance of what James Secord has called "works of scientific reflection," these authors turned science into an engine of social stability, a single function that could operate at the level of the individual and the group. Science was not an apolitical solution, a value-free tool for getting past our differences. Instead, science was supposed to be political, subsuming all that was valuable in human thinking and holding each participant to account.[31]

The idea that science is political seems strange today. Those who style themselves as defenders of science and its method often insist on its *lack* of politics, against those who want to undermine it. It is precisely because science is value-neutral that it is worth protecting, they have argued since the Second World War. But this view of science as apolitical should give us pause. Confusion over whether and how to defend science stems partly from the conflation of politics and partisanship. In the nineteenth century, however, it was easier to pry them apart. From the authors of Darwin's day to their self-proclaimed inheritors later in the century, the inclusion of a range of political views *within* science was a real possibility. These authors included not only British figures who very literally inherited this commitment, but also the American progressives who around 1900 trumpeted the potential of science as an agent of social improvement. What tied these movements together, beyond the shared commitments they avowed, was a goal to use science to secure both mental and social stability. To different degrees, they recast science as political but nonpartisan—as a balance predicated on hearing from all sides.[32]

Reform of mind and method was not conducted in a vacuum. Much like attempts to change political institutions in the same period, reflections on scientific method were spurred by a range of transformations at home and abroad. As we will see, social and political issues shaped views on who was capable of such reasoning and on how it would be best to wield it. In the age of methods, science continued to be identified with traits that cut along lines of gender and class. Increasingly, its ideal method was identified as peculiarly British. The immediate impetus for much of this work was a sense of crisis rooted in unflattering self-reflection by British men of science. Responding to the widespread idea that science in Great Britain had fallen behind, a range of authors proposed a new approach to scientific thinking as a matter of national pride. The first step was to decide how, exactly, science worked—what was essential to it and how that essence could be exported. The scientific study of science was supposed to jumpstart a modern era, an age of methods. And so, it is to those efforts that we now turn.

2

HYPOTHESIS UNBOUND

The age of methods began with a nightmare. Mary Shelley's *Frankenstein: A Modern Prometheus* repurposed Greek mythology as a warning about scientific hubris in an age of upheaval. Published anonymously in 1818, the book expressed a view that was a far cry from the scientific optimism shared by many of Shelley's contemporaries. To her, science was an enemy of the social stability many sought in what felt like turbulent times. This lesson was embodied in Victor Frankenstein, whose struggle with the monster he created serves as a cautionary tale about science's unsettling potential. Beneath this story about the risks of science, Shelley's book also dramatized scientific thinking itself. Its plot tracked Dr. Frankenstein's desperate attempts to pull the world into alignment after knocking it off its course. Frankenstein's give-and-take with his monster stood in for the desperate reactivity of scientific inquiry. The book brings to life not only science's products but also its process. As in the myth on which the story draws, cosmic balance hinged on human limits and the failure to recognize them. Whether regarding a fallen god bestowing a divine skill or a created being denied his proper status, the story made plain the need for equilibrium and the costs of disturbing it. *Frankenstein* is thus science fiction in more than the familiar sense: it not only helped give rise to a genre, but also made scientific thinking into an engine of dramatic excitement and fictional imagination.[1]

Science was not the only pursuit eliciting anxiety in Shelley's lifetime, nor were the members of her social set all anxious about it. The subject of poetry was just as fraught, its essence debated in the same terms: power and identity, imagination and truth. As Samuel Taylor Coleridge wrote in 1817: "What is poetry? is so nearly the same question with, what is a poet? that the answer to the one is involved in the solution of the other." Just like science, poetry was a matter of the mind. "The poet," he continued, "described in *ideal* perfection, brings the whole soul of man into activity, with the subordination of its faculties to each other, according to their relative worth and dignity. He diffuses a tone and spirit of unity, that blends, and (as it were) *fuses,* each into each, by that synthetic and magical power, to which we have exclusively appropriated the name imagination." Poetry meant balancing "the general, with the concrete; the idea, with the image; the individual, with the representative; the sense of novelty and freshness, with old and familiar objects; a more than usual state of emotion, with more than usual order." Its task was to offset revolutionary tendencies with necessary caution, lest Britain fall prey to the turmoil recently observed in France. Poetry, for Coleridge, was the key to progress.[2]

Celebrations of poetry in this era were also celebrations of specific states of mind. To understand poetry's power, critics probed its process. William Wordsworth called poetry "the spontaneous overflow of powerful feelings." Three decades later, in his own essay titled "What Is Poetry?," the young John Stuart Mill pointed into the poetic mind as well: "The poetry is not in the object itself, nor in the scientific truth itself, but in the state of mind in which the one and the other may be contemplated." The meaning of poetry was not in this or that poem, but in the minds that gave rise to them. Poets shared, as Percy Bysshe Shelley put it, "new springs of thought and feeling, which the great events of our age have exposed to view, a similar tone of sentiment, imagery, and expression. A certain similarity all the best writers of any particular age inevitably are marked with, from the spirit of that age acting on all." That spirit was often described in terms of "imagination," a capacity that inspired enthusiasm and suspicion in equal measure—especially when it came to debating the responsibilities of art and science during and after the Enlightenment. Even those who read but did not write the poetry of the period saw it as a way of taming revolutionary impulses within the social order.[3]

Substitute the word "science" for "poetry"—and "scientist" for "poet"—and you arrive at a view of scientific thinking that emerged at the same time and traded on many of the same impulses. Though science and poetry were often opposed, a closer inspection of efforts to define both reveals a shared ambition to balance extremes. When Shelley defended poetry as "that which comprehends all science, and that to which all science must be referred," he suggested a common thread even as he split the two apart. This was what Wordsworth and Coleridge meant when they framed *Lyrical Ballads* as "experiments," which Wordsworth later amended to "*an* experiment." It was a search for balance, an attempt to weigh "vivid sensation" against "metrical arrangement." For Wordsworth, the poet walked side-by-side with the man of science, "carrying sensation into the midst of the objects of the Science itself." Poetry and science were sometimes complementary, sometimes mirror images; the one needed the other, but both contained creative sparks and means of taming them. What they shared, in the minds of these Romantic critics, was that either could be imagined as an individual act, a mode of creative expression rooted in every human mind.[4]

The literary critic M. H. Abrams described this new image of poetry as "the replacement of the metaphor of the poem as imitation, a 'mirror of nature,' by that of the poem as heterocosm, 'a second nature,' created by the poet in an act analogous to God's creation of the world." The same could be said—and was said—of science. The term went from signifying certain knowledge gleaned from facts to the projection of the scientific mind onto the world. An emerging genre of scientific criticism in Britain emphasized the process, not the product, of scientific thinking, just as Romantic critics did for poetry. These scientific critics cared less about science as a body of results, and more about how those results were produced and how to produce more. To them, as to the literary critics writing about poetry, what mattered most was the effort to capture minds at work, producing a living image of human creativity. It was a narrow image, to be sure, based on the minds that were most familiar to them: their own, or those selectively drawn from history. What the scientific critics saw was a window into *all* minds, a universal vision of intellectual progress based on foundations laid close to home. Rather than worry about this obvious narrowness, they saw their inward turn as a source of strength, with consequences that would only become apparent much later.[5]

What scientific critics *did* worry about was the "spontaneous overflow" that literary critics like Shelley and Wordsworth had celebrated. Their worries centered on one outlet for that overflow, a controversial cognitive tool slowly creeping back into British science in the 1830s: the hypothesis. The same qualities that made the imagination so enticing to Romantic poets made hypotheses suspect among their scientific peers. Associated with the French rationalist tradition, especially the work of René Descartes, hypotheses had been seen as speculative, irresponsible, and potentially dangerous for generations. In place of hypotheses, British men of science had long presented a homegrown faith in facts, casting Francis Bacon and Isaac Newton as the figureheads of a native, empiricist tradition. Busts of both graced frontispieces throughout the nineteenth century, their words serving as epigraphic signals of inductive caution. Antagonism to hypotheses was summed up, for many Britons, in Newton's famous dictum, *hypotheses non fingo*—"I frame no hypotheses." The idea was that hypotheses introduced unwarranted (and un-British) speculation into science. Never mind that Newton *did* use hypotheses. The pithy line, along with Bacon's increasing popularity, turned allegiance to facts and opposition to hypotheses into a test of British national character by the nineteenth century.[6]

Hitching method to national identity had an unanticipated effect: as British politics shifted, so did what counted as acceptable scientific thinking. Many of those who in the 1830s were framing science as a dynamic equilibrium were part of a political party—the Whigs—that was pushing a similar social vision. Whigs sought to stave off revolution by meeting popular demands halfway, insisting on the gradual nature of political change. The party had been around in some form for centuries; but in the nineteenth century, reform became the Whigs' watchword. By century's end, long after the Whig party had fallen, the virtues of compromise had come to seem like a basic British approach to almost anything—including natural philosophy. While those pushing the language of balance did not agree on everything, an emerging consensus on the value of equilibrium bordered on a new aesthetics of cognitive practice, a kind of consensus on consensus. Books by men of science framed mind and method as engines for equilibrium, a balance achieved first in the individual and, later, in society as a whole.[7]

Hypotheses were central to this ideological coalescence. Of course, acknowledgment of their role went hand in hand with warnings about the dangers of a "fondness for hypotheses." It was worth the risk. Hypotheses clearly answered a felt need among men of science for a creative spark. Anxious about their status in the international community, and sensing that something was missing from the scientific scene, some British men of science turned to the hypothesis as a way to address their intellectual restlessness. This was not a simple turn to an outcast tool. Instead, what emerged starting in the 1830s was a new kind of hypothesis, remade as an engine for a new, scientific social life. The contested term, long seen as a weapon in fights over methodology, was repurposed as a means of undoing those fights by borrowing a bit from both sides. Those debates persisted, of course, but in hindsight they seem like disagreements on tactics masking agreement on strategy. That strategy, focused on progress, depended on bold leaps and careful testing in equal measure. This is where the hypothesis came in, soon to be set loose on problems like the one to which we now turn.[8]

SIGNS OF THE TIMES

"It is the Age of Machinery, in every outward and inward sense of the word," wrote Thomas Carlyle, "the age which, with its whole undivided might, forwards, teaches, and practises the great art of adapting means to ends." The year was 1829, and the signs were not good. According to

Carlyle, mechanization had hit everyone, from factory workers and their families to members of Parliament. "Not the external and the physical alone is now managed by machinery," he lamented, "but the internal and spiritual also." Science was one victim, a noble enterprise now transformed into something soulless: "No Newton, by silent meditation, now discovers the system of the world from the falling of an apple; but some quite other than Newton stands in his Museum, his Scientific Institution, and, behind whole batteries of retorts, digesters, and galvanic piles, imperatively 'interrogates Nature,'—who, however, shows no haste to answer." It would have been clear where Carlyle laid the blame: the Royal Society. Previously associated with Bacon and especially Newton, the Society had recently come under fire. Carlyle had been among its early detractors, lampooning its topics of discussion as less "the invention of fluxions" and more "the dimensions of a fossil stick," all of which had "very much the look of drivelling."[9]

For Carlyle, the Royal Society and groups like it embodied the folly of mechanical life. "Has any man," he sneered, "or any society of men, a truth to speak, a piece of spiritual work to do; they can nowise proceed at once and with the mere natural organs, but must first call a public meeting, appoint committees, issue prospectuses, eat a public dinner; in a word, construct or borrow machinery, wherewith to speak it and do it." Machinery, in this sense, extended to the social fabric of intellectual life. From his isolated farm in Scotland, far from such societies, Carlyle deemed them not just unnecessary, but pernicious. "This deep, paralysed subjection to physical objects comes not from nature," he concluded, "but from our own mode of viewing Nature." The fault was ours; it was rooted deep in the way we viewed the world. Mechanical science was not inherently evil, but it had to be balanced by a "science of *Dynamics*," one that "rose up, as it were, by spontaneous growth, in the free soil and sunshine of Nature." In its absence, men of science had "lost faith in individual endeavor, and in natural force." Without such a counterbalance, cultivated in ourselves, machines would rule.[10]

Charles Babbage was just the kind of machinist Carlyle had in mind. The "cunningly devised implements" and "pre-established apparatus" Carlyle despised were Babbage's conscious aims. A leading mathematician and one of the founders of computer science, Babbage's lifelong goal was to replace the natural patterns of scientific reasoning with a machine capable of performing arithmetic, algebraic, and even analytical functions. Initially cavalier about his ability to produce this "Difference Engine," Babbage secured a series of grants from the British Treasury on the strength of plans he began drafting in 1822. Part of what he promised to the public was a set of real-world applications such a device could perform, including rapidly and faultlessly computing the numerical tables on which various governmental and industrial projects depended. Matthew Jones has shown how his effort to do so was the climax of two centuries of "reckoning with matter," or attempting to resituate cognitive labor in apparatuses of iron and wood. For Carlyle, the actual construction of such devices mattered less than the goal of building them. The very idea of such a machine, of reducing human thinking to mechanical rules, revealed Babbage's "hidebound wooden irony" and "acridest egoism"—not to mention his "frog mouth and viper eyes." The horror, according to Carlyle, was Babbage's audacity.[11]

Carlyle's instrumental dystopia was the utopia Babbage sought. Babbage's factory mode of inquiry and the intellectual "division of labor" it enabled were the industrial ideas that Carlyle most reviled. Over the course of the 1830s, the two would continue to clash over the promises and pitfalls of the Industrial Revolution, which Carlyle dubbed "the Condition of England question." The position of both could not have been clearer. While Carlyle attacked the effects of industrialization on inner life, Babbage published a strident defense of the same effects in 1832: *On the Economy of Machinery and Manufactures.* Intriguingly, he attributed his own book to machinery, calling it "one of the consequences that have resulted from the Calculating-Engine, the construction of which I have been so long superintending." Babbage argued that Carlyle's "natural force" could actually be produced by natural, human, or mechanical means. When he wrote that "*Knowledge is power,*" he was giving the Baconian maxim a material gloss: literally, knowledge could be made into a powerful machine, and those machines helped spur ideas. "It not merely gives to its votaries control over the mental faculties of their species," he wrote, "but is itself the generator of physical force."

Ideas produce books and books produce ideas. Each could convert the power of the other—like a feat of engineering, a vast system.[12]

Such a system had to be categorized and organized to maximize efficiency. According to Babbage, machines were divided into two classes: "*1st. Those which are employed to produce power;* and as, *2dly. Those which are intended merely to transmit force and execute work.*" The analogue of the "unequal division among machines" in the class struggle of the day would have been hard to miss. And indeed, Babbage extended this logic himself in the book's conclusion: "the division of labour is no less applicable to mental productions than to those in which material bodies are concerned." His effort to develop a "universal language" of machinery was soon used to explain everything from the physiology of animal life to the political economy of the factory system. Babbage's view of machines, in other words, was becoming a view of the world. When he envisioned men of science working together, each "labouring in that department for which his natural capacity and acquired habits have rendered him most fit," Babbage extended industrial efficiency to the world of science—where he fancied himself a manager.[13]

Babbage's book on machinery was a sort of sequel to an attack on the Royal Society he had published in 1830, tellingly titled *Reflections on the Decline of Science in England.* Like Carlyle, he saw the science of the day choked by institutional politics and irrelevant minutia. And as a member of the Royal Society unafraid to name names, Babbage had produced a critique that hit closer to home than Carlyle's. The problem was personal for Babbage: anxious to see his "Difference Engine" funded, he accused the Society of failing to cultivate "difficult and abstract sciences" and "a peculiar class professedly devoted to science" to pursue them. Babbage's self-image and livelihood were both at stake. Where Carlyle had looked to Germany for models to meet his needs, Babbage looked to France, where public funding of science supported a class of practitioners unrivaled closer to home. Given Babbage's own, very material interests in seeing a similar program in Britain, the prospect of the kind of support French science enjoyed was more concrete than the intellectual inspiration Carlyle sought in his German-speaking models.[14]

Justifying public funding meant showing some results, and this proved to be a problem for a machine as complex as Babbage's. Ultimately his emphasis on the value of "basic" and "abstract" science seemed like a form of protesting too much, emerging from a need to show his work was worthwhile even as progress on the actual, physical machine fell behind. This is not to say that he abandoned his earlier calls for application and "practical use"—he simply deferred their realization to some point in the future. The actual materialization of the machine mattered to many critics, but not to all. For Carlyle, after all, mechanical apparatuses mattered less than the dream of reducing human nature to a set of mechanical relations. "When we can drain the Ocean into our mill-ponds," he sneered, "and bottle up the Force of Gravity, to be sold by retail, in our gas-jars; then may we hope to comprehend the infinitudes of man's soul under formulas of Profit and Loss; and rule over this too, as over a patent engine, by checks, and valves, and balances." Whether or not Babbage succeeded in building an engine, it was his conflation of man and machine on paper and in theory that disturbed Carlyle most. And Carlyle was not alone: the *instrumentalization* Babbage represented, has been the object of suspicion ever since.[15]

Decline may have looked different to science's various critics, but all of them agreed on the basic point: British science was in a bind. Despite Carlyle and Babbage's mutual contempt, their critiques converged on this point. Both adopted the withering tone of political tracts of the period, like John Wade's *The Black Book, or Corruption Unmasked!* Published serially starting a decade earlier, *The Black Book* was frequently reprinted as reform peaked during the 1830s. Like Babbage, Wade had made his case by dredging up secretive connections to scandalous effect; like Carlyle, he was waging war on bureaucracy. By adopting a polemical style their audiences would have recognized, Babbage and Carlyle helped make science seem as public as the political debates they mimicked. Their critiques reinforced a shared intuition: science *should be* a matter of public concern. By airing their grievances and turning their foes into enemies of the public interest, both men helped transform natural philosophy from a cloistered, seemingly value-free affair into a public, even political issue, firmly situated in the grubby world of debates over reform and representation in the 1830s.[16]

Carlyle and Babbage also offered similar prescriptions for how to put things right. Along with many of their contemporaries, both promoted versions of what Coleridge called a "clerisy," or a ruling cultural elite. According to Coleridge, an elite was needed to guide society in the direction of orderly progress. This applied as much to science as to politics. Of course, Carlyle and Babbage differed radically on how to compose such an elite. Babbage lobbied for public funding, in the hope that Britain might develop a professional scientific class that would support men of science like himself in full-time research. Carlyle opposed such top-down cultivation, claiming that "an aristocracy of talent" would emerge organically from British society to guide others toward their proper ends. "An actual new Sovereignty," Carlyle wrote in 1843, "Industrial Aristocracy, real not imaginary Aristocracy, is indispensable and indubitable for us." Babbage's scientific bureaucracy was far from Carlyle's intellectual aristocracy, but the rift between their visions masks the fact that, for both, a form of scientific thinking could anchor social stability in society more broadly—if wielded by a select few.[17]

How exactly science would achieve that stability was another question. One way forward, on the model of parliamentary reform, would be to focus on the *social* issues in the scientific community that Carlyle and Babbage had critiqued. This would mean increasing transparency in governance and the awarding of prizes—the precise causes Babbage had championed within the Royal Society, where the recalcitrance of its members stymied any changes. But just as political reform empowered the House of Commons relative to the House of Lords in the 1830s, scientific shifts held out the promise of broader participation in intellectual life as well. The dream of such engagement animated proposals for a new organization, one that would be more democratic than the aristocratic Royal Society. Named the British Association for the Advancement of Science (BAAS), this new group first met in York in 1831—making good on a promise to extend science beyond London and Oxbridge from the very start. An emphasis on participation led to the group being nicknamed a "Parliament of Science," signaling the intention to balance out the elite Royal Society on one side and local scientific societies on the other. Early meetings of the BAAS were different from both. One participant observed: "The most motley literary crowd that ever assembled in this country is thrown together, under all the excitement of a public meeting, to listen to addresses delivered by those who cannot be personally cognizant of the feeling or dispositions of one-tenth of their listeners." A "Parliament" indeed![18]

As elsewhere in the age of reform, realities within the BAAS were more complicated than its democratic rhetoric suggested. While frustrations with the Royal Society and a real fear that other countries were getting a leg up contributed to calls for increased participation, the BAAS was still very much a hierarchical organization. This is signaled perfectly by the title given to the most authoritative history of the group: *Gentlemen of Science.* The early leaders of the BAAS were extremely careful about how they presented the group's identity and mission to the wider public. Even the choice of the term "association" seems to have been calculated. For decades, suspicions sparked during the French Revolution had made it difficult to meet in large groups. So-called Seditious Meetings Acts, for example, imposed restrictions on meetings of more than fifty people, to ensure that such gatherings occurred under government supervision. Against this backdrop, early leaders of the BAAS were wary of links to the radical discussion that followed the Revolution, which partly explains their emphasis on openness and publishing their proceedings. To these gentlemen, the term "association" expressed a spirit of voluntarism that seemed distinctly English. And men of science were not alone. By the end of the century, a "gigantic development of associated life" in England could be called "the great psychological fact in the life of this people, its great characteristic feature."[19]

Early BAAS meetings reflected the liberal (or "Broad Church") Anglicanism espoused by many of its founders. Liberal Anglicans supported relative tolerance in religious observance and backed modest reforms at universities, including at the movement's center of power: Trinity College, Cambridge. The movement also encompassed new modes of scientific and historical inquiry, highlighting progressive principles in the natural and social worlds in ways that echoed the work of Whig contemporaries. Indeed, although Liberal Anglicanism and political Whiggism were not synonymous in the 1830s, both responded to political pressures by carving out space for a benevolent aristocracy that could achieve moderate reform without upsetting groups that were accustomed to unequal distributions of power. Popular resentment at unfulfilled promises would eventually sweep the Whigs from Parliament (in favor of the liberals who soon replaced them entirely), but in the 1830s sympathetic aristocrats could still use economic influence, political agility, and the reputations of institutions like Trinity College to wield authority in Parliament and "the Parliament of Science" alike. Within the BAAS, an "inner core" emerged whose sense of science and society soon became synonymous with scientific practice in Britain.[20]

If domestic politics was a source of reform rhetoric for leaders of the BAAS, imperial politics lent it another set of images. As the group's founding president put it in his first address: "When any science becomes popular, and those who interest themselves in its advancement perceive the necessity of working for it by united exertions, that science is detached from the central body; first one fragment falls off, and then another; colony after colony dissevers itself from the declining empire, and by degrees the commonwealth of science is dissolved." Where parliamentary rhetoric had accommodated the many new practitioners of the sciences in recent decades, this imperial metaphor suggested the perils of the specialization that went along with those changes. The loss of its American colonies—cemented in the War of 1812—sharpened the image of disseverance for Britain, while competition with France on what was now almost a global scale made the issue of continued international influence, and the role that science and technology might play in it, increasingly urgent as the nineteenth century wore on.[21]

The result was a sense of cohesion and fragility, confidence and apprehension, that made its way into the rhetoric of organizations like the BAAS. References to a scientific empire were not (just) expressions of a hope for expansion; they also reflected widely held anxieties about Britain's place in the world. Just a few years after the BAAS's founding, Cambridge professor and cofounder William Whewell warned that the "disintegration" of science was "like that of a great empire falling to pieces." In 1847 another member wrote that "the peripatetic character of the Association constitutes it a scientific institute on the *aggressive* system," one that "takes up its empire piecemeal, by districts and squares." Whewell's friend John Herschel made this more explicit: science was akin to "the occupation and settling of the country under the dominion of the conquerors, quelling the insurrectionary movements of ignorance and prejudice under the new regime, and partitioning out the land in provinces and domains." Anxiety and aggression were the engines of empire, and they were at work in the BAAS as well. Its founders' cautious confidence mirrored political, especially Whig, optimism about imperial expansion. Men of science and members of Parliament alike articulated the hope that such "provinces" might yet be held together and proposed various means for achieving that cohesion.[22]

While individual autonomy may have suggested the language of "association," it quickly revealed a weakness as well. Without a centralized structure, the group's leaders worried that the BAAS would collapse under its own weight. Metaphors of fragmentation expressed real fears for its founders, who began to seek a common principle that could bind them all together. Calls for unity soon devolved into debates over terminology, including, famously, a debate at the group's third meeting, at Cambridge in 1833. The discussion of what to call members of the BAAS was opened by Coleridge, who ruled out the use of "philosopher" on the basis of its being "too wide and too lofty a term" for the sorts of practices pursued by men of science. Some bemoaned the fact that France had "*savans*" and a German could be a "*natur-forscher,*" but these were deemed inappropriate for Britain (the former was "rather assuming" while the latter suggested "such undignified compounds as *nature-poker* or *nature-peeper*"). It was in this context that Whewell coined the term "scientist," crediting it to "some ingenious gentlemen" in an unsigned review the following year. As a word, it was not that unusual. "Artist" was an obvious analogue, after all, and "sciolist" and "atheist" worked, too. But as Whewell noted, the term was "not generally palatable" and was quickly dropped. Men of science in Britain remained skeptical for decades, so much so that when Americans later adopted it, the term "scientist" was labeled "an American barbarous trisyllable" and "an American importation."[23]

Given the preference for the term "*men* of science," it is ironic that Whewell announced the coining of his term in a review of a book by a woman. As its title implies, Mary Somerville's *On the Connexion of the Physical Sciences* addressed the same issue of unity broached at the BAAS meeting the previous year (a discussion that she, being a woman, was not allowed to attend). However, her solution was quite distant from the one attempted by the BAAS. Rather than locate the meaning of science in its practitioners, Somerville sought the "connexion" of her title in the external world, specifically in the principles and forces that give rise to the patterns that are the subject of scientific study. "The heavens," she wrote, "afford the most sublime subject of study which can be derived from science." The cause of this sublimity was gravity, about which she rhapsodized at greater length in the book's second edition: "In it we perceive the operation of a force which is mixed up with everything that exists in the heavens or on earth; which pervades every atom, rules the motions of animate and inanimate beings, and is as sensible in the descent of a rain-drop as in the falls of Niagara, in the weight of the air, as in the periods of the moon." By identifying science with nature, rather than the mind, Somerville was tacitly carving out a space for female practitioners in a male-dominated field. If a scholar's identity mattered less than the laws under study, then nothing—including gender—should preclude her active participation. Science, for Somerville, did not care who you were.[24]

Reviewers failed to see Somerville's attempt at unification for what it was. Whewell, for one, thought the book manifested a "peculiar mental character" of the female mind, enabling her to see what men could not. Women and men were different, he argued, which helped Somerville clarify several thorny issues in a new way for general readers. Herschel, reviewing an earlier, more technical book by Somerville, went the other way, complimenting her on the fact that "beyond the name in the title-page, nothing throughout the work [reminds] us of its coming from a female hand." As both reviews make clear, the default image of a "*man* of science" was not just descriptive, but prescriptive. Whewell's focus on female qualities of mind and Herschel's sense that Somerville had "lost sight of herself" were both *meant* as compliments, but by linking their definitions of science to the identities of practitioners rather than to the external world to which they turned their attention, both reviews missed the political significance of Somerville's definition of scientific unity. They insisted, obstinately, on the role of identity.[25]

And the inward focus of Herschel and Whewell won the day. Indeed, their victory was so resolute that by 1874 Somerville was recast in their image. In that year, James Clerk-Maxwell opined that the "unity shadowed forth in Mrs Somerville's book is therefore a unity of the method of science, not a unity of the processes of nature." Nevermind that Somerville seemed to say the opposite. The period's focus shifted inward, eventually into the mind itself, and swept her right along with it. Proponents of this approach erased or reworked dissenting views, bringing them into line with the emerging consensus that method was what held the sciences together. For some, unity had been "out there"—either in a universal mind, as the German Romantics held, or in the structure of knowledge, as the followers of Auguste Comte suggested. But it became "in here," a feature of individual minds. No matter how collaborative, science had become something you could do yourself—with the help of a long-lost tool: the hypothesis.[26]

THE SPIRIT OF THE AGE

Signs may have been inauspicious to Carlyle, but not everyone was gloomy. Just two years after his lament, in 1831, another anonymous barometer appeared on the scene, with a far cheerier outlook. A series of articles entitled "The Spirit of the Age," published in five parts over seven issues of a journal with radical sympathies, found hope where Carlyle felt horror. Signed "A.B.," the piece was written by John Stuart Mill, the son of the well-known radical James Mill and presumptive heir to Jeremy Bentham's utilitarianism, of which his father was a vocal champion. Like Carlyle's essay, Mill's reflected a feeling of transition, of being caught up in change. And indeed, when Carlyle first read it, he saw its author as a kindred spirit, another soul unsettled by the precipitous moment in which they lived. But Carlyle was wrong. Unlike his lament about the destructive force of change, Mill's essay foresaw a fusion of past and present. In this, he had less in common with Carlyle's Germans than with Auguste Comte, whose *Cours de Philosophie Positive* was already changing how Mill saw science's potential.[27]

Where Carlyle railed against "the age of machinery" for blunting our mental lives, Mill saw it the other way around: our minds shaped the material world. "A change has taken place in the human mind," Mill announced in the essay's first installment, and that change was destined to alter the course of politics and society. Mill's task was to predict where all this might lead, but after seven installments it remained unfinished. The conclusion of the series called for "a careful survey of the properties which are characteristic of the English national mind, in the present age—for on these the future fate of our country must depend." Written amid the "bustle and tumult" leading up to the passage of the Great Reform Act, this pronouncement was also a promissory note for more work. Mill's idea was that such a study would help everyone grapple with the political changes that were already afoot. "The revolution which had already taken place in the human mind," he concluded, "is rapidly shaping external things to its own form and proportions." This faith in the power of mind grounded much of Mill's later writing, from his *System of Logic* in 1843 to *On Liberty,* which he published in 1859.[28]

Mill's faith made a certain amount of sense. With his father bent on "manufacturing" a new champion of Bentham's views, Mill was taught Greek at age three and political economy as a teenager. His *Autobiography* details his ups and downs, including a mental breakdown in 1826 from which he was still recovering when he wrote "Spirit of the Age." During his recovery, Mill realized that Bentham was not enough on his own and began supplementing his teachings with Romantic impulses he found in Wordsworth and especially Coleridge. From Bentham, Mill took what he called "his *method,*" or "those habits of thought and modes of investigation, which are essential to the idea of science." From Coleridge, he took "the spirit of philosophy," a search for underlying truths. The goal was balance: liberal and conservative, radical and reactionary. "We hold," Mill argued, "that these two sorts of men, who seem to be, and believe themselves to be, enemies, are in reality allies. The powers they wield are opposite poles of one great force of progression." To achieve this, you had to keep both in your mind. "Whoever could master the premisses and combine the methods of both," Mill predicted, "would possess the entire English philosophy of his age."[29]

In the 1830s such synergy seemed distant, if not impossible. Past and present were at war, in Mill's own mind and in the nation's. The "spirit of the age" was discord, not harmony. One way to bring these sides together, Mill thought, was to look to another arena: "While these two contending parties are measuring their sophistries against one another, the man who is capable of other ideas than those of his age, has an example in the present state of physical science, and in the manner in which men shape their thoughts and their actions within its sphere, of what is to be hoped for and laboured for in all other departments of human knowledge; and what, beyond all possibility of doubt, will one day be attained." Unity could be found in the practices with which men of science "shape their thoughts." What it offered was "a compact mass of authority" that ensured stable, progressive growth of the sort those in the political reform movement wanted. He saw in science a solution both to his personal quest for stability and to national progress. No wonder Mill was enchanted by it.[30]

Mill's enchantment had a very specific source. While serializing "Spirit of the Age," he set himself the task of reviewing a new book on the nature of the physical sciences by John Herschel, an acknowledged expert: *Preliminary Discourse on the Study of Natural Philosophy.* Herschel's book appeared in *The Cabinet Cyclopaedia,* an accessible series run by the eccentric editor Dionysius Lardner. The recent success of the Society for the Diffusion of Useful Knowledge and its publishing enterprise, the Library of Useful Knowledge, provided a model for Lardner's own series. Founded in 1826, that Society fit in with other programs of self-education and improvement under the banner of the reform movement. The title of the Library's first volume, by Whig politician Henry Brougham, made the Society's scientific assumptions clear. *Objects, Advantages, and Plea-*

sures of Science framed the nature of science in that order: by the things it studied, the (material) benefits of such study, and the enjoyment to be had in doing it. Science, according to Brougham, "in its most comprehensive sense only means *Knowledge,* and in its ordinary sense means *Knowledge reduced to a System;* that is, arranged in a regular order, so as to be conveniently taught, easily remembered, and readily applied." On this view, science sounded almost material, like something you could reach out and touch.[31]

Starting in the 1830s, however, this materialist sense of science began to change. In a conclusion on the "pleasures of science," Brougham focused on the subjective side of scientific thinking: what it felt like to entertain grand thoughts, and how one might mimic the discovery process oneself. The certainty of scientific results was unconfirmed, he argued, until you learned to "trace the steps by which those doctrines are investigated." Herschel appreciated this idea of "steps," and set out to sketch how individuals working in the physical sciences thought through their problems. He also felt compelled to publish on these more general matters by the public reaction to Babbage's *Decline of Science.* Though Herschel agreed with a lot of what Babbage wrote, the book's tone struck him as so intemperate that, as he told Babbage in a letter, it merited "a good slap in the face." Even if Herschel agreed that the Royal Society was due for reform, he thought that divisive books like Babbage's were more likely to exacerbate existing problems than further the cause of progress.[32]

In his own book Herschel worked hard to avoid the kind of polemical topics and tone to which Babbage gravitated, steering clear of questions of decline or the governance of scientific societies. Instead, Herschel sought to provide an account of ways of thinking that could transcend these differences, one that would ring true to practitioners and make the nature of their practices clear to those without formal training in the sciences. Unity was Herschel's organizing principle, a fact he reiterated toward the end of the book: "Natural philosophy is essentially united in all its departments, through all which one spirit reigns and one method of enquiry applies." Even though the whole must still be separated into parts for analysis, methodological unity was both a useful tool for inculcating a scientific "habit of seeing" and a way to sidestep the hot-button issues of content and governance that were still roiling the sciences under study. The Royal Society was conspicuous by its absence, while an account of the "well constituted mind" took center stage. By replacing the social world of scientific strife with an individual search for stability, Herschel turned the dynamic equilibrium he and others sought into a matter of self-improvement, if not self-help.[33]

Herschel's focus on method, rather than results, is what attracted Mill. His claim about "the revolution which had already taken place in the human mind," about which he was writing when the *Preliminary Discourse* appeared, seemed to gather strength from Herschel's evidence. Mill's review appeared exactly halfway through his "Spirit of the Age" series (also in the *Examiner*) and captured his hopes precisely. "If the utility of the *very* modern physical inquiries were to be estimated solely by the intrinsic value of their *results*," he began, ". . . we know not that the labours of a *savant* would be deserving of much higher commendation than those of a bricklayer." Science aids progress not so much "by the truths which it discloses, but by the process by which it attains to them." Method, not this or that matter of fact, was the mirror that the sciences held up to the mind, the solution to the "Spirit of the Age" Mill was trying to define. What is more, this allowed Mill to turn his lack of familiarity into a strength. Soon after reviewing Herschel, he wrote to a friend that "[if] there is any science which I am capable of promoting, I think it is the science of science itself, the science of investigation—of method." The problem for someone like Mill, without an education in the physical sciences, was to learn enough about science to see how it worked from the inside.[34]

Not that men of science were much better. They were, according to Mill, "usually as little conscious of the methods by which they have made their greatest discoveries, as the clown is of the structure of his eye." Not just that: such self-study was downright suspect. "An inquiry into the nature of the instrument with which they all work, the human mind, and into the mode of bringing that instrument to the greatest perfection, and using it to the greatest advantage, has usually been treated by them as something frivolous and idle." This is what made the *Preliminary Dis-*

course so valuable for Mill: it offered up examples from inside scientific practice, and did so in a way that did not shy away from their mental or cognitive components and limitations. For Herschel, these went hand in hand: his decision to explain science to outsiders, and the role he granted to hypotheses in particular, required him to delve into the delicate balancing act that made those controversial tools safe to use. To do that, Herschel turned the hypothesis into a kind of boundary object, the tipping point between induction and deduction, rather than an engine of wild speculation. In his hands, the hypothesis was no longer (merely) fanciful, foreign, or—worst of all—French.[35]

Herschel's *Preliminary Discourse* domesticated the hypothesis. Good science, according to Herschel, could never be *merely* inductive or deductive. The two approaches depended on one another, he argued, and not just in the long run or as a matter of social equilibrium. For Herschel, the balance between induction and deduction was embodied in individual practitioners and their judicious testing of hypotheses. "The successful process of scientific enquiry," he wrote in the middle of the *Preliminary Discourse,* "demands continually the alternate use of both *inductive* and *deductive* method." It was the latter approach—deduction—that helped natural philosophers *produce* hypotheses and design ways to test them. This was where the spark came from. But such a spark meant nothing without the induction, which ensured that ideas, arrived at in advance, fit with experimental or observational evidence. Every practitioner, indeed every theory, was going to need both modes in order to advance. In including both, Herschel was offering a model of the scientific process that was, in a very specific sense, *neutral.* Science encompassed opposites.[36]

Herschel underscored this neutrality, and undercut potential criticism, by appealing to the very figures whose rules he seemed to be flouting. Bacon's bust appeared on the frontispiece of the *Preliminary Discourse,* and his famous warning about the limits of cognition served as an epigraph: "Man, as the minister and interpreter of nature, is limited in act and understanding by his observation of the order of nature: neither his knowledge nor his power extends farther." If Bacon was seen as authoritative by most readers, Herschel sought to corral that authority for an unlikely cause. Citing Bacon approvingly throughout the text (twenty times in all), Herschel held up his "immortal countryman" as a hero who, along with Galileo, had "dispelled the darkness" of previous ages. Only Newton was mentioned more often (around forty times), and both men were praised for the rigor and caution that had made them famous. Herschel's aim here was a common one: invoking authorities like Bacon and Newton was a standard way for men of science in these years to establish their bona fides, even in the service of radical new theories.[37]

Name-dropping Newton and Bacon was supposed to get Herschel's audience on his side. But it also created a problem. As we have seen, both forbears explicitly precluded hypothesizing, which was precisely the practice Herschel had set out to support. Newton's famous *vera causa* principle—"No more causes of natural things should be admitted than are both true and sufficient to explain their phenomena"—was almost as clear on the issue as his line about not framing hypotheses. Causes, Herschel saw, had to be "recognized as having a real existence in nature, and not [as] mere hypotheses or figments of the mind." He would have to tread carefully. The ghostly specters of guesswork and speculation haunted critiques of deduction and rationality at the time. Indeed, Herschel himself knew all too well the risks entailed by such flights of fancy. If you introduced false hypotheses, for example, they could easily poison subsequent work that depended on their truth. As science expanded, bad hypotheses formed cracks in the foundation. What Herschel needed, in this context, was to come across as staunchly Baconian, as careful and inductive, while making room for hypotheses at the same time. Doing so required a performance of caution that would render his critics more amenable to small risks.[38]

Herschel's plan was to link the hypothesis to a more familiar tool: the analogy. Insisting that analogies were necessary for establishing Newton's *verae causae,* Herschel went on to argue that one could only concoct appropriate analogies by using hypotheses. The line of thought was almost irresistible. If "the analogy of two phenomena be very close and striking, while, at the same time, the cause of one is very obvious, it becomes scarcely possible to refuse to admit the action of an analogous cause in the other, though not so obvious in itself." Through a clever rereading of Newton, Herschel made analogies the essence of induction. Linking areas of thought that had previously seemed separate required spotting unseen connections—which, by definition, required you to take a leap: that is, to hypothesize. Hypotheses, for Herschel, "afford us motives for searching into analogies; grounds of citation to bring before us all the cases which seem to bear up on them, for examination." They were, in other words, "a scaffold for the erection of general laws," a kind of mortar shoring up the edifice of science. "Regarded in this light," Herschel concluded, "hypotheses have often an eminent use: and a facility in framing them, if attended with an equal facility in laying them aside when they have served their turn, is one of the most valuable qualities a philosopher can possess; while on the other hand, a bigoted adherence to them, or indeed to peculiar views of any kind, in opposition to the tenor of facts as they arise, is the bane of all philosophy." As Herschel saw it, "the disposition of the mind to form hypotheses, and to prejudge cases" was unavoidable. Rather than try to stamp out this basic human element, he urged an acknowledgment of hypotheses—in order to control them.[39]

Once again, balance was key. Rather than *either* "examining all the cases" *or* "forming at once a bold hypothesis," Herschel suggested we do both. Good science was "a process partaking of both of these, and combining the advantages of both without their defects." On the one hand, such evenhandedness seemed to render the origins of hypotheses moot. If a theory explains the facts, "it matters little how it has been originally framed. However strange and, at first sight, inadmissible its postulates may appear," all that mattered was their explanatory power. Or, as Herschel put it when explaining inductive leaps: "In the study of nature, we must not, therefore, be scrupulous as to *how* we reach to a knowledge of such general facts: provided only we verify them carefully once detected, we must be content to seize them wherever they are to be found." On the other hand, hypothesizing was risky, and had to be done with care. Herschel rejected the idea that, in "the formation of theories, we are abandoned to the unrestrained exercise of imagination, or at liberty to lay down arbitrary principles, or assume the existence of mere fanciful causes. The liberty of speculation which we possess in the domains of theory is not like the wild licence of the slave broke loose from his fetters, but rather like that of the freeman who has learned the lessons of self-restraint in the school of just subordination." It was a grim metaphor, and a telling one. As a centrist Whig, Herschel supported the emancipation movement that finally succeeded in 1833—but only up to a point. Young children were immediately free, but anyone older was renamed "an apprentice." It was another decade before this exception was ended. Compromises like these were the bread and butter of Whig politics in the 1830s, which helped the party build coalitions but also gives Herschel's choice of image a chilling specificity that would not have been lost on his eager readers in the 1830s.[40]

Whatever else it implied, Herschel's troubling metaphor did *not* endorse a division of labor like those envisioned by Babbage or Carlyle. If we take him at his word, Herschel's "school of just subordination" was an *internal* one, enacted by the individual mind on itself. The dialectic of speculation and control was meant to be private. This is not to say that science was antisocial: much the opposite. According to Herschel, "there is no body of knowledge so complete, but that it may acquire accession, or so free from error but that it may receive correction in passing through the minds of millions." In the language of the liberal platitudes of his time, progress emerged from the coordination between individuals. But ultimately the minds involved were separate, the idea being that each person can and should use the same methods and observe the same phenomena as Herschel's men of science. As they did, thinkers would be reproducing, in the individual equilibrium of guessing and checking, the social order so many of them were seeking. When Mill pointed to unity in the physical sciences, this mirroring of individual and social dimensions was precisely what he had in mind. Society was not a big machine, full of interlocking parts with different roles to play. Rather, every individual used the same method, with coordination occurring both inside and outside. Political stability was not a matter of social opposition; it was achieved when individuals incorporated opposition and did the work of equilibrium themselves. This internalization was what science was all about.[41]

The emphasis on individual balance in the *Preliminary Discourse* is what caught Mill's eye—and he was not alone. Herschel's friend Whewell applauded exactly the same thing in his own anonymous review of the book. According to Whewell, Herschel's authority to speak on the matter resided in his formal scientific accomplishments, which made him exceptional in English (as opposed to Continental) discussions of science. "Having been thoroughly disciplined in all that can form the character of an accomplished man of science," Whewell wrote, Herschel was the perfect man to start a British conversation about the methods of science. For "while volumes upon volumes have been written upon the nature of human knowledge and the laws of human thought," no one had focused on "those mental processes which have been exemplified in the progress of modern science—the most undisputed and the most admired assemblage of truths which has ever yet been obtained." Whether or not this claim of originality was true—other works already pointed in this direction by Herschel's day—Whewell's judgment was generally accepted. As a result, a book Herschel had originally written for a general audience became a founding document in the expert study of the history and philosophy of science.[42]

Whewell dwelt at length on Herschel's attempt to balance induction and deduction. It was this "combination of induction and deduction, the complete cycle of ascent and descent, which forms the entire scheme of a perfect science, from its first origin to its final development." With this balance as an ideal, Whewell set out to test existing sciences to see whether or not they lived up to it. Results were mixed: botany and chemistry leaned inductive, whereas mechanics and astronomy had moved on from induction to deduction. The hinge between the two, in both the individual mind and in the development of sciences over time, was the hypothesis. Herschel, in a passage Whewell quoted approvingly, put it this way: "Such is the tendency of the human mind to speculation, that on the least idea of an analogy between a few phenomena, it leaps forward, as it were, to a cause or law, to the temporary neglect of all the rest." Like Herschel, Whewell then echoed the usual Baconian warning about the dangers of "anticipation"—before defending it, at least for mature sciences. These were safe, Whewell argued, because "the necessity of verifying our generalizations, or of abandoning them, is irresistibly felt and acted upon." In other words, the "spontaneous impulse" to hypothesize was checked in the "school of just subordination." The result was that Herschel's hypotheses were naturalized as a function of the ordinary human mind. Performed in the right setting, and constrained in the right ways, this natural mode of projecting forward in time was fundamental to scientific thinking. Like hypotheses, science could be read as a natural response to external conditions—which secured it new forms of authority in time.[43]

REFLEX SCIENCES

Knowing how science worked had not always depended upon, or even related to, how mental life evolved. Nor does it still. Many scientists and philosophers today would argue that the imperfect, adaptive mechanisms of the human mind fail to account for what makes science special. Turning science into embodied cognition meant naturalizing what had seemed before, and has seemed since, like the ultimate artifice: human reason. Changes in the study of the mind and the study of science converged in this emerging naturalistic account of scientific thinking, with psychology assuming the role of explaining scientific progress. And the admission of hypotheses was a key step in this transformation. Hypotheses explained how theories exceeded evidence and yet stayed true to it—they were the justified leaps required to make any sort of advance. This was as true of accounts within science as it was of accounts *of* science. If science was mental, it was possible to study the mind in order to study science. And if the mind was a part of nature, then the study of nature was a natural process; scientific scrutiny of the mind put the very method being used under the microscope. The reflexive potential of these changes was, and is, almost dizzying. The terms start to tighten on themselves as they are used to explain more and more.[44]

Hints of such things were rare in Herschel's book, with its steady focus on physical science. But the few suggestions that remained excited his acolytes. Mill, for one, concluded his review of the *Preliminary Discourse* by quoting a long passage that hinted at new applications of the scientific method he described. "The successful results of our experiments and reasonings in natural philosophy," Herschel wrote, "tend of necessity to impress something of the well weighed and progressive character of science on the more complicated conduct of our social and moral relations." This view of the social impact of natural philosophy led to an optimistic view of science's more general effects: "It is thus that legislation and politics become gradually regarded as experimental sciences; and history, not, as formerly, the mere record of tyrannies and slaughters, which, by immortalizing the execrable actions of one age, perpetuates the ambition of committing them in every succeeding one, but as the archive of experiments, successful and unsuccessful, gradually accumulating towards the solution of the grand problem—how the advantages of government are to be secured with the least possible inconvenience to the governed." Mill's excitement about this passage is understandable given his emphasis on reform. It was this same eagerness about future prospects that led Mill to cut off his "Spirit of the Age" series early in order to engage in reformist politics directly. Though Herschel steered clear of such engagement in the *Preliminary Discourse,* Mill was able to turn it to political account.[45]

Something similar happened to Whewell, who saw his review of Herschel as a chance to finally get his ideas on method into print. In a letter to his friend Richard Jones, he called his review "as good an attempt as I could make to get *the people* into a right way of thinking about induction." That "right way of thinking" was to see hypotheses as central to induction, rather than anathema to it. Whewell was even more intent than Herschel on locating hypotheses in the functions of the mind. Doubling down on the idea that speculative thinking came naturally, Whewell argued that the organic occurrence of hypothesizing meant it was best to channel rather than extinguish the impulse. Like Mill, Whewell concluded with a lengthy extract from Herschel, one that framed science's advantages in mental terms: "By cherishing as a vital principle an unbounded spirit of inquiry and ardency of expectation, [science] unfetters the mind from prejudices of every kind, and leaves it open and free to every impression of a higher nature, which it is susceptible of receiving, guarding only against enthusiasm and self-deception by a habit of strict investigation, but encouraging, rather than suppressing, everything that can offer a prospect or a hope beyond the present obscure and unsatisfactory state." On this view, science is a product of minds that are unfettered *and* self-disciplined; or rather, it is the process of give-and-take that unfurls "boundless views of intellectual and moral, as well as material, relations."[46]

These "boundless views" hint at a further aim of Whewell's, one that went beyond a new description of scientific inquiry toward a prescription for other forms of work. In the same letter in which he had imagined a new account of induction, he suggested a further goal: "a popular exposition of the matter applied mainly to moral, political and other notional sciences." Like Mill, Whewell planned to use his review of the *Preliminary Discourse* as a platform from which to announce the political and social consequences of methodology. Unlike Mill, he did not intend to rush off and join the reform movement (much less Parliament). A decade Mill's senior, born not into intellectual royalty but to a Lancaster carpenter, Whewell's eye was on making a name for himself at Trinity College, Cambridge, where he would soon take up a professorship in moral philosophy. If Mill was exploring boundaries between philosophy and politics in a very literal sense, Whewell was after something different: an account of the sciences that was both unified and plural. The politics of Whewell's vision were more academic, centered on the BAAS and its need for a common thread as well as his own position at the margins of a set of disciplines that were slowly drifting apart. For Mill and Whewell, method was a common cause that solved very different problems and anchored political projects that were increasingly at odds.[47]

Whewell's project came in two parts, each embodied in a different book. His *History of the Inductive Sciences,* published in three volumes in 1837, was followed up three years later by a two-volume *Philosophy of the Inductive Sciences.* Though it was not the "popular exposition" he had floated in 1834—both sides of the project were quite technical—Whewell's books carved out a space where not only history and philosophy but also logic, psychology, and even social analysis of science could complement ongoing efforts in the sciences themselves. Though he had published on a variety of scientific topics, including minerology and what he called "tidology," Whewell was anxious about his status in a world of increasing specialization. Already in 1818, he had described himself in a letter to Herschel as one of the "lookers on, who, not making a single experiment to further the progress of science, employ ourselves with twisting the results of other people into all possible speculations mathematical, physical, and metaphysical." His books were meant to make room for a kind of outsider-specialist, someone who could turn a wide view of the sciences into a scientific project of its own.[48]

It was not always easy. By 1840 Whewell came to see even his crowning achievements as hindrances. "My *History* and *Philosophy* of Science," he wrote by way of refusing to lead the BAAS, "are disqualifications, not qualifications, for my being put at the head of the scientific world." Those members who "laboured hard in special fields" would feel "indignant at having a person put at their head, recommended only by what they think vague and false general views." And certain critics agreed. One even argued that the *Philosophy* proved Whewell's "singular want of acquaintance" with the sciences which it claimed to explain. Though others would come to see Whewell's work as scientific in just the sense to which Whewell aspired, his position as an enthusiastic outsider both enabled and constrained his participation in debates over how best to define "science," much less practice it. The idea of studying science, *qua* science, as an object in its own right was far from an obvious or even permissible pursuit in the 1830s.[49]

Yet pursue it Whewell did, from this opening salvo to the end of his career. Part of what emboldened him to do so, and made some of his contributions counter-intuitive or seemingly out of step with his moment, was his obvious debt to the critical philosophy of Immanuel Kant. From his earliest sketches of a scientific philosophy, Whewell sought science in the inner workings of the mind. In 1832, for example, Whewell defined scientific thinking in terms derived from Kant and at odds with British orthodoxy: "Science is formed by acts of mind, in addition to which there may or may not be employed, as materials, special facts given by experience." To a good Baconian, such "facts given by experience" were all science was—the idea of "acts of mind" was potentially problematic. From early on, Whewell was formulating what he would come to dub a "fundamental antithesis" between facts and ideas as central to scientific practice. Facts were the products of experience and observation, while ideas emerged from the mind at work. Interactions between the two, according to Whewell, "give rise to scientific truths." As he put it in the *Philosophy:* "Inductive truth is never the mere *sum* of the facts. It is made into something more by the introduction of a new mental element; and the mind, in order to be able to supply this element, must have peculiar endowments and discipline." Science was *mental.* And not only that. Different fields were anchored by different "fundamental ideas," which meant the sciences could be organized by the mental constructs on which they depended. Geometry, for example, grew out of the idea of space. The mind gave rise to the diverse pursuits that made up modern science. By studying the mind, you could account for what gave science its special power.[50]

Whewell was no Whig, but he cast his philosophy in terms that would have been familiar to them. For both, reform was a way of avoiding revolution. This was true for individuals and for society, for one mind and for a discipline. Writing about science, Whewell argued that progress was never "formed by one single act" or "completed by the discovery of one great principle." Rather, science was "a long-continued advance; a series of changes; a repeated progress from one principle to another, different and often apparently contradictory." Older stages were "taken up into the subsequent doctrines, and included in them." In the end, "the history of each science, which may thus appear like a succession of revolutions, is, in reality, a series of developments." Through a "series of transformations," every theory was "penetrated, infiltrated, and metamorphosed by the surrounding medium of truth, before the merely arbitrary and erroneous residuum has been finally ejected out of the body of permanent and certain knowledge." Bizarre imagery aside, Whewell was adapting the gradualist rhetoric of his reformist contemporaries in order to justify including hypotheses (even *erroneous* hypotheses) in science. As far back as his review of Herschel, he had insisted that the slow back-and-forth of hypotheses and testing was a natural feature of the mind and the normative heart of scientific practice.[51]

Intriguingly, Whewell's *History* and *Philosophy* were not the whole project—or were not supposed to be. His plans make it clear that there was to be a third part, one that pointed forward into the future. "Book 3," as he put it in a letter, would address "*Prospects* of Inductive Science. The question of the possibility and method of applying Inductive processes, as illustrated in the philosophy of Book 2, to other than material sciences; as philology, art, politics, and morals." Whewell soon pushed this further, saying that he planned "to try to discover the nature of the analogy which exists between these sciences and our knowledge respecting morals, taste, politics, language, and generally, all hyperphysical knowledge." This was a tall order, he admitted, but lucky for him he had built a justification for this way of thinking into the theory that would make up these works: "I must take advantage of my own philosophy, which, as it points out that all knowledge comes by induction and that induction is guessing, allows us to publish guesses, acknowledgedly imperfect, as contributions to knowledge. In short, I shall write reviews of the progress of morals, taste and the rest, much as any other critic would, only keeping in view the analogy of the true type of the progress of science." Whenever we do science, Whewell said, we guess. And if we guess when we are conducting scientific inquiry, what is stopping us from guessing *about* science—including its next steps?[52]

Guess Whewell did—but not at first. He had begun with sincere doubts about whether a method gleaned from the physical sciences would be applicable to these other areas. The same year in which he reviewed Herschel and began to map out his larger project, he confided to Jones that he had "no removal yet of my doubts as to the identity of the scientific method (that is, the method of making a science) in physical and moral sciences." There was something separating these sciences, a gap that prohibited simply transferring lessons from one field to another. "Every attempt to build up a new science by the application of principles which belong to an old one," he later insisted, "will lead to frivolous and barren speculation." Avoiding this fate led Whewell to reflect, historically, on "the process by which Science is formed" and "the method by which scientific discoveries have really been made." Privately, he referred to such studies as "the reflex sciences," inviting science to turn inward and, through reflection, identify common principles to be extended. This was, he went on, "the *usual* method by which men have arrived at the principles" of inquiry. Seeking a tool to speed up science's advance, he turned to the sciences themselves for the resources with which to accelerate their progress.[53]

Even in his *History*, we find Whewell looking ahead, seeing the roots of future research in past and present sciences. In the chapter "Physiology," for example, he asserted that recent studies of the nervous system had pushed the field up against its own limits. "In tracing the phenomena of sensation and volition to their cause," he wrote, "it is clear that we must call in some peculiar and hyperphysical principle." Here we see echoes of his earlier reference to "hyperphysical knowledge," where the terms of philosophy collide with the scientific practices they describe. This collision was precisely what Whewell was driving at:

> We have here to go from nouns to pronouns, from things to persons. We pass from the body to the soul, from physics to metaphysics. We are come to the borders of material philosophy; the next step is into the domain of thought and mind. Here, therefore, we begin to feel that we have reached the boundaries of our present subject. The examination of that which lies beyond them must be reserved for a philosophy of another kind, and for the labours of the future; if we are ever enabled to make the attempt to extend into that loftier and wider scene, the principles which we gather on the ground we are now laboriously treading.

Referring back to this discussion in his *Philosophy*, Whewell remarked on "how irresistibly we are led by physiological researches into the domain of thought and mind," how biological questions seem to prompt psychological ones. The "reflex sciences" he imagined were just that: reflexive, unconscious, an unavoidable by-product of organic scientific development.[54]

By projecting forward, even as he looked back, Whewell imagined not only new subjects for scientific inquiry but a new account of the mind as well. The *History* ends with a move "from matter to mind, from the external to the internal world." This inward turn inaugurated fields that Whewell variously called "immaterial," "notional," "hyperphysical," "moral," and "reflex" sciences. Physics and psychology blurred together, though their methods "must assume different aspects in cases where a mere contemplation of external objects is concerned, and where our own internal world of thought, feeling, and will supplies the matter of our speculations." What history revealed was "a unity and harmony throughout all the possible employments of our minds." This elision of external and internal modes of study became an obsession for Whewell, eager as he was to claim for his grand project the imprimatur of scientific status. He spent much of the rest of his career defending his view of the value of hypotheses against a variety of enemies.[55]

Enemy number one was none other than John Stuart Mill, with whom he clashed publicly from the moment of the publication of Mill's *System of Logic* in 1843 until Whewell's death in 1866. The back-and-forth of two of England's most famous philosophers, soon known as the "Mill-Whewell Debates," fit into the larger political clashes of the period. According to Laura Snyder, both Mill and Whewell were intent on "reforming philosophy"—they just disagreed, in profound ways, about how to do it. Each worked criticism of the other into ongoing projects, intent on defending his personal understanding of induction as the way forward both for British science and for social and political life. Although it is true that Mill was the radical and Whewell the conservative, the two had much more in common than they or their champions let on. And indeed, the result of their conflict—it would be too much to call it a "resolution"—was a focus on the methods of science that borrowed from both views. Like many debates, the polemical statement of their disagreement masked an underlying consensus, one that was broader than their two particular philosophical projects but was cemented, paradoxically, by serving as the foundation of their more spectacular statements about the errors and even bad faith of the other.[56]

Mill's main line of attack against Whewell was to cast him as an "intuitionist"—someone who appealed to innate ideas as a means of justifying the status quo, rather than appealing, as Mill did, to experiential learning as a means of achieving progress. In framing Whewell in these terms, Mill misrepresented matters in ways that seem both intentional and accidental. Whewell's "Fundamental Ideas" were innate, to be sure, but they were not antisocial. These ideas needed to be developed through a socio-individual process that Whewell called "explication" before they could be used to organize (or, as Whewell put it, "colligate") scientific facts. Whewell's ideas were not intuitive in any simple sense, but were instead products of the confrontation between the mind and the world—a view not all that distinct from the one Mill explicitly opposed to it, wherein science is a product of shared human experiences. Whewell, for his part, accused Mill of cherry-picking his examples and of a general lack of familiarity with the scientific practices on which he drew for his evidence. While there was no love lost between the two, close inspection reveals more commonalities than they would ever have admitted.[57]

Where Mill and Whewell did diverge was on the origins of the mental elements that organize sensory evidence. Mill was an adherent of what was called "associationism," a view of the mind dating to John Locke and of which Mill's father was the period's best-known champion. For associationists, discrete ideas "associate" to form more complex ideas and the train of one's thoughts generally. Grounded in Locke's image of a "blank slate," the theory proposed that the senses wrote onto the mind ideas that were subsequently associated. James Mill's *Analysis of the Phenomena of the Human Mind* had appeared in 1829, and although the younger Mill disagreed with certain aspects of his father's argument, he worked hard to defend it against the book's many critics. Whewell disagreed. To him, associationism might explain what happened to ideas once we have them, but it did not (and *could* not) tell us where ideas came from in the first place. While Whewell did not ignore the problem of organizing facts, he was committed to articulating the origins of ideas outside of sensory experience, not least because those ideas gave shape to separate scientific fields. Whether or not this view is best called "intuitionism," it was distinct from the associationism touted by the elder Mill and marshaled against Whewell by his son. But associationism's dominance in the era meant that Whewell was fighting an uphill battle, which helps explain the tenor of the disagreement on both sides.[58]

On the surface, the dispute between Mill and Whewell was about induction. But at a deeper level it was also a debate over the nature of the mind and the politics of method. At issue was the origin of our ideas, including scientific ideas: where they came from and what kinds of people could produce or use them. For associationists like the Mills, the origin of every idea was in our senses, which meant they were open to everyone. As James Mill put it: "Our ideas spring up, or exist, in the order in which the sensations existed, of which they are the copies." With five senses and a brain for processing their data, you had what it took, for an associationist, to be a full participant in scientific discovery and political discourse. The "associations" from which the theory got its name had to do with the order and relations of your sensations. Our ideas associate with ideas that occur before, after, or alongside them. If associations like these were the key to ideas of every sort, then education amounted to learning and reinforcing the proper associations. This was a radical theory of the mind, one that held out the promise of empowerment for anyone willing to put in the time (or, in Mill's own case, to be guided from a young age) to develop a set of associations from which to draw. The blank slate was, in this sense, emancipatory.[59]

But it was also fleeting. Even though Mill seemed to defeat Whewell in their celebrated debates, and thus earned the right to bestow on posterity his own definition of induction, the associationism he had inherited from his father was soon on the decline. This is not to say that Whewell's own theory of mind was winning out. Instead, new accounts of the mind, including others adapted from Romantic notions of imagination, rose to challenge associationism in British and European psychology. One competitor was phrenology, which purported to map the mental "faculties" onto the physical body, specifically the brain and its visible case, the skull. Britain's phrenological bible—George Combe's *The Constitution of Man in Relation to External Objects*—was first published in 1828 and became one of the nineteenth century's best-selling books. Combe's phrenology was progressive, but it also outlined the limits of education. Like Whewell's view of science, Combe's account combined the possibility of achievement with an acknowledgment of received nature, as embodied in the brain. There was room to progress at the individual level, but its boundaries were set by the size and shape of one's skull. Despite such warnings, Combe's popularity stemmed from what phrenology shared with associationism: the fact that it lent scientific authority to the aspiration for personal improvement.[60]

Phrenology's own fortunes were limited, however. Responses from established men of science were swift and withering. Already by the 1840s, phrenology was "a popular science" in the backhanded sense in which that phrase is still used today. The fate of associationism was thus sealed less by a direct theoretical challenge, given phrenology's travails, and more by issues with its preferred method: reflecting on one's own mind. Gradually, over the nineteenth century, this practice of introspection gave way to a set of experimental procedures most often associated with the laboratories of the research universities in Germany. And rightly so. One of the consequences of Kant's critical philosophy was the separation of psychology from philosophy, on the grounds that the former could study the minute phenomena of the mind and leave the latter to address the larger questions related to truth and falsehood. Disputes among German practitioners over these boundaries and the methodologies appropriate to specific studies helped give rise to a new discourse of objectivity within the mind sciences. The data on which Mill and Whewell had relied, more than their specific theoretical commitments or their emphasis on method, was what paved the way for new developments in the human sciences later in the century.[61]

Crucially, German experimentalism's rise to international prominence largely postdated these British debates over the nature of mind and method. The back-and-forth between Mill and Whewell reveals that both men sought data by other means in order to support their positions. Recall Mill's exasperation at the inability of men of science to reflect on their own methods, or Whewell's efforts to define a role for those who were drawn more to broad engagement than specific studies. Shifting social organization and scientific specialization led Mill and Whewell to turn their relative lack of distinction in the sciences into a new foundation for the pursuit of philosophy. For Mill, this meant a sweeping survey of science—including, to Whewell's chagrin, relatively recent fields whose scientific status the latter doubted. Historical analysis served the same role for Whewell, who saw in past accomplishments evidence of the sort a present-day practitioner might encounter through introspection. Both men focused on a familiar subject—science—but their ways of doing so modeled a new form of scientific reflection, new ways of thinking about science and its method. Despite their differences, they both sought a living account of the scientific mind at work—what Whewell had called "the reflex sciences." Mill and Whewell, through separate projects and divergent politics, set those sciences in motion in ways neither would have predicted.

WHIG INTERPRETATIONS OF SCIENCE

Until recently, "whiggish" was the worst thing you could call a historian of science. This was an ironic turn for the political term, not least because it was the capital-W Whigs themselves who put something like the modern history of science on the map. Starting in the early twentieth century, however, calling a historian "whiggish" was a way of calling out their presentism, their reliance on an unacknowledged assumption about how history—and especially the history of science—worked. We owe "whiggish," as an insult, to the historian Herbert Butterfield, whose *The Whig Interpretation of History* introduced its negative meaning in 1931. Whiggish historians, according to Butterfield, were guilty of trying to "produce a story which is the ratification if not the glorification of the present." It was no secret whose sins Butterfield had in mind: Whig historians like Macaulay and Mackintosh, whose approach had already come in for criticism along these lines. By giving political histories like these a name, Butterfield was simply shoring up a widely held assumption that knowledge should be apolitical, unbiased, pure. The irony is this: A century earlier, in the period Butterfield critiqued, it was becoming clear that such knowledge was a fantasy. All thinking was for a purpose, all methods have a politics. Science was not apolitical; it was (and remains) deeply political, but in a very specific sense. At least initially, methodologists sought to *incorporate* society, not keep it at bay.[62]

As we have seen, this politics of method aligned with the platform of the Whig party. It was an age of reform, though not all Whigs were as fond of science as this limited cohort of men of science were, nor were all these scientific critics Whigs. The overlaps between the two are telling, but the overall effect of analyzing the emergence of methodology in the 1830s against the backdrop of parliamentary reforms and the expansion of the franchise (among other shifts) is one of family resemblance rather than identity. In some movements, such as calls to reform British universities, the two worlds came closer together. But even where they did not, an emphasis on progress tempered by a respect for tradition (or, as their critics would put it, having it both ways) characterized political and scientific reforms across a wide range of contexts. The champions of the movement recast both politics and science, far into the past and through into the future, in a set of terms they shared with one another—sometimes consciously, sometimes not. More than a zeitgeist (though some saw it that way), this was a matter of specific problems being addressed with materials that were close to hand. In the 1830s, among those engaged in the sciences and their reform, those materials often came in the form of hypotheses and the language of equilibrium.

Whewell was not a Whig, but he was whiggish in just this sense. Much like those Butterfield called out a century later, Whewell's parliamentary leanings are less important than a general view of progress he shared with his would-be opponents. He opposed certain issues that were central to reformist Whig politics in the 1830s, but he saw in science and its method a means of achieving a balance not unlike the one Whigs were advocating. Indeed, Whewell's views of science were ecumenical enough that a critic called his definition of induction "the whole box of tools," a permissive approach that belied his curmudgeonly political leanings. This permissiveness, the willingness to see science as a stabilizing force, held Babbage and Carlyle, Herschel and Somerville, Mill and Whewell together. All saw scientific thinking as a potential source of harmony and progress in society. Though they disagreed about specific policies and quarreled over basic features of human cognition and the history of science, these critics helped make method central to how scientific progress was imagined. Those whiggish interpretations of science were articulated by Whigs and by Tories, by those promoting parliamentary reform and those opposing it. The return of hypotheses and the celebration of balance were, like Macaulay's Whig histories, responses to what many felt were the inevitable transitions of the age.[63]

In the same year in which he published his *History,* Whewell came out forcefully against calls to change the university system. His argument against reform drew an analogy between political and intellectual life that would have been right at home in the arguments of those who were for it. Whewell pilloried philosophical teaching, for example, as too revolutionary. Too often, he wrote, "the old system is refuted; a new one is erected, to last its little hour, and wait its certain doom, like its predecessor. There is nothing old, nothing stable, nothing certain, in this kind of study." The student thus "lives among changes, and has not the heart to labour patiently for treasures that may be ravished from him by the next revolution." Philosophy in Germany was the worst, where it was "as unfavorable to the intellectual welfare of its students, as the condition of the most unstable government of the East is, to the material prosperity of its subjects." How could anyone, subjected to such an education, "cultivate his own thoughts, and possess in a tranquil and even spirit the knowledge and the habits of mind which he has acquired?" Whewell inferred, hyperbolically, that pursuing such a system would put "in serious and extensive jeopardy the interests of the civilisation of England and of the world."[64]

But England was different. Its "admirable combination of active and original intellect" and "unequalled practical sagacity and force of character" meant, according to Whewell, that England "constantly impelled the progress of thought and of institutions in Europe, while at the same time, she has held back from the extravagances and atrocities to which the progressive impulse has urged more unbalanced nations." He used a brief, somewhat spurious history to make his case:

> The bold and vigorous metaphysicians of England first put in action the speculative movement of modern Germany; but England refuses to follow the wild onward whirl of system after system, to which this movement has led. English teachers of political freedom, and the free institutions of England, called up the spirit which has broken the bonds of more than one despotism; but England (thank God!) was never hurried into the democratic madness of her nearest neighbor. England had a Reformation of religion without the abolition of her Church; she had a Revolution of dynasty without the destruction of her Loyalty; she has had a Reform of her Parliament, we trust, without any fatal wrench to her Constitution.

In philosophy and politics, in science and religion, England adopted a third way. Rejecting systems in favor of tools, emphasizing reform instead of revolution—in this, if not in much else, a wide range of British authors, including Whewell, came to seem like scientific whigs.[65]

In predicting what came next, however, the scientific critics diverged. Whewell defended the status quo while Mill demanded further reforms; some underlined science's reform potential while others followed specialization away from public engagement into cloistered research. The authority of the BAAS increased over the ensuing decades, but unity among its sections and its integration into British life remained concerns. The causes of this fracturing were complex, with opposing sides each claiming a victory of foresight. Carlyle's anxieties about mechanization's effects came true, yet so did Babbage's prediction that machines would not replace human labor. But even as these critics were talking past each other—as, to a certain extent, critics always do—their visions of science shared as much as they split. In particular, their accounts inaugurated a discourse, if not a reality, of unified method that shaped generations of scientific practitioners. It is easy to see the discursive divisions in this age of scientific criticism, but it is just as important to identify their common features, the methodological center that increasingly held.

Two immediate effects of these years stood out: the twin convictions that science was a product of actual minds and that probing those minds would shed light on science. Studying science in mental terms and using science to study the mind were the concrete consequences of this first methodological moment—but the two need not have gone together. Science *might* have persisted as a philosophical topic, but the advent of psychology turned it into a scientific object of its own. The tools used to study it—the tools of psychology—might have been captive to a narrow meaning of psychology; that is, to the laboratory measurement of senses and observable phenomena. But psychology was not so limited, at least not for everyone. Averting these fates was partly the result of a new way of viewing both mind and method in *evolutionary* terms. But it is crucial to see that the goals of those who achieved this adaptation were different from the goals of those on whose work they drew. Turning science into hypothesis testing solved one problem in the age of methods; adapting that method into a theory of the natural world solved another. This history was not a smooth journey from one end to the other, as we shall see. When Charles Darwin turned the politics of method into a theory of evolution, he did so for reasons—and with consequences—that set him and his moment apart from what had preceded them.

3

NATURE'S METHOD

In the spring of 1831, Charles Darwin found himself with free time on his hands. Three years earlier, he had "gone up" to Cambridge late, which meant he now had to wait a few months in order to receive his degree after completing his exams in January. Darwin passed the time in relative leisure, riding horses and collecting insects like he had throughout his college years. But some things did change. For one, he grew closer with his mentor John Stevens Henslow, a professor of botany who had fueled Darwin's passion for natural history through both lectures and private conversation. Becoming "the man who walks with Henslow" opened doors for Darwin, and invitations to parties provided a window into the scientific life and how to live it. Social opportunities came along with a new intellectual freedom. No longer under the pressure of formal study—such as it was, given that he was never a very dedicated student—Darwin could turn his extracurricular passion for the naturalist tradition into a reading program as well. Taking up new books along with old favorites, he began to imagine a future as a scientific practitioner. For the short term, Darwin envisioned tropical travels in pursuit of new objects for his collections; in the longer term, he foresaw a life in the church.

A country parsonage could provide the space and time for pursuing his passion: collecting and describing the natural world.[1]

But it was not all parties and books. At Henslow's urging, Darwin soon put into practice the approach he had been reading about in John Herschel's *Preliminary Discourse,* Alexander von Humboldt's *Personal Narrative,* and William Paley's *Natural Theology.* Geology seemed the most promising path forward, and Henslow "crammed" Darwin on its concrete components: how to use instruments, what fieldwork might be like. But Henslow was a botanist, and to really learn the tricks of the trade Darwin would need a new mentor. Happily, the geologist Adam Sedgwick agreed to let Darwin tag along that summer as he set out to rework the geological maps of Wales. Though Sedgwick disagreed with the views that Darwin would eventually pick up from Charles Lyell's *Principles of Geology,* he gave him an education that proved essential once Darwin struck out on his own in the field. From the skills needed for surveying landscapes to the confidence required to make inferences about what lay beneath them, Sedgwick equipped Darwin with a practical toolkit he would soon take with him into contexts he could only imagine. When, toward the end of their time together, an opportunity arose to join HMS *Beagle* on a trip around the world, Darwin was ready. Though not yet "a *finished* Naturalist," as Henslow put it, Darwin was poised to benefit immensely from just such a voyage.[2]

The trip's impact on Darwin is famous—but often exaggerated. We know, for example, that there was no "Eureka" moment in the Galápagos, no single bird or plant that revealed how the world had evolved to the young naturalist. Darwin's views on most things remained constant during his years at sea. Yes, he gave up the prospect of a parsonage (owing to his declining belief) while he was away. But upon his return his ideas about how a man of science should think echoed Herschel, Humboldt, and even Paley as much as they had when he embarked. Indeed, as Darwin turned his attention to the specimens he had gathered while away—farming many out to experts as he drafted a contribution to the voyage's official publication—he leaned on his old textual exemplars more than ever. These books were models for how to handle evidence, how to publish, and, increasingly, how to buttress a burgeoning itch to theorize about all the things he had observed. The notebooks in which Darwin scribbled his ideas begin to reveal a man of science on the make, his efforts in geology informed by a deep attention to the methodological strictures of his day. As he began participating in the scientific community he had glimpsed during his days walking with Henslow, Darwin quickly turned to scientific mentors and manuals for guidance on how to think theoretically and rigorously at once.[3]

Just as it had been for his mentors, Darwin's attraction to methodology was about more than getting his facts straight. He found himself drawn to Herschel's hypothetical method, which he saw in both Sedgwick's geological lessons and Lyell's *Principles* (as well as in the letters the two began exchanging). As Alistair Sponsel has shown, Darwin's anxieties during and after the *Beagle* voyage centered on his status as a theorizer and the instability of that status in an age so obsessed with Bacon's anti-hypothetical virtues. Self-consciousness about method led Darwin to seek out patrons and patterns: first for his theory of coral reefs, and only later for his speculations about transmutation. Darwin was looking for a way to safeguard himself against the withering criticism he knew men of science were capable of. It was this simultaneous interest in theorizing about the data he had gathered and desire to protect his reputation that made Darwin a careful student of scientific methodology. As prevailing ideas about how science should work shifted from orthodox induction toward tolerance of hypotheses, Darwin gradually gained confidence in his own fascination with the changes he was observing and his ability to say something new about them. It was a precarious time to take up such inquiries—but an exciting one, too. Method was the terrain on which disputes over evolution and extinction were fought, the language in which Darwin levied his sternest critiques and the source of the worst rebukes he received in turn. A lot was at stake in methodology, in how you thought you thought.[4]

This methodological engagement rubbed off on Darwin in some unexpected ways. For one, it led him down certain experimental rabbit-holes that seem obscure in hindsight. His five-year study of barnacles, his careful attention to breeding, and his eventual fascination with worms and mold were not just attempts to shore up his speculative theory of species change. They also index a desire to adhere to the methodological precepts that were shifting around him. Another effect of this methodological fixation, both subtler and more profound, was that Darwin came to see life itself as *scientific.* The theory he eventually formulated—evolution by natural selection—seemed a lot like the science he was trying to practice. Natural selection was nature's *method* for making species. This analogy between science and nature rested, in part, on the tradition of natural theology, the search for God's purpose and, indeed, God's *mind,* in the material world. When Darwin famously likened evolution to how human breeders "selected" animals and plants, he was drawing on a deeper likeness according to which both nature and humans were imbued with a power previously reserved for the creator. It was this ability to see mind in nature that paved the way for what historians have called "Darwin's metaphor," the idea of nature "selecting" from available varieties. For some, this metaphor implied divine guidance; for others, it was proof of pure chance. Either way, the idea of a "selector" preserved the ancient image of a mind behind the natural world—an idea that proved crucial for the development of Darwin's theory and its eventual application to human affairs.[5]

The theory toward which Darwin worked was not only methodological in the general sense it shared with natural theology. Nature's method was more specific than that. As he formulated new ideas about how species emerged and changed over time, Darwin saw in these processes something familiar: the actual method to which he aspired, with its careful balance of hypotheses and tests. In the 1830s and 1840s, the whig interpretation of science and the theory of evolution converged on a shared mechanism of violence and stasis. Choosing wisely between alternatives, producing novelty through growth and pruning, came to describe science and nature at once. This fusion was helped along by Darwin's methodological anxiety and his moderate politics, rendering it not just plausible but *natural* that mind and matter shared a single, stable mode of operation. Nature was more than mental: it was a "man" of science, the natural philosopher *par excellence.* Variations, like hypotheses, were tested in the realm of experience; the life of the mind was as vibrant and brutal as life in the bush. Darwin's anxieties and ambition, his assumptions about a divine mind and his ability to translate them into a secular view of nature's method, paved the way for him to see his own techniques in nature's constant activity. Natural theology and Humboldt's *Personal Narrative* helped him see the world as a kind of mind, while his use of Herschel and Whewell turned it from a divine mind into one that was more familiar: his own. Writing his chosen method onto the world he described, Darwin made both nature and science into systems that evolved through a version of trial and error.[6]

THE DIVINE MIND

What was it like to be God, or an animal, or a pebble? Charles Kingsley had an answer. "Ages and Æons since," Kingsley told an audience in 1846, "thousands on thousands of years before there was a man to till the ground, I the little pebble was a living sponge, in the milky depths of the great chalk ocean." Holding aloft a rock he said he had "picked up out of the street as I came along," Kingsley exhorted his audience to suppress their "likes and dislikes, fancies, and aspirations" and hear what it had to say. "Remember," he told them,

> that while England is, and ever will be, behindhand in metaphysical and scholastic science, she is the nation which above all others has conquered nature by obeying her; that as it pleased God that the author of that proverb, the father of inductive science, Bacon Lord Verulam, should have been an Englishman, so it has pleased Him that we, Lord Bacon's countrymen, should improve that precious heirloom of science, inventing, producing, exporting, importing, till it seems as if the whole human race, and every land from the equator to the pole must henceforth bear the indelible impress and sign manual of English science.

Kingsley's pebble was part of an empire, echoing the imperial imagery of the BAAS. But it was also a sign of something internal, an invitation to think speculatively. "Imagination is a valuable thing," Kingsley insisted, "and even if it were not, it is a thing, a real thing, a faculty which every one has, and with which you must do something. You cannot ignore it; it will assert its own existence." The pebble, he suggested, reminded us that we *always* imagine. We cannot help it.[7]

Today Kingsley is best remembered for another act of imagination: a children's book called *The Water-Babies: A Fairy Tale for a Land Baby.* Published in 1863, the book was a satirical supplement to Darwin's *Origin of Species,* which Kingsley read (with great enthusiasm) before it was published in 1859. *Water-Babies* presents an evolutionary narrative, buttressed by racialist claims and, intriguingly, a naturalization of learning as a kind of evolution. Following Tom, a chimney-sweep who falls into a river and becomes a "water-baby," the plot is structured around lessons taught by underwater fairies, each representing a developmental stage through which Tom must pass. Toward the end, his lessons become explicitly evolutionary. A fairy named Mother Carey tells him: "If I can turn beasts into men, I can, by the same laws of circumstance, and selection, and competition, turn men into beasts. You were very near being turned into a beast once or twice, little Tom. Indeed, if you had not made up your mind to go on this journey, and see the world, like an Englishman, I am not sure but that you would have ended as an eft in a pond." Evolution goes both ways, according to Mother Carey (and Kingsley). Our direction—up or down, progressive or regressive—depends, literally, on how we make up our minds. If evolution contains an element of choice, there is also the prospect of blame.[8]

Tom's lessons reveal that seeing "like an Englishman" meant thinking like Darwin. A turnip, for example, tells Tom, "My mamma says that my intellect is not adapted for methodic science, and says that I must go in for general information." Science started where the turnip failed. Kingsley parodied naive Baconianism, as he had in his Reading lecture, for its lack of imagination. This meant purposefully conflating Tom's individual learning process and the history of science, suggesting that minds—like embryos—follow a progressive path blazed by larger groups over historical stretches of time. While this theory of "recapitulation," the mirroring of micro- and macro-level processes, has a vexed history in the life sciences, it was certainly possible to read it into Darwin's work in the 1860s. Indeed, many did. Recapitulation theory structured the application of evolutionary theory to political and social topics in the period, such as when Herbert Spencer claimed that "education should be a repetition of civilization in little." Spencer and others used the logic of recapitulation to extrapolate from historical patterns to individual potential and back again, from single children to the future of races. *Water-Babies* translated this move into a children's story, a curious form that belied the book's serious engagement with evolutionary self-reflection in the mid-nineteenth century.[9]

Kingsley has been called "Darwin's other bulldog" for good reason. The original bulldog, Thomas Huxley, came to Darwin's defense as a man of science—most famously in a debate with the bishop Samuel Wilberforce at the 1860 BAAS meeting. Kingsley did so in Anglican robes. The task he set himself was to make room for natural selection within the tradition of natural theology, which framed the study of nature in divine terms. Kingsley hit on a solution right away, telling Darwin in a letter that "it is just as noble a conception of Deity, to believe that he created primal forms capable of self development . . . as to believe that He required a fresh act of intervention to supply the lacunas which he himself had made." The insight appeared in *Water-Babies,* when Mother Carey tells Tom that "any one can make things, if they will take time and trouble enough: but it is not every one who, like me, can make things make themselves." And eight years later, in an address titled "The Natural Theology of the Future," Kingsley restored the idea's explicit religiosity: "We knew of old that God was so wise that He could make all things; but behold, He is so much wiser than even that, that He can make all things make themselves." Like Darwin's devout friend Asa Gray, Kingsley accommodated himself to the theory by insisting that it bolstered, rather than undercut, the idea of a benevolent creator. After all: Why would God design a world that required constant attention? Evolution made room for grace.[10]

This is the first idea in the *Origin*—but the words are not Darwin's. They are Whewell's, who is quoted in the first of two epigraphs to the first edition. Like any author, Darwin chose his epigraphs carefully, hoping to suggest specific precedents and parameters for his argument. Just as Herschel and Whewell claimed Baconian patrimony this way, Darwin did so too—quoting a call for scientific progress from *The Advancement of Learning*. But before quoting Bacon, he used Whewell to set the tone for the book. Darwin's choice of epigraph was telling: "But with regard to the material world, we can at least go so far as this—we can perceive that events are brought about not by insulated interpositions of Divine power, exerted in each particular case, but by the establishment of general laws." The line was chosen as much for its content as for its source, which was Whewell's volume in a series of theological texts from the 1830s known as the *Bridgewater Treatises*. The series, endowed by a bequest from the eighth Earl of Bridgewater, was meant to reveal "the Power, Wisdom, and Goodness of God, as manifested in the Creation." The *Treatises*' "theology of nature" influenced subsequent efforts to popularize scientific ideas. Whewell's volume, on the topic of astronomy and physics, was the first to appear and the most successful, and it was partly this popularity on which Darwin was leaning when he quoted it.[11]

But popularity was not all Darwin was after. Whewell could also help buffer him against charges of atheism or materialism he knew to expect. Indeed, Darwin was concerned enough that he asked Kingsley if he could share his letter (anonymously) with readers, which in the second edition he credited to "a celebrated author and divine." Along with a new epigraph from Joseph Butler's *Analogy of Revealed Religion*, Darwin's use of Kingsley has been cited as evidence of his concerns about religious readers and his effort to assuage them. Kingsley's response suggests he succeeded, at least in part, as does Gray's in the American context. The reactions of some of Darwin's devout readers show that the marriage between natural theology and natural selection was not as difficult to achieve as is often remembered. Contrasted with the infamous debate between Huxley and Wilberforce, the book's beautiful closing line seemed to leave space for just such an interpretation: "There is grandeur in this view of life, with its several powers, having been originally breathed into a few forms or into one; and that, whilst this planet has gone cycling on according to the fixed law of gravity, from so simple a beginning endless forms most beautiful and most wonderful have been, and are being, evolved." When Darwin made creation explicit ("breathed *by the Creator*") in the second edition, he lent credence to this reading—no matter what his personal views were on the matter.[12]

Indeed, the roots of Darwin's attraction to Whewell ran even deeper. Beyond popularity and propriety, the idea of "Wisdom . . . manifested in the Creation" encapsulated a key feature of how Darwin had learned to see the natural world. During his college days, natural theology was most closely associated with William Paley, whose *Natural Theology* was published in 1802 and, by the 1830s, had come to stand in for the movement as a whole. Paley was required reading for Cambridge undergraduates, which Darwin described as "the only part of the Academical Course . . . of the least use to me in the education of my mind." Looking back, he added: "I did not at that time trouble myself about Paley's premises; and taking these on trust I was charmed and convinced by the long line of argumentation." The natural theology of Darwin's curriculum was reinforced outside the classroom, where he collected plants and surveyed landscapes with devout mentors like the botanist John Stevens Henslow and the geologist Adam Sedgwick. It was on long walks with the former and postgraduate excursions with the latter that Darwin put Paley into practice, learning to see the path of the naturalist as a calling, if not a career.[13]

Central to the natural theology of Darwin's mentors was the idea that studying animals and plants was a way of "reading the book of Nature." This metaphor comes through clearest in Paley's famous watchmaker analogy, from the opening lines of his *Natural Theology:*

> In crossing a heath, suppose I pitched my foot against a *stone,* and were asked how the stone came to be there; I might possibly answer, that, for any thing I knew to the contrary, it had lain there for ever: nor would it perhaps be very easy to show the absurdity of this answer. But suppose I had found a *watch* upon the ground, and it should be enquired how the watch happened to be in that place; I should hardly think of the answer I had before given, that for any thing I knew, the watch might

> have always been there. Yet why should not this answer serve for the watch, as well as for the stone? Why is it not as admissible in the second case, as in the first? For this reason, and for no other, viz. that, when we come to inspect the watch, we perceive (what we could not discover in the stone) that its several parts are framed and put together for a purpose.

Stones do not imply a maker, at least not how watches do. Paley's point was that they *should.* When we contemplate the natural world, we should see the evidence of its creation as easily as we see that of the watch and other human contrivances. There is a guiding purpose everywhere, which explains the order of the natural world and is, in turn, evidence for God's existence.[14]

This was, as best we can tell, how Darwin understood scientific inquiry when he sailed. But natural theology in the mold of Paley was not Darwin's only source for seeing the world as a result of deliberate (and divine) mental activity. Alexander von Humboldt's *Personal Narrative,* which drew on and represented the same Romantic impulses that Mill got by way of Coleridge, was another. Just how much Humboldt's ideas shaped Darwin's thinking is debated, but the book's importance in framing his *Beagle* experience, especially in South America, is beyond dispute. His diary is full of references to Humboldt, making clear that Darwin's obsession with the Prussian naturalist was early and ardent. "As the force of impression frequently depends on preconceived ideas, I may add that all mine were taken from the vivid descriptions in the Personal Narrative which far exceed in merit anything I have ever read on the subject." The book had been a gift from Henslow, himself a natural theologian, and both men thought it bolstered Paley's views. On the voyage, Darwin praised Humboldt for "the rare union of poetry with science which he so strongly displays," even gushing that "he like another Sun illumines everything I behold." Humboldt shaped not only Darwin's ideas, but his way of *formulating* ideas. "If you see him again," he wrote to his friend Joseph Hooker in 1845, "pray give him my most respectful & kind compliments, & say that I never forget that my whole life course is due to having read & reread as a Youth his Personal Narrative."[15]

The *personal* dimensions of Humboldt's *Personal Narrative* affected Darwin most. He noted "with what intense pleasure he appears always to look back on the days spent in tropical countries." Before the voyage, he "read and reread Humboldt," so excited he could "hardly sit still on my chair." Once aboard the ship, he needed the *Personal Narrative* even more. "Nothing could be better adapted for cheering the heart of a sea-sick man" than Humboldt's "glowing accounts of tropical scenery," Darwin wrote from his bunk. The book's "enthusiasm" was so great, Darwin was afraid the actual Amazon would disappoint him! "I have written myself into a Tropical glow," he told his sister Caroline, who saw how right he was as soon as she read Darwin's journal: "I thought in the first part (of this last journal) that you had, probably from reading so much of Humboldt, got his phraseology & occasionly made use of the kind of flowery french expressions which he uses, instead of your own simple straight forward & far more agreeable style. I have no doubt you have without perceiving it got to embody your ideas in his poetical language & from his being a foreigner it does not sound unnatural in him." From rainforest reveries to family letters, the *Personal Narrative* shaped Darwin's immersion in the natural world. Humboldt alone, he told his mentor Henslow, "gives any notion, of the feelings which are raised in the mind on first entering the Tropics." It was this affective dimension that drew Darwin in and sustained both his interest and his theorizing over a long career.[16]

Humboldt's "rare union of poetry with science" was such an inspiration to naturalists in the period that their work has been dubbed "Humboldtian science." As Darwin's praise implied, such science was a hybrid endeavor. Part Enlightenment rationalist, part Romantic idealist, Humboldt was obsessed with exact measurement and given to flights of fancy at once. He took the chemist's precision laboratory instruments and the artist's watercolors into the field, insisting that both were required to fully comprehend the universe. Natural unity demanded methodological unity, which meant the naturalist could became one with the world only if his mind was open to all it had to offer. As Humboldt later reflected: "Such a result can, however, only be reaped as the fruit of observation and intellect, combined with the spirit of the age, in which are reflected all the varied phases of thought." The unity of nature was not only *mirrored* in the unity of mind—it was, in a sense, a consequence of it: "man has laboured, amid the ever-recurring changes of form, to recognise the invariability of natural laws, and has thus by the force of mind gradually subdued a great portion of the world to his dominion." For Humboldt, our inner lives are not just dim reflections of the outside world, but instead help to constitute that world. The experience of this link was evidence of a divine mind and the source of science's sublimity.[17]

The nineteenth century was full of such appeals to unity, to ideas of nature expressing "the mind of the creator." In relation to Darwin, this approach was closely associated not only with Gray but also with Gray's colleague and nemesis, Louis Agassiz. A Swiss-born naturalist who helped turn Harvard—and thus the United States—into a new center for scientific research, Agassiz made the contemplation of a divine mind into a prerequisite for doing science. "There will be no *scientific* evidence of God's working in nature," he wrote, "until naturalists have shown that the whole Creation is the *expression of a thought,* and not the *product of physical agents.*" Though this might sound like Agassiz saw science and religion as separate, he meant something like the opposite. God's work in nature was the proper object of science, and being a good naturalist meant learning to see the world that way. As Agassiz had put it in a textbook a few years earlier: "To study, in this view, the succession of animals in time, and their distribution in space, is therefore to become acquainted with the ideas of God himself." And in the preface to his monumental *Contributions to the Natural History of the United States* (published in the same year as the *Origin*), Agassiz turned classification into mimicry. Distinct species were "instituted by the Divine Intelligence as the categories of his mode of thinking," with naturalists "only the unconscious interpreters of a Divine conception." Like Humboldt, Agassiz saw our categories as pale reflections of the divine mind, the best an imperfect species could do.[18]

Another of Darwin's eventual critics, Richard Owen, expressed similar ideas about the order behind natural occurrences. Borrowing, like Agassiz, from German *Naturphilosophie,* Owen argued that each organism was based on an "archetype," or *Urbild,* representing a unified divine plan in the mind of the creator. This is clearest in the area of his expertise: vertebrate anatomy, especially of the limbs. An archetype, though embodied differently in every bone, was shared between individuals, species, and even broader taxonomic groups. Owen was explicit that archetypes were ideas: a bat's wing articulated a divine *idea* of a wing. Like Agassiz's species, Owen's archetypes were ideal, but also *real.* Sorting through bones, Owen built a unified vision out of the incredible diversity of the British Museum's collections, which gave his study purpose and linked it up to other branches of natural history. Always careful to define his terms, Owen began his lecture *On the Nature of Limbs* with a discussion of its title. His use of the term "nature" was meant to approximate the German *Bedeutung,* which was less like "structure" and more like "the 'idea' of the Archetypal World in the Platonic cosmogony." Or, as he put it in his conclusion: "The Divine mind which planned the Archetype also foreknew all its modifications." In nature, the human mind met its divine inspiration.[19]

Though Owen ended up as one of Darwin's fiercest critics, his theory of archetypes balanced British natural theology and German *Naturphilosophie* in much the way natural selection would. And when he boarded HMS *Beagle* in 1831, Darwin felt he and Owen were after the same thing: elucidating the contours of nature's mind. Owen even helped him do it, interpreting Darwin's fossil specimens for him when he got back. Reading them as proof of an archetypal relationship and, thus, of divine plans in nature, he emboldened Darwin to see a different kind of unity in diversity. While Darwin's interpretation led in a different direction—naturalizing the "divine mind" as a selective process—it is important not to see their views as completely opposed. Given how easily Darwin's friends Gray and Kingsley accomodated natural selection to the tradition of natural theology, there was clearly room for assimilation. Darwin may have told himself that "the old argument of design in nature, as given by Paley, which formerly seemed to me so conclusive, fails, now that the law of natural selection has been discovered." But he was probably protesting too much. His desire to gain a fair hearing for his theory from both devout practitioners like Gray and agnostics like

Huxley (who had coined the term "agnostic" himself!) suggests a degree of ambivalence—strategic or otherwise—on the matter of natural theology's influence on his own thought.[20]

The antithesis of science and religion, though articulated by many at the time, was more of a polemical construct than a lived reality for most Victorians. The idea of a divine mind—of natural theology, in the fullest sense—persisted, in both implicit and explicit ways. It is there on the surface, in the agency implied by "Darwin's metaphor" and seized upon by his readers. But it is also present in subtler ways. Take, for example, a curious admission Darwin made in 1871, reflecting on these early years: "I was not able to annul the influence of my former belief, then widely prevalent, that each species had been purposely created; and this led to my tacitly assuming that every detail of structure, excepting rudiments, was of some special, though unrecognised, service." Darwin suggested that his attention to adaptation was a consequence of his natural theology. That same year, Kingsley picked up on these connections from the other end. Defending Darwin from Anglican attacks, as he had promised to do, Kingsley urged his fellow priests to meet men of science halfway: "For if we are ignorant," he told his audience, "not merely of the results of experimental science, but of the methods thereof, then we and the men of science shall have no common ground whereon to stretch out kindly hands to each other." Mind and method, science and religion—Darwin stretched easily, even eagerly across such binaries. And doing so left "special, though unrecognized" stretch marks in his work.[21]

MYSTERY OF MYSTERIES

Natural theology helped make the world seem like the product of a mind—specifically, God's mind, with science a kind of spiritual mind-reading. But there were other analogies, furnishing different ways to read nature's book. Some, such as the study of "natural laws," emphasized its universal qualities, revealed through laboratory instruments. The fact that we could contemplate those laws shored up the idea, for some, that

God was knowable. For others, nature seemed even closer to home, less universal and more familiar. Sedgwick, for example, explicitly used "the language in which we describe the operations of intelligence and power" to account for natural processes. Nature worked on the same principles as men of science—that is, both were modes of "slow and toilsome induction." Building from the ground up, accumulating facts in Baconian fashion, was "the only path which leads to physical truth" because it was the path taken by the objects under study. In a famous "Commemoration Sermon" of 1832 and its published version, *A Discourse on the Studies of the University,* Sedgwick went further: induction not only accessed truth, but also embodied it, mirroring the virtue of Christian sacrifice, if not Christ himself. The scientific mind stood in for religious truth, just like Tom's lessons in Kingsley's *Water-Babies.* What one historian has called "the moral dignity of inductive method" was a working principle for Sedgwick, transforming his description of how we think into a prescription for how *to* think. The "laborious but secure road of honest induction" was both the road to scientific and Christian truth and a symbol of that truth's moral depth and spiritual significance.[22]

Science was a mirror that reflected the divine mind. Like any mirror, it was imperfect. Whewell had insisted, after all, that "the Ideas of a Divine Mind must necessarily be different in kind, as well as in number and extent, from the Ideas of the human mind." Science attempted to reflect something perfect, but the mortal efforts of men of science could never live up to it. We see the divine mind only "dimly," according to Whewell, and as a result our progress will always be "scanty and incomplete." And yet: we progress. For Whewell, this fact was evidence of God's mind and justification of science as a way of studying it. The human mind was "the image of God in its faculties," reflecting in its best moments a divine universality. "When it has arrived at a stage in which it sees several aspects of the universe in the same form in which they present themselves to the Divine Mind," Whewell concluded, "we cannot suppose that the Author of the human mind will allow it and all its intellectual light to be extinguished." This was an argument for immortality—not of the individual practitioner, but of the larger scientific process. Whewell sounded a hopeful note about the progress of science that could be achieved beyond the bounds of any individual's life, so long as the scientific method mirrored the divine mind that was its ultimate object of study. The method was individual, but scientific immortality was social.[23]

The mirroring relationship between science and God worked in both directions. Whewell saw "the image of God in its faculties," but the scientific mind was also written onto nature (and God). Natural theology's discourse of a mind in nature made it easy to project theories of mind and method onto the world one sought to describe in scientific terms. Sedgwick's defense of the inductive method was more than an effort to practice Christian virtue, in other words. It worked in reverse, naturalizing and even deifying his preferred approach to science by writing it onto the powers of the Creator. Whewell, too, defended induction in this way, though as we have seen his view of the method—and thus his interpretation of the divine mind—differed from Sedgwick's. Still, he agreed that one's approach framed one's facts. As Whewell put it in his *Bridgewater Treatise,* entitled *Astronomy and General Physics Considered with Reference to Natural Theology:* "Deductive reasoners, those who cultivate science of whatever kind, by means of mathematical and logical processes alone, may acquire an exaggerated feeling of the amount and value of their labours." The problem was not the method, so much as the adherence to one method over others. In other words, it was a problem of balance. As he would elaborate in his *History* a few years later, a sense of the past can help men of science avoid the pitfall of writing erroneous methods onto nature.[24]

Whewell's friend Babbage responded directly with his own account of the divine mind. He published his interpretation as an uninvited contribution to the same series, calling it *The Ninth Bridgewater Treatise.* This was a sign of his commitment to a divine interpretation of deductive method despite lacking a formal affiliation with the Bridgewater project. In the book, Babbage extended Whewell's meditation on how we tend to write our own methods onto the natural world, but did so in defense of his own way of thinking: "We take the highest and best of human faculties, and, exalting them in our imagination to an unlimited extent, endeavour to attain an imperfect conception of that Infinite

Power which created every thing around us. In pursuing this course, it is evident that we are liable to impress upon the notion of Deity thus shadowed out, many traces of those imperfections in our limited faculties which are best known to those who have most deeply cultivated them." Though he sounds critical, Babbage was not acknowledging the "imperfections in our limited faculties" to extinguish them. After all, doing so was impossible: our imperfections make us human. Seeing God in the phenomena we observe, we project the limits of our own minds onto what is ostensibly perfect. The task, then, is to know those limits, to turn our analysis backward on ourselves from time to time. Science only achieves its aims when it acknowledges the mind's limits. As Babbage concluded: "All those discoveries which arm human reason with new power, and all additions to our acquaintance with the material world, must from time to time render a revision of that notion a necessity." Our sense of God changed with our sense of science—as it should.[25]

For Babbage, this meant that we see the divine mind in light of our own. God was akin to an inventor, if not to a calculating engine, a perfect match for Babbage's interests. "Conscious that we each of us employ, in our own productions, *means* intended to accomplish the objects at which we aim, . . . we are irresistibly led, when we contemplate the natural world, to attempt to trace each existing fact presented to our senses to some precontrived arrangement." Working on his engines seemed to present "matter for reflection on the subject of inductive reasoning." Like Paley, Babbage saw a machine that governed itself, according to law, as superior to one needing constant interventions. From there, it was a short hop to see something similar in how rocks were formed—or, for that matter, how the lives of animals and plants changed over time. In the book's second edition, Babbage expanded on these points by quoting a letter from Herschel to Lyell that defended a view of God as an inventor of just this sort, operating by "intermediate causes" rather than by constant, direct interference. Herschel's letter was a thank-you note for a copy of a new edition of Lyell's *Principles of Geology,* made famous by a passage in which Herschel deemed the question of species change "that mystery of mysteries." To Babbage, the idea that new species could emerge according to natural laws only bolstered the identity between calculation and the divine mind. And he was not alone.[26]

Around the time that Babbage published his *Ninth Bridgewater Treatise,* Darwin also turned to Herschel for support. Having read Thomas Malthus's *Essay on Population* in the fall of 1838 ("for amusement," by his own account), Darwin very quickly assembled what we now see as the major pieces of his theory of evolution by natural selection. In this heady context, Darwin turned again to Herschel, whose *Preliminary Discourse* he had read both before and during the voyage of the *Beagle.* In a notebook listing his "Books Read," Darwin wrote "Whewell inductive History {References at end}" and "Herschel's Introd to Nat. Philosophy {do 2d time of Rdg}" on the page after he wrote "Malthus on Population" next to the date "Oct 3." Whether it was Whewell's dedication of the book to Herschel, or his own memory of its prior value, Darwin would recall the *Preliminary Discourse* in terms he had, until then, reserved for Humboldt. "This work," he later said of the *Personal Narrative* in his autobiography, "and Sir J. Herschel's Introduction to the Study of Natural Philosophy stirred up in me a burning zeal to add even the most humble contribution to the noble structure of Natural Science. No one or a dozen other books influenced me nearly so much as these two." As with Humboldt, Herschel did more than inspire the young naturalist. Both helped him see the world around him in new ways.[27]

Like Babbage, Darwin had read Herschel's line about "that mystery of mysteries," and it too had made an impression on him. "Herschel calls the appearance of new species the mystery of mysteries," Darwin scribbled in a notebook in 1838, "& has grand passage upon the problem.! Hurrah—'intermediate causes.'" Beyond the motivation of seeing an idol wax poetic about his secret obsession, Darwin also leaned on Herschel's defense of hypotheses to authorize his own guesswork. In a notebook devoted to metaphysical speculations, Darwin wrote: "Arguing from man to animals is philosophical, viz. (man is not a cause like a deity, as M. Cousin says), because if so ourang outang,—oyster & zoophyte: it is (I presume—see p. 188 of Herschel's Treatise) a 'travelling instance' a—'frontier instance,'—for it can be shown that the life & will of a conferva is not an antagonist quality to life & mind of man." These lines are hard to parse. Darwin was thinking rapidly, even sloppily, across lines of species and topic. But the basic point was simple: by a "frontier instance," Herschel meant that a theory should be able to extend to the limits of what was known, to reach out and "grab" some far-flung phenomenon and help it make sense. This idea licensed the connections Darwin saw and the boldness he wanted so desperately to embody. He was *using* Herschel, just as Herschel had used Newton: seeking not only guidance but also authority when he felt (intellectually) at risk.[28]

Darwin needed the help. When he began speculating about transmutation (not to mention its application to the human species), such trains of thought risked being labeled "materialist," an insult tantamount to "atheist." The threat of such censure was powerful enough that Darwin, even in his private notes, worried where his theory might lead him. In one of his "Transmutation Notebooks," for example, he imagined just such a critique of his theory as it applied to the origins of religion: "love of the deity effect of organization, oh you materialist!" Soon, Darwin cleaved his thinking in two: he continued to pursue the origin of species but migrated his ideas about humans and the mind to a series of what he called "Metaphysical Notebooks." There, he pursued his ideas on religion and other cases that invited dangerous thinking. "It is an argument for materialism," he wrote in one, "that cold water brings on suddenly in head, a frame of mind, analogous to those feelings, which may be considered as truly spiritual." Later, he plotted ways to "avoid stating how far, I believe, in Materialism." Even before the cautionary tale of the controversial (and, not coincidentally, anonymous) *Vestiges of the Natural History of Creation,* Darwin was wary of the effect that speculating about human origins could have both on his own thinking and on his reputation among colleagues and the general public.[29]

It was not just taboo topics like human religiosity that got Darwin's heart racing. His famous two-decade delay in publishing on natural selection was about more than the retribution that was sure to be visited upon a speculator on such topics. As Darwin had learned from early on, the risks of speculation were everywhere—even in geology, where his well-known "blunder" on the parallel roads of Glen Roy and his work on corals had taught him caution. "All young geologists," he later recalled in a letter to a young acquaintance, "have a great turn for speculation; I have burned my fingers pretty sharply in that way, & am now perhaps become over cautious." He did not forbid "the sin of speculation" entirely, however: "I can have no doubt that speculative men, with a curb on make far the best observers." Hypothesizing, for Darwin, was a natural tendency—indeed, all too natural—but young practitioners had to earn their rights to engage in it. It was crucial to scientific advances, but without proper attention to facts it was more harm than help. As Darwin put it, years earlier, to another colleague with a tendency to speculate: "such zeal as yours, is a main element in discovery." What he had learned, the hard way, was that such speculation was not the *only* element. One needed caution as well.[30]

In his volume of 1839's *Journal of Researches,* Darwin included a vexed view of hypotheses from his *Beagle* diary that is worth quoting in full:

> In conclusion, it appears to me that nothing can be more improving to a young naturalist, than a journey in distant countries. It both sharpens, and partly allays that want and craving, which, as Sir J. Herschel remarks, a man experiences although every corporeal sense be fully satisfied. The excitement from the novelty of objects, and the chance of success, stimulate him to increased activity. Moreover, as a number of isolated facts soon become uninteresting, the habit of comparison leads to generalization. On the other hand, as the traveller stays but a short time in each place, his descriptions must generally consist of mere sketches, instead of detailed observations. Hence arises, as I have found to my cost, a constant tendency to fill up the wide gaps of knowledge, by inaccurate and superficial hypotheses.

Here, speculation was rooted in the language of "habit" and "tendency." Just as in his letters to young naturalists and in admonishing himself, the effect of this way of framing was to *naturalize* scientific thinking itself, to turn it into something like an instinct. Herschel had suggested as much, when he began a review of Whewell by noting that "man is a speculative as well as a sentient being." The point struck Darwin, who noted: "From Herschel's Review Quart. June (41) I see I *must study* Whewell on Philosophy of Science.—Speculates on Instinct." Here, Whewell's speculation was doing double duty: Darwin was hungry for new ideas about instinct, to be sure, but also for models of how to speculate appropriately on psychological matters.[31]

Darwin's anxieties about speculation made him averse to certain analogies. For example, he pushed back repeatedly against what sounded like inappropriate analogizing between human and natural phenomena in his copious reading. Once he got around to reading Whewell, he found too much of this sort of reasoning for his taste. Reading another book that organized "animals according to varieties of man," Darwin wrote that it "confounds, *like Whewell,* affinity with analogy." Later in the same set of notes, he returned to Whewell: "Mayo (Philosoph. of Living) quotes Whewell as profound because he says length of days adapted to duration of sleep of man!!! Whole universe so adapted!!! & not man to Planets.—instance of arrogance!!" Around the same time, but in one of his "Metaphysical Notebooks," he revisited the topic of universal laws to push back against a similar anthropocentric tendency: "This unwillingness to consider Creator as governing by laws is probably that as long as we consider each object an act of separate creation, we admire it more, because we can compare it to the standard of our own minds, which ceases to be the case when we consider the formation of laws invoking laws." The habit of seeing ourselves in nature, as a source of metaphor and analogy, potentially obscured the very patterns that naturalists sought to elucidate. We could be our own worst enemies.[32]

Ironically, the solipsism Darwin feared was part of his biggest breakthrough. Had he been totally opposed to anthropocentrism, he might have missed it. Luckily, Darwin was a realist about the ways that we filter the world. It mattered less *that* we made such analogies—after all, he thought, we did so naturally—than *how* we made them. In his first secret notebook, begun in the summer of 1837, Darwin was already filtering nature through mental language. "Generation here," he scribbled, "seems a means to vary, or adaptation. Again we believe (know) in course of generations even mind and instinct become influenced.—child of savage not civilized man.—birds rendered wild through generation acquire ideas ditto." Flitting back and forth between morphology and mental development, Darwin blurred the boundaries between the human and natural worlds. Of course, this blurring is what his theory is most famous for today—but he was blurring things in a more specific sense, and much earlier, than is often attributed to him. Ideas, like birds, changed over time—and, Darwin speculated, according to the same sort of process. A series of staccato lines a few pages later underscored the point: "Each species changes.—does it progress—Man gains ideas. The simplest cannot help—becoming more complicated: & if we look to first origin there must be progress." Changing species and changing minds appear side by side, each progressing from simple origins to the complexity observed in nature—or science.[33]

Soon Darwin gave the first public hint of his private work, revealing just how close he was bringing method and evolution in his theorizing. In a letter to Lyell, he asked: "Has your late work at shells startled you about the existence of species? I have been attending a very little to species of birds, & the passages of forms, do appear frightful—every thing is arbitrary; no two naturalists agree on any fundamental idea that I can see." It is easy to assume that the "frightful" thing was the idea of species change, which Darwin had likened to "confessing a murder." But such ideas were all over in those years (after all, Herschel had written about it to Lyell the previous year). It seems that, instead, Darwin was talking about the state of science, not nature. His use of "arbitrary" hints at this reading: the disagreement among specialists had him just as worried as the arbitrariness of life, if not more. Darwin's affective response to scientific stakes was blurring into existing worries about what a dynamic, chance-riddled world might mean—for religious and irreligious naturalists alike.[34]

The blending of these anxieties, about science and about nature, made sense in Darwin's day. It was a common belief that scientific thinking reflected order and stability in the natural world. This was especially true of systems of classification, seen as the indexes to the book of nature within and beyond natural theology. Disorder, for someone like Darwin, could go either way: it might suggest a human failure to classify things correctly—or deeper instability, which was being accurately captured by a disordered taxonomy. As Darwin tried to get his thoughts in order, the disarray of scientific opinion would have been as frightful as his jumble of specimens. The diversity that "startled" him could be either scientific or natural, and indeed it was likely both; the complexity he complained about to Lyell was proof that something more was needed. This did not shake Darwin's faith that some lawlike pattern *would* be discovered—far from it. His anxieties about science and nature revealed that he was thinking through both at once. New strategies for collecting and publication were as important as finding a way to account for change over time. In other words, he was running the problems he encountered in his own mind together with those he detected in what, until recently, had been a divine mind.

Darwin continued this line of thinking a year later, linking it back to the theological concerns that had helped frame his theory originally. "I look at every adaptation," he wrote in a separate notebook, "as the surviving one of ten thousand trials.—each step being perfect . . . to the then existing conditions.—An adaptation made by intellect this process is shortened, but yet analogous, no savage ever made a perfect hinge.—reason, & not death rejects the imperfect attempts." Selecting, in the mind, was a way of short-circuiting the bloodier events in nature. But it was the same pattern, as he made clear a few pages later: "In the Mollusca / Bees / the nervous system is endowed with the knowledge of trying a hundred schemes of structure, in the *course of ages* / step by step / .—in man, the nervous system, gains that knowledge before hand, & can in idea (with consciousness) form these schemes." Material links between body and mind formed a bridge for the application of his theory. In his "Metaphysical Notebooks," he reinforced the point by reframing these "ten thousand trials" in more material terms. Thinking through the evolution of the mind, Darwin proposed a thought experiment to explain what can seem to be the random train of our thoughts: "Shake ten thousand grains of sand together & one will be uppermost:—so in thoughts, one will rise according to law." That law, riddled with chance and operating on a vast scale, helped link mind and nature to human artifice, illustrating the randomness with which his skeptics so struggled at work in both the natural world and the scientific mind.[35]

The "savage" incapable of making a hinge was no idle example. Darwin was fascinated by what he saw as intermediate stages between the artificial selection of breeders and natural selection in the nonhuman world. In an 1857 letter to Gray—later excerpted as part of the announcement of the theory in a joint publication with Alfred Russel Wallace—he called this "unconscious selection," suggesting that the step predated what he called "methodical" selection by thousands of years. The idea found its way into the *Origin,* too, where even an unintentional preference between two animals was enough, over generations, to produce differences that modern husbandry could intentionally exploit. Darwin used the example of dogs to make the case for the plausibility of such selection: "If there exist savages so barbarous as never to think of the inherited character of the offspring of their domestic animals, yet any one animal particularly useful to them, for any special purpose, would be carefully preserved during famines and other accidents, to which savages are so liable, and such choice animals would thus generally leave more offspring than the inferior ones; so that in this case there would be a kind of unconscious selection going on." Here, the gap between natural and artificial selection almost closes. Unconscious selection linked the divine to the artificial, paving the way for imagining "a being, who did not judge by mere external appearance, but could study the whole internal organization." Humans were imperfect approximations of a grander process.[36]

And "savages" were not the only ones who worked in unconscious ways. Naturalists proceeded unconsciously, too. As Darwin put it in the *Origin:* "Community of descent is the hidden bond which naturalists have been unconsciously seeking, and not some unknown plan of creation." What made all of this unconscious was the fact that naturalists *thought* they were building the case for natural theology even as they gathered evidence that undermined it. Seeking data to prove that the living world was a perfect reflection of a divine plane, these naturalists inadvertently stocked the evidence room for their enemies. Like Darwin's "savages," defenders of natural theology "selected" for traits that inadvertently revealed transformations that either did not matter to them (in the case of "unconscious selection") or would actually come to seem anathema to their original motivations (in the case of the naturalists who opposed Darwin's theory). "Unconscious selection," in this sense, was the scientific equivalent of dog breeding, as Darwin saw it: evolution occurred through human action, in both cases, but not for the reasons imagined by those who were doing the selecting. Whether the products of husbandry or theory, whether it was an idea or a physical attribute that was being selected, the result was the same sort of winnowing over time. Naturalists came to seem, well, natural, as their own work came to mirror the process of natural selection that Darwin and others were soon describing.[37]

SELECTION BIAS

Darwin was torn. On the one hand, his notebooks record a zeal for hypothesizing over his whole career. On the other, he tempered this zeal, right from the start, with self-directed warnings about the risks of speculation. For example, while thinking through the problem of atavistic organs, Darwin found himself on the threshold of explaining improvement in terms of progressive change. "Such law would explain every thing," he gushed, before quickly cautioning himself: "*Pure hypothesis* be careful." Elsewhere, he underscored a similar warning, in Spanish: "Cuidado." Such admonitions were partly the careful note-taking strategy of a potential author. Warning an acquaintance that if "facts are so mingled with speculation" you could "injure your reputation," Darwin was expressing in his correspondence the kinds of anxieties he directed at himself in the notebooks. Explicit accounts of his own methods served a similar purpose, linking wilder ideas to the hard work he was doing and the evidence he was so eager to assemble. "The line of argument often pursued throughout my theory," he wrote at one point, "is to establish a point as a probability by induction, & to apply it as hypothesis to other points, & see whether it will solve them." This way of thinking was pure Herschel, an approach to "travelling instances" lifted straight from the *Preliminary Discourse* and explored elsewhere in Darwin's notebooks. As he supposedly said quite often: "I love fools' experiments. I am always making them." Speculation was essential (and fun)—but he prided himself on keeping things from getting *too* foolish.[38]

At the same time, Darwin went beyond Herschel's methodology. Just a few pages after these reflections on his "line of argument," Darwin pivoted from explaining his scientific work to speculating about mental evolution. "So with the mind," he mused, "the simplest transmission is direct instinct & afterwards enlarged powers to meet with contingency." Linking science to his theory of evolution was more than Herschel seemed to license. Darwin was connecting the dots, drawing a line from scientific thinking to natural selection. What he was pointing toward, in his notebooks, was the idea that mental evolution was analogous to methodological development, a way of turning scientific method into evidence of how the mind had changed over time. The best practices of science, gleaned from Herschel and carefully put to work in Darwin's notebooks, were now a template for his discussion of how ideas, like species, had evolved. At one point, earlier in the same notebook, Darwin brought the two together explicitly. "Mine is a bold theory," he announced, "which attempts to explain, or asserts to be explicable every instinct in animals." Instincts were a bridge between scientific thinking and organic evolution.[39]

Darwin's followers would later turn scientific thinking into an instinct, using the theory of natural selection to explain how the capacity for science had first emerged. But at this earlier moment, in the 1830s, Darwin had not even committed his theory to paper. Yes, he had hinted at it in letters and to himself in notebooks for half a decade, but other tasks intervened: his coral reef theory, the *Beagle* book, a volume on geology. This changed in May 1842, when Darwin visited his in-laws, left his notes behind, and set about outlining his theory from beginning to end. Words flooded in. What he called a "pencil sketch of my species theory," thirty-five cramped and crisscrossed pages, was both a relief and a motivation. In January 1844 he wrote the famous letter to Hooker in which he "confessed" his belief that species might be mutable. Insisting that he had tried to "collect blindly every sort of fact which c^{d} bear any way on what are species," Darwin worked to distance his approach from philosophers who put theory first. But he was not far off. That summer, he expanded the sketch into a longer "essay" that would eventually form the bulk of his contribution to the 1858 joint publication with Wallace. Indeed, Darwin set enough store by this new version that he asked his wife, Emma, to "devote 400£ to its publication" in the event of his untimely death. The money was meant to entice both an editor and a publisher. Lyell was ideal for the former job, Darwin thought, as "the Editor must be a geologist, as well as a naturalist." The task was to gather together books and scraps of paper, to transcribe his many scattered thoughts and to present them in an organized fashion; the goal was to give his theory a chance of survival in the harsh environment of Victorian science.[40]

Of course, Darwin did not die before he could publish, and so a great deal changed between the "Essay" of 1844 and what was eventually printed in 1858. For example, the earlier version blurred variation and selection together, while the published one kept them apart. His ideas about progress changed, too: what he later decried as "Lamarck nonsense" played a role in his early work. Huxley took note of these differences: "Much more weight is attached to the influence of external conditions in producing variation," he wrote to Darwin's son Francis, "and to the inheritance of acquired habits [in the "Essay" of 1844] than in the *Origin.*" In its earliest articulation, Huxley said, Darwin's theory seemed more like Lamarck's "inheritance of acquired characteristics." The 1842 sketch even starts with the idea: "habits of life develope certain parts. Disuse atrophies. [Most of these slight variations tend to become hereditary.]" Though seeking a "contrast with Lamarck,—absurdity of habit" later in the same piece, Darwin could not shake the idea that differences were "partly habit, but the amount necessarily unknown, partly selection." The speculations were even clearer in 1844: "habits of body or consensual movements, habits of mind and temper, are modified or acquired during the life of the individual, and become inherited." Darwin would eventually hedge these claims in the *Origin,* but his earliest efforts made natural selection more purposeful, more progressive, and more *mental.*[41]

Darwin never abandoned these ideas entirely. About a decade after publishing his theory, he proposed a mechanism for the interaction between environmental pressure and the production of variation—closing the circle on his search for a means of inheritance by offering a means for passing acquired characteristics to one's offspring. Darwin introduced this idea, which he called "The Provisional Hypothesis of Pangenesis," in 1868's *Variation of Animals and Plants under Domestication.* It was a simple proposal, but one Darwin recognized required a long leap of faith. As part of the process of division, he hypothesized, our cells "throw off minute granules or atoms," tiny structures that he called "gemmules." These collect in bodies and pass from parent to offspring, providing a mechanism for inheritance. And they could change. If "certain cells or aggregates of cells had been slightly modified by the action of some disturbing cause, the cast-off gemmules or atoms of the cell-contents could hardly fail to be similarly affected, and consequently would reproduce the same modification." Darwin thought gemmules might explain how "any habit or other mental attribute, or insanity, is inherited," when "gemmules derived from modified nerve-cells are transmitted to the offspring."[42]

There was only one problem: no one had seen a gemmule. Darwin knew this made the proposal risky, which is why he packed *Variation* with some of his most explicit methodological reflections. In the book's introduction, he insisted that "[in] scientific investigations it is permitted to invent any hypothesis, and if it explains various large and independent classes of facts it rises to the rank of a well-grounded theory." Darwin used natural selection as an example: it "may be looked at as a mere hypothesis," he admitted, "but rendered in some degree probable by what we positively know of the variability of organic beings in a state of nature,—by what we positively know of the struggle for existence, and the consequent almost inevitable preservation of favourable variations,—and from the analogical formation of domestic races." The question is whether the idea "explains several large and independent classes of facts." If it fails to do so—if "we gain no new knowledge; we do not connect together facts and laws; we explain nothing" by it—then it should be discarded. Natural selection lived or died, in other words, by its own sword: survival of the fittest theory. Anxious about evidence, Darwin here applied the theory to itself. "Man," he wrote early on in the book, "may be said to have been trying an experiment on a gigantic scale; and it is an experiment which nature during the long lapse of time has incessantly tried." Species and theories alike had to struggle for life.[43]

The pangenesis episode casts some key Darwinian issues into relief. As we have seen, Darwin's early attention to methodology shaped his theory in surprising ways. For one, anxiety about proper procedure led him to see methods everywhere, in the faults of scientific colleagues and in nature's twittering, tangled bank. Danger lurked around every corner, in science and in the field, its presence heightening the need for caution even when boldness was demanded. Yes, equilibrium was a kind of regulative ideal in how Darwin and others thought about scientific and natural communities. But it was fear—of error, of extinction—that motivated Darwin's yearning for a method and the intense caution that characterized his public persona (if not his private notebooks). When, thirty years after he allowed himself to formulate the theory for the first time, he put it at risk with the idea of gemmules, it might seem like an aberration, a rare moment of rashness brought on by success or late-blooming confidence. And it certainly was, in part. But in a certain sense, the idea of pangenesis and the willingness to air it were of a piece with his earlier theorizing, a bold idea that helped explain a variety of phenomena. As with variations, so with gemmules: great leaps required imagination. When laboratory experiments were impossible, mental ones were called for.

Pangenesis also helps reconcile a confusion in Darwin's reputation. Although we now think of Darwin as having *separated* the cycles of variation and selection—of having, in a sense, turned them into interlocking, dialectical processes—his early notebooks and the hypothesis of pangenesis blur the two together. And both ways of thinking about Darwin, as lumper and as splitter, were true. He repeatedly emphasized the gap between variation and selection in order to distance himself from theorists of directed evolution like Lamarck (and later, Herbert Spencer), something that followers like William James gleefully celebrated in his theory. But Darwin also continually toyed with concrete explanations for how chance variations were influenced by the conditions that governed selection. If we take both perspectives seriously, it can turn what seems like a problem for understanding natural selection into a solution. Beneath Darwin's statements about method and his evolving approach to the natural world ran a persistent worry, perhaps even a hope, that the two spheres ran together. Like the testing of hypotheses, the selection of variations was part of a complex whole; nature, like the mind, seemed to hold these processes apart—until it did not. Hypothesis testing, natural selection, pangenesis: all were attempts to have it both ways, to integrate error and truth, violence and peace, inheritance and acquisition. Clean divisions, pure dualisms, were ideals, not reality—in science and nature alike.

In defending his "hypothetical chapter" on pangenesis, Darwin reached for familiar sources of strength. Just as he had in his notebooks when he extended the theory to encompass mind and instinct, Darwin drew on both Herschel and Whewell to shore up his latest hypothesis. "I am aware that my view is merely a provisional hypothesis or speculation, but until a better one be advanced, it may be serviceable by bringing together a multitude of facts which are at present left disconnected by any efficient cause. As Whewell, the historian of the inductive sciences, remarks:—'Hypotheses may often be of service to science, when they involve a certain portion of incompleteness, and even of error.'" He sounded this theme again while reflecting on his "well abused hypothesis of Pangenesis" a few years later. "An unverified hypothesis is of little or no value," he admitted, "[b]ut if any one should hereafter be led to make observations by which some such hypothesis could be established, I shall have done a good service, as an astonishing number of isolated facts can thus be connected together and rendered intelligible." Despite the abuse his idea had sustained, he held out hope that it might still bear the fruit of Herschelian speculation: that is, that it might prove fit in the environment of a new set of experimental results or observational desiderata.[44]

As with the scientific method, so with the natural world: what mattered was the result of the test, not the origin or even the veracity of the hypothesis being tested. While pangenesis was roundly ridiculed, Darwin had said the same thing about natural selection. In the letter in which he revealed his speculations to Gray (almost twenty years after he had hinted at them to Lyell), Darwin framed the theory in the same way: "*Supposing* that such hypothesis were to explain general propositions, we ought, in accordance with common way of following all sciences, to admit it, till some better hypothesis be found out." If you "*assume* that species arise like our domestic varieties," then you have to "test this hypothesis by comparison" with all evidence bearing upon it. Balance was the key to avoiding what today we call "selection bias": cherry-picking data to suit your best guess or mixing your hypothesis with what it purports to explain. "I have steadily endeavoured to keep my mind free," he recalled, "so as to give up any hypothesis, however much beloved (and I cannot resist forming one on every subject), as soon as facts are shown to be opposed to it." It was part confession, part method: speculating came naturally, and so he was always on the watch. Though he would protest (perhaps too much) against it, this kind of cognitive blending was there when the ice beneath Darwin's feet was thinnest and when his speculation was boldest. Anxious to seem scientific, he turned the world into an experiment.[45]

After reading Malthus in September 1838, Darwin seems to have shifted toward chance as the explanation for most variations. Analogous to differences in human populations, variation in natural populations arose by the work of mysterious forces. What mattered was the pressure that resulted from fecundity and differential survival, however the relevant distinctions between traits or organisms might have arisen. But the question of chance variations is ambiguous, even at the origin of Darwin's analogy from the *Essay on Population* to the natural world. Darwin imagined nature as a surface into which a thousand wedges were driven, each pushing out another as they competed for space. "The final cause of all this wedgings [*sic*]," Darwin wrote in his initial notes on Malthus, "must be to sort out proper structure, & adapt it to changes.—to do that for form, which Malthus shows is the final effect (by means however of volition) of this populousness on the energy of man." At this crucial turn, Darwin hedged: the role of volition in human and natural change is blurry at best. His invocation of directed variation in 1842 and 1844—not to mention his later defense of pangenesis—suggest that Malthus's impact on his sense of progressive change in the natural world was a complicated one. Chance mattered, but there were always ways for the selective pressure of environmental conditions to spur, and even shape, variations.[46]

Something similar underlay Darwin's view of science. Hypotheses and tests were separate, but they had ways of shaping one another. Looking back on his "Malthusian moment," for example, Darwin recalled: "Here, then, I had at last got a theory by which to work; but I was so anxious to avoid prejudice, that I determined not for some time to write even the briefest sketch of it." Darwin thus allowed even the *recording* of his ideas to be shaped by what might become of them. It is possible to see "prejudice" in two different lights, given his methodological anxieties and those he harbored about the reception of his theory. Of course, Darwin knew how angrily an audience could respond to fanciful speculation, especially as it bore on human origins. But he also might have had his *own* prejudice in mind. Having "a theory by which to work," though essential, was also risky: one's eagerness to see the theory confirmed made observation, experiment, and even articulation a constant battle *with oneself.* Internal strife of this sort was the first guarantor of truth; anguish kept men of science honest. Ideas struggled for existence in two stages: first, in the mind—against the twin impulses to abandon or enshrine them prematurely—and then second, in the wider world—in letters and journals and public debates. Darwin feared the latter most, but the former was where psychology and methodology first met.[47]

Darwin's use of military metaphors makes clear just how agonistic he thought intellectual life could be. He anticipated "a long uphill fight" in defense of his theory, but assured Hooker that "the battle is worth fighting" and that he would "resolve to buckle on my armour." Asa Gray heard the same thing: Darwin vowed to "fight my best," anticipating "a long fight." Darwin did little of the fighting himself, famously staying home for the brawls and goading his friends from afar. But his bulldogs were happy to help: "now that I hear that you & Huxley will fight publicly," he wrote to Hooker, ". . . I fully believe that our cause will in the long run prevail." Some lieutenants extended the military metaphor into Darwin's intellectual strategizing. Gray, for one, called some of Darwin's later work a "beautiful flank-movement" against evolution's opponents, which Darwin echoed later as "a flank movement on the enemy." Even though the enemy's identity could be unclear, Darwin's feeling of danger spurred both offensive and defensive efforts. As one scholar has pointed out, the project Darwin began immediately after the *Origin* seems to have been of just this sort: that is, "a 'flank movement' to capture the attention of a variety of readers." One battle was over, but the war raged on.[48]

Why? Importantly, it was not simply in *defense* of his theory that fighting was needed. The production of science was "a long battle, before it is known how much truth & how much error there is in my Book." Darwin was not a protective parent, confident of his theory's veracity and keen to keep it safe from jealous critics. Nothing less than truth was at stake, and it depended on a process of open, competitive criticism to be worked out in detail. Babbage had said much the same thing: error, "although it may be unfavourable or fatal to the temporary interest of an individual, can never be long injurious to the cause of truth." If we see our flaws for what they are (and correct them), then "the fiercest warfare of intellectual strife" produces accuracy. It may hurt, of course, but that pain was a necessary evil in pursuit of progress. Such views were the rule, rather than the exception, among the kinds of critics urging the use of hypotheses and reforming the social structure of the scientific enterprise in the mid-nineteenth century. Darwin was one of them, as were Herschel and Whewell. And their ideals of combat and competition stuck, at least for some. Indeed, such views are common down to the present day.[49]

What set Darwin's work apart from these other celebrations of competition was what he did with this image of scientific struggle: he projected it, piece by piece, onto the natural world. If Babbage's language of individuals and the collective sounds familiar, this is partly because it blurs into Darwin's eventual description of evolution's reliance on violence. By turning his experience of competition in science into a theory of natural competition, Darwin made the animals and plants he studied reflect those same values back at him. The progressive striving, the arc of individual violence and collective equilibrium that Babbage and others saw in science, Darwin found in nature, too. Both species and theories emerged and changed according to this logic of progress through competition. With Darwin's view of nature complementing his view of science, the theory of natural selection was just one more variation, its fitness determined only through experiment and debate. What struck some readers as an amoral depiction of nature, a detached description of what might seem horrific or terrifying, was in fact a naturalized account of a deeply moral commitment to testing speculative ideas through rigorous experimentation. In other words: the scientific process, like nature, was "red in tooth and claw."[50]

Darwin was not alone in blurring science and nature in this way. Gray, the natural theologian, worked hard to render the theory of natural selection compatible with his religious views. The image of a selector—or, as we might say, an experimenter—helped here, retaining some of the teleological flavor that characterized views of the Creator in this period. Defending this view of the teleological core of natural selection led Gray to mimic Darwin, drawing on the theory to explain its own methodological prospects. In a review of Darwin's book, he wrote: "To many, no doubt, Evolutionary Teleology comes in such a questionable shape as to seem shorn of all its goodness; but they will think better of it in time, when their ideas become adjusted, and they see what an impetus the new doctrines have given to investigation. They are much mistaken who suppose that Darwinism is only of speculative importance and perhaps transient interest. In its working applications it has proved to be a new power, eminently practical and fruitful." Ideas, like organisms, "become adjusted," their fitness determined by their ability to explain far-flung phenomena and, by doing so, prove "eminently practical and fruitful" in just the way that an adapted limb or organ might be for an animal.[51]

In the same review, Gray highlighted another aspect of Darwin's approach that spoke to his methodological self-awareness and the close relationship between evolution and experiment as the two friends understood them. Darwin, according to Gray, "takes his readers into his confidence, freely displays to them the sources of his information, and the working of his mind, and even shares with them all his doubts and misgivings, while in a clear and full exposition he sets forth the reasons which have guided him to his conclusions." This invitation into his own mind resulted partly from Darwin's vigilant attention to methodological matters, adhering to the Herschelian principles he saw as proper scientific theorizing. But it was also substitution: in lieu of the kind of evidentiary smoking guns other theories boasted, natural selection depended on his own mind to fill in the gaps left by nature. "In order to make it clear how, as I believe, natural selection acts," Darwin wrote in the *Origin,* "I must beg permission to give one or two imaginary illustrations." Natural selection exceeded the familiar modes of observation, and Gray appreciated how up-front Darwin was about his limits and the tools he needed to overcome them.[52]

Evolutionary theory, according to Gray, depended on a version of what Steven Shapin has called "virtual witnessing." The idea is that, to prove the existence of any phenomenon, you must enroll your readers in the experiment too, as though they had seen it themselves. Virtual witnessing achieved what geographical, chronological, or social distance seemed to preclude: the affirming presence of viewers, testifying to the veracity of what was claimed. But while Shapin described experiments an audience *could* have seen, Darwin's witnesses were virtual in a deeper sense: his experiments were imaginary, which means Darwin himself was a witness, too. What James Lennox calls "Darwinian thought experiments" were not just helpful aids to those who could not attend the original demonstration—they *were* the demonstration, for experimenter and audience alike. More than proxies for private events or future facts, his "imaginary illustrations" blurred the line between fact and fiction, between the mind and nature. This was more than a "cognitive strategy" or a rhetorical ploy. Darwin's use of the imagination, his dependence on what he could not observe, indicates the impact of methodology on the content of theories and foreshadowed new, imaginative uses to which natural selection would be put.[53]

THE COGNITIVE TREE

The collapse of Darwin's vocabulary of method into his theory of evolution can be detected at levels large and small. His use of the theory to explain its own prospects, the new readings he gave to old sources, and evolution's implications for human affairs all stemmed from the blurring of mind and nature traceable throughout Darwin's work. It even gives new meaning to one of the very icons of evolutionary theory: Darwin's famous "tree" sketch of 1837. Scribbled above its forking branches is a prefatory "I think," which has generally been read as a slippage between linguistic and imagistic thinking, as though the theory could only be expressed as a drawing. But "I think" can be read in a different way, reflecting the interpenetration of Darwin's theoretical and methodological anxiety. On this reading, the tree represents not just how nature works, but *how we think*. From stem to branch, in violent clashes and abortive attempts: the tree is a symbol of Darwin's theory and of his process, the transmutation of species and ideas. A few pages later, Darwin shifts from nature to method in a new set of notes: "difficult for man to be unprejudiced about self, but considering power, extending range, reason & futurity it does as yet appear clim . . ."—but here the page is cut, the rest of his train of thought missing. It is a tantalizing absence. While we cannot know where Darwin's reasoning was going, it is safe to say—based on his "Metaphysical" notebooks and his later publications—that this was a convergent moment, the terms of mind and nature traveling back and forth at speed.[54]

Experiment and evolution do more than parallel one another. Their balance between tests and hypotheses or selection and variations grew from the same root. The conditions of possibility for both emerged from anxiety about scientific authority and the whig interpretation of science that resulted from it. Darwin integrated them, consciously and unconsciously, as he worked to adhere to novel methodological precepts and win over skeptics who wanted to see nature as evidence of a divine mind. His ambiguity seems to have stemmed from a sense of something shared. You can call this a hedge, especially when it came to courting religious readers, but the move was more than rhetorical. At the end of his orchid book, for example, Darwin returned to comparing human and divine creation. His studies had convinced him that the actual flowers he had studied "transcend in an incomparable degree the contrivances and adaptations which the most fertile imagination of the most imaginative man could suggest with unlimited time at his disposal." On the surface, natural and human creations were "incomparable." The former awed Darwin. But below awe was wonder of another kind: that of a student paling before his teacher. The book's title emphasizes the flowers' "contrivances"; the human imagination was "fertile." These were potent metaphors that ran both ways: nature modeled perfect invention, and the mind was a branching tree of its own.[55]

Imagination, as we have seen, was both prized and feared among men of science. Partly, this was because it seemed beyond their grasp. The invisible roots of imagination led critics to naturalize it on the one hand, while holding it up as peculiarly human on the other. As Kingsley had put it, imagination was "one of the faculties which essentially raises man above the brutes, by enabling him to create for himself." Like so many others, Kingsley cautioned against letting things get carried away, while insisting that the imagination had a role to play, and a vital one at that. Kingsley's prescriptions for how to think drew upon and amplified the lessons arrived at by those arguing over the proper role of hypotheses in the work of theorizing. Imaginative ideas were the products of "a faculty which every one has," a fact that supported Kingsley's calls for broader science education even as he cautioned that it was what you *did* with your imagination that mattered most. When, in *The Water-Babies,* Kingsley taught Tom a lesson about foresight, this was the balance he struck: warning him about "the fanatics, and the theorists, and the bigots, and the bores, and the noisy windy people, who go telling silly folk what will happen, instead of looking to see what has happened already," Kingsley was not forbidding the use of imagination as such. Rather, he saw the imaginative use of the past as the key to the future.[56]

Kingsley divided imagination between "the How" and "the Why." Physical science could reveal *how* the world worked, but only natural theology saw *why*. He pressed the point to Darwin in 1865: "It is better that the division of labour sh^{d}. be complete, & that each man should *do* only one thing, while he looks on, as he finds time, at what others are doing, & so gets laws from other sciences w^{h}. he can apply—as I do—to my own." Kingsley was increasingly eager to "keep the How and the Why more religiously apart from each other," as he urged in "The Natural Theology of the Future" a few years later. After all, Darwin's theory led some to feel that "there is no Why; the doctrine of evolution, by doing away with the theory of creation, does away with that of final causes." Kingsley denied such conclusions: "Let us answer, boldly: Not in the least. We might accept all that Mr. Darwin, all that Professor Huxley, has so learnedly and so acutely written on physical science, and yet preserve our natural theology on exactly the same basis as that on which Butler and Paley left it. That we should have to develop it, I do not deny. That we should have to relinquish it, I do." As he put it in another children's story around the same time: "We must talk first with Madam How, and perhaps she may help us hereafter to see Lady Why. For she is the servant, and Lady Why is the mistress." Both depended upon "imagination, wonder, awe"—it was a matter of finding a point of equilibrium between the two.[57]

Darwin saw things in much this way. Looking back on his accomplishments around the time Kingsley was writing *Madam How and Lady Why,* he attributed his impact to "complex and diversified mental qualities and conditions," each of which balanced out another. "Of these," he concluded in his *Autobiography,* "the most important have been—the love of science—unbounded patience in long reflecting over any subject—industry in observing and collecting facts—and a fair share of invention as well as of common-sense." Part humility and part braggadocio, the location of a meeting place for invention and common sense was full English. It was the whig interpretation of science, written into a memoir of scientific achievements. Darwin was both a creature of his times and a Whig. In the 1870s, this meant that he emphasized "common sense" over "invention." Looking back on his early theorizing, he chose to memorialize his habits in familiar terms: "My first note-book was opened in July 1837. I worked on true Baconian principles, and without any theory collected facts on a wholesale scale, more especially with respect to domesticated productions, by printed enquiries, by conversation with skilful breeders and gardeners, and by extensive reading." This was, at best, a partial account. But the fact that it persisted, alongside his introspective sense of the role of imagination in his work, suggests that old habits were hard to lose.[58]

Even as he protested too much about his Baconian origins, Darwin was not above relying on the powers of imagination to fill in evidentiary gaps when he needed to. In the letter he sent to Gray in 1857 (the one that was reprinted the following year with Wallace's paper), he invited imaginative labor to aid in the acceptance of his theory. "This sketch is *most* imperfect," he admitted, "but in so short a space I cannot make it better. Your imagination must fill up many wide blanks." Like the fossil record, the theory was patchy; you had to think imaginatively to make it whole. Darwin projected this work into the future, too: "I believe," he wrote in 1863, "however, from what I see of the progress of opinion on the Continent, and in this country, that the theory of Natural Selection will ultimately be adopted, with, no doubt, many subordinate modifications and improvements." In language that shades into anatomy, Darwin pointed to hypothetical tests as proof of the theory's utility. This gives the "extensive reading" Darwin credited for his success a new meaning: the use of imagination extended beyond his own mind, into those of his readers, both present and future. The extension of science secured its fitness, proving its utility and thus its truth. Its experiments were literary, in that they operated at a distance and between the minds of those who entertained its theories. What counted most was whatever proved successful in the real world. Evolution's truth was in the testing, and test is exactly what Darwin and others did in the ensuing decades.[59]

4

MENTAL EVOLUTION

"Darwin, by the way, whom I'm reading just now, is absolutely splendid." Two weeks after the *Origin* appeared, Friedrich Engels was already urging his friend Karl Marx to read it. "Never before has so grandiose an attempt been made to demonstrate historical evolution in Nature," he continued, "and certainly never to such good effect. One does, of course, have to put up with the crude English method." It took Marx a year to get to it, but he liked what he found. He agreed with Engels about "the crude English fashion," but went even further in his praise: "This is the book which, in the field of natural history, provides the basis for our views." As Marx put it to another friend: "Darwin's work is most important and suits my purpose in that it provides a basis in natural science for the historical class struggle." Though his evolutionary enthusiasm remained, Marx soon found certain aspects of Darwin's argument off-putting. A year later, he told Engels: "I'm amused that Darwin, at whom I've been taking another look, should say that he *also* applies the 'Malthusian' theory to plants and animals, as though in Herr Malthus's case the whole thing didn't lie in its *not* being applied to plants and animals, but only—with its geometric progression—to humans as against plants and animals. It is remarkable how Darwin rediscovers, among the beasts and plants, the society of England with its division of labour, competition, opening up of new markets, 'inventions' and Malthusian 'struggle for existence.'" Natural selection seemed to invert Hegel, for whom "civil society figures as an 'intellectual animal kingdom,' whereas, in Darwin, the animal kingdom figures as civil society."[1]

Marx's reading was truer than he realized. Darwin had found more than "the society of England" in its beasts and plants—he had found the *science* of England there, too. In the natural world, hypotheses were refashioned as variations, subjected to selective pressures not unlike experiments. Pointing out this analogy between natural processes and social (or scientific) ones was supposed to undermine it; Marx's remark was meant as a critique. But something else happened instead. Linking nature and society actually bolstered Darwin's claim, by making evolution seem familiar and helping social practices seem *natural.* If both the commercial marketplace and the marketplace of ideas were reflected in nature, then they could draw on the fact that nature often felt inevitable, unavoidable, even beneficial. In other words, it was *because* Darwin had written the logic of science (and society) onto natural history that his theory, or new variants of it, could be adapted to so many intellectual and political purposes. Marx recognized something others had missed: Darwinism had *always* been social. These social roots made social applications of the theory that much easier in the decades ahead.[2]

And Marx was right. Natural selection could be "the basis for our views," as he had put it to Engels. But it could be the basis for other views, too. Evolutionary theories could support not only liberal, revolutionary projects but also conservative, reactionary ones. What has been called "Darwin's dangerous idea" was dangerous in many ways; its intentional applications and unintended consequences veered in many directions, attached to a wide range of political aims. Thomas Huxley saw the theory as an antidote to the static universe proposed by natural theology; Asa Gray saw it as just the opposite: a piece of evidence for divine order. Famous fights over these issues, refracted through such combative dyads as "science versus religion," have tended to obscure the flexibility of evolutionary thinking in the years after Darwin published the *Origin.* Its applications were underdetermined by the theory itself, responsive instead to the divergent needs of those who took it up. By assuming the priority (and prominence) of division, a framework of ideological "warfare" misses many creative readings and applications of evolutionary theory to a range of surprising issues and projects during the nineteenth century.[3]

If the theory's social roots paved the way for its broad applications, so did what a critic called its "strange inversion of reasoning." Natural selection asks us to act counterintuitively: we are supposed to interpret what seem like signs of design as random chance, not purpose. The idea, the critic argued, was that "in the formation of a perfect and beautiful machine it is not requisite to know how to make it." This was supposed to be a fatal critique, but some Darwinians wore this "strange inversion" as a badge of honor. It was not enough to banish intelligent design: they had to flip it on its head, turning the divine craftsman into an unthinking fumbler. Huxley seized the bull by the horns: "an apparatus thoroughly well adapted to a particular purpose might be the result of a method of trial and error worked by unintelligent agents." Rather than drop the idea of design, he doubled down on it, mocking Paley's natural theology by proposing imperfect, childlike processes in its place. To accomplish this, he reached for language from his own field: the study of nonhuman animals, including the great apes. Huxley was busy putting the "natural" in "natural selection" and "natural science," insisting that the two worked according to the same (unintelligent) principles. As they had for a generation, matters of method—of how minds had an impact on the material world—turned on concrete studies of human and animal nature.[4]

Marx had noted this reorientation in his mocking appraisal of Darwin. Evolution was built on social theory, which paved the way for applying natural selection to social issues. But it was not Darwin who pursued that application most aggressively. As Marx well knew, the contours of nature and nurture, of science and society, were being probed from many directions and with multiple tools in the middle decades of the nineteenth century. Darwin was part of this shift; his *Descent of Man* (1871) and *Expression of the Emotions in Man and Animals* (1872) brought his theory of plants and animals to bear on human affairs. But Darwin's attempt to naturalize human behavior as biological functions was far from original. Such efforts were part of a longer story about the rise of "scientific naturalism" or just "naturalism," as it came to be called in the Victorian era. Scientific naturalists saw the world only in terms of their preferred modes of observation. Whatever their tools did not reveal did not exist. To them, nature's hidden processes unfolded according to laws discernible by science alone.

But work on mental evolution, inspired by Darwin and others, altered the parameters of this scientific confidence. If the mind was a part of the natural world, and science was rooted in the mind, then scientific naturalism had two meanings. Most familiarly, it meant studying nature using science's tools. This is what we usually mean by "scientific naturalism": a commitment to being scientific about nature. When the story of Victorian science is told, this is the meaning of naturalism at center stage. That story tends to revolve around replacing the natural theology of Paley (and the younger Darwin) with a narrower sense of what could be inferred from specimens or equations. As science's power expanded, more and more people saw the world around them in the terms set by scientific discourse. In other words: the world, for some naturalists, was *merely* natural. Scientific interpretation gradually crowded other modes of analysis out of the room.[5]

The second meaning of naturalism folded science inward. This sense of the naturalistic turn, transforming scientific thinking into an object of its own inquiry, is less familiar but more profound than the simple commitment to scientific study. Why? Because turning science into a natural phenomenon changed not just how science was imagined to work, but also the authority claimed on its behalf. Science, embedded in the human mind as an adaptive process conferring fitness advantages, came to seem like the obvious tool for addressing an ever-widening range of theoretical and practical problems. Ongoing transformations such as the rise of the research university and increasing secularization were bolstered by science's naturalization, which in turn was shaped by the elaboration of evolutionary theories starting in the mid-nineteenth century. Unfolding against clashes over democracy and cultural authority, the embodiment of science took some unexpected forms. Neurological studies of learning and the anthropology of innovation contributed to this shift in science's imagined location. In mid-Victorian Britain, to be a naturalist meant seeing science in the world around you—and treating it as you would treat any other natural process: by studying it scientifically. The result, over the course of the 1850s and 1860s, was a science grounded in the messy, organic, evolving functions of the mind.

EMOTIONS OF INTELLECT

At the end of the *Origin,* Darwin allowed himself a bit of forecasting. "In the distant future," he predicted, "I see open fields for far more important researches. Psychology will be based on a new foundation, that of the necessary acquirement of each mental power and capacity by gradation. Light will be thrown on the origin of man and his history." This was the only glimpse Darwin gave in the book of his long-standing interest in the mind and morality, and it was controversial for at least two reasons. The first is the familiar one: implying that a book that was (mostly) about fossils had any bearing on the exalted human mind. Any such applicability was far from generally accepted, as Darwin's decades of hesitation make clear. But his prediction was also controversial for what it left out. According to many of his peers, the new foundation to which Darwin alluded was already set—it just was not an evolutionary one. Contemporaries already working in psychology saw their future, not in natural selection, but in the language of association, the careful study of our ideas and the taxonomy of how they connected with one another to form trains of thought. In their view, association—not adaptation—was going to be the foundation of a new science of mind in the nineteenth century.[6]

A month before the *Origin* appeared, this alternative foundation was being celebrated in nationalist terms. "The sceptre of Psychology," an anonymous reviewer wrote in 1859, "has decidedly returned to this island. The scientific study of mind, which for two generations, in many other respects distinguished for intellectual activity, had, while brilliantly cultivated elsewhere, been neglected by our countrymen, is now nowhere prosecuted with so much vigour and success as in Great Britain." Though it may sound like it, this was not a claim for the importance of the laboratory. It was instead a plea for associationism, a plea authored by none other than John Stuart Mill. The stakes for Mill were personal, as his father James had helped turn associationism into a point of national pride. The elder Mill argued, as John Locke had, that "elements" of the mind were "compounded" from experience by mental activity, organizing his analysis chapter by chapter in accordance with this view. Unlike his father, John Mill was not committed to experiential phenomena alone; he wanted to have it both ways, to include prior or innate ideas in a broadly associationist scheme. As he did in his political philosophy (including *On Liberty,* published in the same year as the anonymous review), the younger Mill struck a balance between competing forces at the level of the individual mind.[7]

To do that, Mill was seeking a disciple, someone to whom he could hand the "sceptre" and who could help modernize his father's approach. The 1859 review was an announcement that he had found such a protégé: Alexander Bain, a personal friend and follower of both Mill and his father. Bain, who had come from Scotland to London two decades earlier to sit at their feet, was the perfect embodiment of modern associationism. His *Senses and the Intellect,* published in 1855, and *The Emotions and the Will,* which had just appeared, were the subject of Mill's review. Calling them "the most advanced point which the *a posteriori* psychology has reached," Mill was in essence praising himself and his father. But Bain went even further than either Mill, highlighting the physical (that is, physiological) causes of mental states. By pointing inward, to the brain, rather than outward to the state as the younger Mill had done, Bain carried the mantle of psychology into a period in which the organic substrates of mental phenomena were increasingly emphasized. This was associationism for a neurological age.[8]

Bain's background was a long way from Mill's. Born poor in Aberdeen, he had to teach himself the things Mill had learned from his father, reading books when not working at the loom. Remarkably, he found his way to the utilitarianism and associationism Mill himself had learned, a convergence Bain made official by moving to London. His early essays betray a psychological bent, including a fascinating piece on toys from 1842 and a glowing review of Mill's *System of Logic* the following year. The two young men traded compliments early on, with Mill's passing of the "sceptre" mirroring Bain's claim that Mill had modeled "a philosophy by a systematic generalization of the methods and processes of modern positive science." The relationship between the two remained close and complex until Mill's death, as each attempted to improve upon the ideas of Mill's father and to meet the challenges of new accounts of the human mind and human nature. Mill may be better remembered today, but it was Bain whose psychological work had the most immediate impact on scientific discussions. The two books Mill reviewed were standards in British psychology for half a century, and *Mind*—the journal Bain founded in the 1870s—was the first of its kind in the world.[9]

A quick glance at the table of contents for *The Senses and the Intellect* and *The Emotions and the Will* shows how similar Bain's approach was to James Mill's. Both broke their books down by trait, starting with the senses and proceeding toward more complex phenomena. The very idea of a mind split into four "divisions" (senses, intellect, emotions, and will) mirrored Mill's taxonomic approach, while Bain's insistence that his own work "proceeds entirely on the Laws of Association, which are exemplified with minute detail and followed out into a variety of applications" perfectly echoed James Mill's *Analysis of the Phenomena of the Human Mind.* Bain "reduced" (in his words) the mind to a set of laws governing associations, another nod to his predecessors. Even though Bain tried to distance himself from the elder Mill in some ways—claiming, for instance, that Mill's "subdivision into faculties is abandoned"—these efforts were questionable at best. Insisting that the "various faculties known under such names as Memory, Reason, Abstraction, Judgment, &c. are modes or varieties of Intellect," he nevertheless granted each a distinct role—thereby echoing associationist themes even as he complicated them.[10]

In Bain's scheme, every thought is reducible to a set of associations. This was meant to be a continuation of John Mill's own defense of his father. As Bain put it near the end of *The Senses and the Intellect:* "Mr. John Stuart Mill, in his *System of Logic,* has shown, I think, conclusively, that the basis of all inference is a transition from particulars to particulars, and not, as usually supposed, the application of a general affirmation to the special affirmations included in it." In other words, we might *think* we have inferred a conclusion ("Socrates is mortal") from a general rule ("All men are mortal") and an instance ("Socrates is a man"), when in fact the *rule* was actually an inference, one in which we imagined "all men" instead of applying an abstract test. We do not discover Socrates's mortality at the end of a chain of reasoning so much as import it at the beginning. Or, as Bain put it in a discussion a little further on: "Much of the acquisitions of a strong intellect is in reality the re-discovery of what is already known," not least because "it is the nature of an advanced science to contain innumerable identifications summed up in its definitions and general laws." Insofar as it was "by a vigorous similarity that these were first formed; by the same power they are rapidly acquired." Individual learning recapitulated the history of the field; associations that were hard-won through experiments were picked up almost unconsciously after the fact by youthful learners.[11]

Science, on Bain's view, is a kind of natural process, an extension of normal ways of thinking to their logical conclusions. It was not always easy, of course. For one thing, science demanded certain unnatural acts of the mind—including, intriguingly, associationist method. "It is always to be remembered," Bain noted halfway through his first book, "that although scientific method requires us to take the different aspects of mind apart, yet in the mind itself they are always working together; Emotion, Intellect, and Volition, concur in almost every manifestation." Bain was arguing that we split pieces of consciousness up without meaning to, in order to analyze them. In doing so, he continued, we violate not only our experience of thinking or reasoning but also the nature of the relations between our thoughts. In certain sciences, these difficulties border on the absurd—but have, at their root, the same natural tendencies. Take what Bain had to say about the thinking required in mathematics:

> The self-denial that enables us to dwell among algebraical symbols, and to concentrate the whole force of the brain upon

> these, to the exclusion of all those things that gratify the various senses and emotions,—this abnegation, so to speak, of human interest, is the moral peculiarity of the mathematician. To be able, for the sake of the ends of Science,—the attainment of truth and certainty as to the causes of things,—to force the mind to entertain willingly conceptions so meagre as the diagrams of Geometry or the symbols of Algebra and Chemistry, proves that the cerebral currents go naturally towards the fixing of mere visible forms, such as have no interest in themselves, but serve as the instruments of our practical ends.

Spoken like a man who had to teach himself arithmetic between shifts at work, Bain's account of scientific experience emphasizes the *labor* of making mental associations, as well as the motives for doing so in both science and everyday life.[12]

This labor was linked to what Bain called, in his second book, "emotions of intellect." Even in our most abstract modes of reasoning, affective dimensions held sway; emotions played a causal role in every act of cognition. When we face a paradox, for example, "a certain suffering is caused by the attempt to entertain contrary accounts of one thing." Puzzles are never logical dilemmas alone—we *feel* them. The "intellectual revulsion" that accompanied confusion was, for Bain, a natural response, an impulsive reaction that shaped the reasoned reflection to follow. One such response was scientific inquiry, which could be seen as a natural way of processing adverse affective states. According to Bain, this was a fundamental feature of high-level reasoning: we *want* to solve the problems we confront, experiencing doubt and uncertainty in visceral, charged ways. This view of scientific method grounded in feeling and reasoning, affect and judgment, was one of Bain's most important ideas, judged by the reception it got from his peers and the uses to which it was put. It made the most detached modes of inquiry into deeply human, even animal impulses—and did not reduce them in the process. Far from it, in fact. By making science into something natural, Bain's physiology of rationality helped turn the alienating activity of high theory into something inherently worthy of pursuit.[13]

Emotions were not just there at the start; they were also involved in how experiments ended. When we solve some puzzle, Bain argued, we experience "a rush of delight, the delight we feel when we are relieved of some long-standing burden." As with the beginning, so with the end: in the laboratory and outside of it, the emotions of intellect help us decide when an answer is good enough, when we have solved something. Whether theoretical or practical, questions and answers are *embodied,* wrapped up with the gut feelings of doubt and satisfaction. "When new knowledge has a practical bearing," Bain elaborated, "the emotion is that produced by the more easy fulfilment of practical ends, and is of the same character of deliverance from labour or enlargement of effect." If you resolve an intellectual dilemma, the feeling is the same as after a hard day's work. Indeed, the two were impossible to separate. "The attainment of Truth is best proved in the region of practice," Bain argued, and a bad theory is bad not only because it "lacerates the sensitive intellect," but also because "the operations of daily life would be frustrated" by its use. Science was physical, a matter of feelings and—by extension—of matter. We clench our fists and rub our temples, proof that the stuff of thought is really, truly *stuff,* just like we are. If science was cognitive in a new way, cognition was newly embodied.[14]

For Bain, higher-order processes like hypothesizing and rational deliberation were a part of our bodies, the same as instinctive reactions to scary situations or reflexive responses to stimuli. While the Mills' associationist approach was exclusively mental, Bain thought that associations linked the mental and material worlds. Not just our senses but our *bodies* were psychological, as subjects to study and tools for studying them. In his notes on Mill's *Analysis,* Bain wrote: "Every mental state can be proved to have its counterpart physical state; joy, sorrow, fear, are each embodied in a distinct group of physical effects in the nervous system, the muscular movements, and the organic processes." Later, in a critical review of the younger Mill, Bain argued that his greatest error was his "disregard of the physical conditions of our mental life." Neither the father nor the son denied these connections, but both failed to account for their role in thinking. Both men "did not allow what every competent physiologist would now affirm to be the facts," instead leaning their accounts of subjective experience on "the *fixity of order* in our sensations." For the Mills, then, psychology was the study of patterned associations, accessed through introspection; for Bain, by contrast, a natural science of the mind would include a greater range of phenomena, including facts about our bodies to complement those about our minds.[15]

Emphasizing the bodily dimensions of inquiry highlighted what inquirers had in common rather than what set them apart. Or at least—it could. Bodies could be used to divide people, but they could also bring them together. Shared properties were the special province of physiology, a field aimed at common processes. By insisting on an embodied approach, then, Bain emphasized the links between our inner selves and "an External and Independent World." Those links could be elucidated only by "dwelling upon the 'common to all,' in contrast with the 'special to me,' to use one of Ferrier's forms of phraseology." In other words, a true theory of how we think had to start with the sensory and motor physiology we share, as humans and even as mammals, rather than the subjective experiences that make us who we are and distinguish us in private from one another. Internally, thinking is what Bain called "the singling out of one among many trains" of thought. But to get it right, to justify the claims of a new physiological psychology, those trains were not enough. What was needed was a study of how trains in general worked. Or, put another way: We need to study not only the trains, but the tracks on which they run. According to Bain, those tracks are laid by our physiology, and specifically by its development over the life-course as a reflection of our shared inheritance. The tracks are what bind us together.[16]

The importance of physiology comes through immediately in Bain's work. *The Senses and the Intellect* begins with a huge section on nervous physiology and the role of the senses as a link between the body and the mind. In *The Emotions and the Will,* anatomical issues are once again at the heart of Bain's analysis, with a special emphasis on *spontaneous* nervous activity. In both books, the claim is that all thinking—and indeed, all human activity—begins with reflexive movements that arise from the spontaneous action of nervous energy. We blink, swing our arms, cry out—and over time we develop sensory and intellectual associations with bodily motions that enable us to achieve our aims. At the end of a chain of bodily and mental associations, extended out over individual and collective learning, was the scientific process. Though many theorists made it seem disembodied, objective *because* it was removed from everyday concerns, Bain was pushing the other way. Science and its method had their roots not in abstract considerations but in the same passions and frustrations as all human and even nonhuman behavior. The physiology of mental functions pointed, ultimately, to the root causes of scientific achievements.

Our associations also *end* in the active dimension. Take the theory for which Bain was (and is) best known: his account of belief. We believe something, Bain argued, only when we are prepared to act on it. This applied to all beliefs, including scientific ones: "When I believe in the circumference of the earth being twenty-five thousand miles, if I am not repeating an empty sound, or indulging an idle conception, I give it out that if any occasion shall arise for putting this fact in practice, I am ready to do so. . . . If there were any hesitation in my mind as to running those risks, my alleged belief would be proved hollow, no matter how often I may have heard the statement, or repeated it, with acquiescence." What begins in bodily movement ends in putting our bodies on the line; the only meaning of belief, for Bain, was that upon which we were prepared to act. It is not that disembodied beliefs do not exist or are less important. Rather, when we struggle to identify the practical bearings of a belief, we are called to just the kind of inquiry in which Bain was engaged. Seeking out the embodied causes and effects of mental phenomena was the point of reflexive inquiry. Science was about finding the foundations of belief.[17]

Bain called this process "trial and error." His use of the phrase is where we get our sense of the term, as (potentially) applying to a range of mental and natural phenomena. Before Bain, "trial and error" described a technique for the quick solution of arithmetic puzzles, named as such in the sorts of elementary textbooks out of which Bain taught himself mathematics in the 1820s. It was a trick you could use to solve a problem by guessing its solution. In Bain's hands, that technique was remade into the core of all thinking, human and otherwise. Right down to our brain, we are always guessing at solutions and checking to see if they fit. What he called "the grand process of trial and error" in *The Senses and the Intellect* was the key to all mental activity, not just simple things like solving puzzles. "In all difficult operations for purposes or ends," he continued, "the rule of 'trial and error' is the grand and final resort." For Bain, "trial and error" was a balancing act: we guess and check; we try, fail, and repeat. At its foundation, the process was governed by pleasure and pain—Bain was, after all, still loyal to the utilitarianism of his youth—but it was also more than that. Linking observable behavior to the "imaginary activity" of abstract cognitive functions marked a new step in the *embodiment* of human thinking in the nervous system. Trial and error started with physiology; with it, the old ideal of equilibrium burrowed deep into the brain. At its simplest, science was at work in our cells.[18]

Likening all thought to "trial and error" had some surprising effects. If all our behaviors, for example, followed "the primitive course of trial and error," then we had a lot to learn about the links between our conscious and unconscious actions—and our agency in general. If much of what we took to be intentional was (partially) accidental, then we would have to rethink learning as a purposive process and the educational system that encourages it. On Bain's theory, school is another word for developing a "specializing, or selective, consciousness" with which to control the barrage of sense data that hits us from the day we are born. "The cognitive process," Bain argued, "is essentially a process of selection, as the mind is moved to special, or monopolized, consciousness of certain portions of its various experience." The task of the educator was split. We needed to create space for trials (and errors) in the classroom, and then train students to distinguish between success and failure when the tasks were more abstract or the evidence was harder to read. The skill of abstraction required an empirical foundation to ground it; science had to start by recognizing failure. "Trial and error" meant both freedom and control.[19]

Written in the year in which Darwin published the *Origin,* Bain's account of "trial and error" was not so much influenced by evolutionary theory as a kind of complement to it. Drawn from the same sources as Darwin's thinking, Bain's psychology operated on similar principles of natural balance and the mirroring of science and nature. While Bain and Darwin were aware of one another's work, their theories of the mind and of transmutation ran in parallel in the 1850s, not together. A few years later, when Huxley argued that natural selection was "something which may fairly be termed a method of trial and error," the two theories were unified in a way neither man had done for themselves. Their connection, at least partly, had to do with the dual nature of the processes both men described: the seemingly random production of variations or trials, on the one hand, and the "process of selection" through which some were repeated or deemed erroneous, on the other. Bain's theory would *become* evolutionary, but it was not there yet.[20]

The major roadblock was methodological. If Bain's midcentury books bridged early- and late-nineteenth-century psychology, as many have claimed, they did *not* do so on the level of practice. Like those who came before him, including the Mills, Bain was mostly an introspective thinker: he gathered up the evidence for his theories by analyzing his own mind, presenting his findings as though they were shared and asking his readers to examine themselves. In the world in which Bain came of age, this kind of reflection was just what it meant to do mental science. But around the time Bain published, thanks in part to the physiological work he highlighted, the hegemony of introspection was beginning to crumble. Soon, self-proclaimed psychologists began to distance themselves from methods they deemed philosophical (which is to say, metaphysical). In Germany, the rise of laboratory teaching at universities enabled psychologists to replace their own senses with those of their colleagues, using new instruments to test the thresholds of experience. Among British psychologists, for whom such instruments played a more minor role, field observations of other minds gained prominence. This included watching other humans, but also—significantly—observing nonhuman animals. Just as Bain took on the mantle of British psychology, what it meant to do so was evolving.[21]

EXTENDING THE SENSES

Not everyone was a fan of Bain—or of associationism. In another anonymous review from around the same time, we start to see why. Though the reviewer praised Bain's "natural history of the mind," he argued that it failed to meet the standards for scientific work. The problem was again one of method. "It is not in itself a system of mental philosophy, properly so called," the reviewer complained, "but a classified collection of materials for such a system, presented with that method and insight which scientific discipline generates, and accompanied with occasional passages of an analytical character." The books were a mere "collection," and a partial one at that. A true theory of mind had to be more than a litany of traits or an accounting of what you could find in a mind at any given moment. There was a larger story, a broader tapestry of which instances were just that: instances. By "confining himself to an account of the emotions as they exist in the adult civilized man," Bain had missed a key aspect of mental life: it evolves.

The reviewer, Herbert Spencer, argued that any psychology worthy of the name had to be more than a table of contents or a series of snapshots. Mental evolution was a necessary topic for a field that was itself evolving. That meant comparing across lines of species and race but also tracking development over individual lives. Spencer had published his own account in the year of *The Senses and the Intellect,* and he was busy recasting it as a piece of a larger project geared toward bringing the theory of evolution to a wide audience. This explains his critique of Bain's book: it lacked an account of how mental states were related, historically and developmentally. "Ascertaining by induction the actual order of evolution of the emotions," Spencer argued in his review, "we are led to suspect this to be their order of successive dependence; and are so led to recognise their order of ascending complexity, and by consequence their true groupings." The kind of classification Bain provided, an organization of a familiar set of traits along new lines, was only the beginning. It would take a grand theory like Spencer's to make sense of its many disjointed parts—at least according to the person authoring that theory.[22]

Self-promotion made sense for Spencer. Born in 1820, his childhood was a lot like both Bain's and Mill's: heavy on industry, light on affection. He shared with his fellow associationists an early interest in logic and a nonconforming approach to religion, which for Spencer seems to have extended into general distrust of tradition for his whole life. What his biographer termed a "longing for passion" seems to have informed Spencer's view of the mind as an active force, not unlike Mill's effort to "balance" his father's philosophy with the emotional zeal he gleaned from Romantic literature. For Spencer, life was less a balancing act and more a constant struggle—an experience he turned into an account of how *all* organisms contend with their environments. He built a system of philosophy out of a lifelong need, turning his own struggles growing up into a picture of struggle all over the universe. Disappointment and ill health gave Spencer's work an increasingly bleak cast over the course of his career, but in the 1850s he was busy building up a system that would make him one of Britain's most famous scientific figures. Part physiology (in the spirit of Bain) and part philosophy, Spencer's approach was predicated on a tension between self and other that he claimed was evident in firsthand experience—as in his own biography.[23]

At its heart, Spencer's project envisioned a new place for science in society. Like the younger Mill, Spencer thought Britain had gone dark "for a generation or two"; unlike Mill, he placed Bain in the dark age. Likening contemporary psychology to the transformations that had recently roiled the field of geology, Spencer cast himself as kind of torch-bearer: "Reasoning without adequate data having led to nothing, inquirers went into the opposite extreme, and for a length of time confining themselves wholly to collecting data, relinquished reasoning. The Geological Society of London was formed with the express object of accumulating evidence; for many years hypotheses were forbidden at its meetings; and only of late have attempts to organize the mass of observations into consistent theory been tolerated." Such attempts led naturally to his own work, Spencer thought. Selling subscriptions to what would be a twelve-volume *System of Synthetic Philosophy,* he felt the need to reject Bain's approach and make something more of psychology.

The mind became part of a universal puzzle, the key to which was an evolutionary worldview Spencer sold to those who yearned for the bigger picture that Darwin, at least, seemed hesitant to provide. Spencer made his name by meeting that desire (and indeed, his books sold more than Darwin's for the rest of the century).[24]

Spencer was not alone in seeing Bain's "natural-history-method" as a partial solution to the problems psychology faced at midcentury. Bain himself described his work in similar terms in the preface to *The Senses and the Intellect* in 1855: "A systematic plan has been introduced into the description of the conscious states in general, so as to enable them to be compared and classified with more precision than heretofore. However imperfect may be the first attempt to construct a Natural History of the Feelings, upon the basis of a uniform descriptive method, the subject of mind cannot attain a high scientific character until some progress has been made towards the accomplishment of this object." Though Bain aspired to provide psychology's "high scientific character," it is possible to read this pronouncement as an acknowledgment of the book's preliminary status. The experience of reading Bain is not unlike that of reading a museum catalogue or travel narrative: here and there, arguments shine through, but the overall impression is of a march through parts of body and mind. Its organization was assumed more than asserted. Spencer's criticisms may have been self-serving—but they also stuck.[25]

The preliminary nature of Bain's account was not its only methodological issue, or even its main one. What hit closer to home was the charge that natural history, in his sense, miscast the mind. It was the Oxford philosopher F. H. Bradley who put the finest point on this problem: "Mr Bain," he wrote in a pithy footnote in his polemical *Ethical Studies*, "collects that the mind is a collection. Has he ever thought who collects Mr. Bain?" Collecting can take you only so far, but what is worse is if it takes you in the wrong direction, by inventing the object it purports to describe. If the mind is a collection, the question is: Who (or what) collected it? As Bradley noted in 1876, "this collection is aware of itself; it talks about itself as if it were simple." Picturing a collection places you outside it, which means the model is missing something: you.

Bradley satirized the problem with an image of onions hanging from a rope: "Talking about 'self,' we (i.e. the onions) fall into the belief that there is something there under the onions and the rope, and on looking we see there is nothing of the kind." This is where associationism got you: "we ourselves, who apprehend the illusion, are ourselves the illusion which is apprehended by us." Bradley's circularity was intentional. The associationist theory of mind-as-a-collection was as coherent as self-conscious onions or a wisp on a river theorizing its own existence. By situating the psychologist as both collector and collected, the theory preserved a tension that its adherents could not resolve in the language of associations alone.[26]

Collecting was the problem. Bain was bullish on psychology's scientific potential, which led him to model it on what he saw as the most scientific methods of his day. A lengthy footnote in the first chapter of *The Emotions and the Will* makes clear just what these were: "The whole plan of the present book can be best explained by quoting the method of a good treatise on one of the Natural History sciences, namely, Mineralogy, Botany, and Zoology." What follows borders on a parody, as though Bradley or Spencer themselves had written it into the book. Summarizing the taxonomic practices across a range of fields, Bain argued that the emotions, similarly, "are to be described in families with genera and species, so to speak, included under them." Citing a textbook on botany and "Mr. Charles Darwin's Monograph on the Cirripedia," he insisted that a science of the mind must treat mental states like so many trees or beetles, to be collected using introspection and pinned to a sheet for further study. The effect of Bain's view of science was to write his preferred method onto the mind; as Bradley said, he "collects that the mind is a collection." It was not enough to say that, to a collector, everything is like a collectible (though this was no doubt true). In the case of someone like Bain, eager to build up a complete mental science, he saw the mind as a collection just like the one he was amassing, held together by both material and mental associations. The model could get messy in a hurry.[27]

But if not collecting, then what? Spencer was not proposing that we abandon natural history, but rather that we do it better. "Alike in classing separate organisms, and in classing the parts of the same organism," he wrote in his Bain review, "the complete natural-history method involves ultimate analysis, aided by development." Psychologists, like taxonomists, needed a set of guiding principles in order to arrange their objects accurately. Without physiology or history, psychologists might confuse external signs for internal reality, just as a zoologist without the proper tools might look at a whale and call it a fish. However, while the interpretive risks were analogous, their solutions seemed quite distinct. Digging around inside a dead marine mammal might be intimidating, but you could imagine doing it. How were you supposed to "dig" into the human mind? And what were you digging for? On the surface, Spencer seemed to study the mind just like Bain did: introspecting on his personal experiences and reading widely in relevant fields. But quietly, Spencer's reflection and reading were in the service of a specific hypothesis, one much larger than psychology alone. His idea was that a universal form of evolution governed everything from gravity to cognition. It was in pursuit of evidence for this theory that Spencer probed the mind. Far from collecting for its own sake, he was doing the opposite—collecting evidence for a conclusion of which he was already certain.[28]

Spencer always said that he had been an evolutionist since childhood. A loss of faith led directly to belief in natural causes. "Doubtless, too," he recalled in his *Autobiography,* "a belief in evolution at large was then latent; since, little as the fact is recognized, anyone who, abandoning the supernaturalism of theology, accepts in full the naturalism of science, tacitly asserts that all things as they now exist have been evolved." What was tacit became explicit when, at the age of twenty, he read Lyell's criticism of Lamarck—and was converted to the views of the latter, not the former. Spencer only publicized this conversion a decade later, in a short (anonymous) piece, "The Development Hypothesis," in 1852. In it, Spencer lampooned religious demands that supporters of transmutation give evidence for their interpretation without requiring any evidence at all for their own beliefs. Asking not for any proof of divine creation, but just for "a *conceivable* mode" of its having occurred, Spencer set the issue of evolutionary development alongside the questions of mind and method that Darwin was tackling in the privacy of Down House. Framed in this way, the superiority of the development hypothesis over a theory of individual creations was cognitive: You can imagine various ways in which creatures evolved over time, whereas the view that they were created separately required a mental leap. It was this leap, according to Spencer, that undermined the idea.[29]

The difference was not one of intelligence, Spencer insisted, but of habit. Whether or not you thought evolution was possible had more to do with what you were taught as a child and the way you learned to see the world than with your ability to entertain propositions as an adult. Or, to put it in a more Spencerian idiom, your ability to entertain propositions as an adult was tied up with the modes of conception into which you had been trained as a young person, and indeed those into which your *ancestors* had been trained. "Habitually looking at things rather in their statical than in their dynamical aspect, they [the creationists] never realize the fact that, by small increments of modification, any amount of modification may in time be generated." A geologist who believed in special creation would "be forced to confess that the notion was put into his mind in childhood as part of a story which he now thinks absurd." But habitual views were not just the stuff of religious believers or Spencer's enemies. He, too, saw the world through his habits—and admitted it. Even though Spencer insisted that "the acceptance of things simply on authority was not habitual" for him, habit had made him speculative from a young age. Even his biggest idea had habitual roots. "From this time onward," he recalled of the year 1852, "the evolutionary interpretation of things in general became habitual, and manifested itself in curious ways." Habits seemed to explain a great deal—including Spencer's own interest in habit.[30]

A year after publicizing his development hypothesis, Spencer applied it reflexively to the very method he used to articulate it. Published as "The Universal Postulate" in *The Westminster Review,* his argument was that our minds had evolved exactly like organisms. Both species- and individual-level changes had equivalents in mental evolution. As evidence, Spencer invited his readers inside his own head, to consider his mind at work in the act of writing the very essay they were reading:

"Regard the philosopher objectively. Is it not clear that the faculties he is now employing in reasoning about consciousness and ideas are the same faculties with which in childhood he drew his simplest inferences? Must not the action of these faculties follow throughout, the same law? Must not the results of their action be therefore congruous? And when they are not congruous, does not the fact indicate something abnormal—some nonconformity with the laws of their action—some error, as we say?" For Spencer, our theories accord with instinctive ways of seeing the world—with our beliefs, that is, and specifically with "beliefs that invariably exist." This correspondence between a view of the world and our deepest beliefs had become, for Spencer, "the universal postulate." It was the key to evolutionary history.[31]

To illustrate the postulate and its importance, Spencer turned to a familiar disagreement in which he had grown interested: the dispute between Mill and William Whewell over method and the mind. Where Whewell defended a version of Spencer's view on "inconceivability," Mill lampooned it—but, in doing so, seemed to support Spencer's evolutionism. According to Mill, our inability to conceive something was "an affair of accident, and depends on the past history and habits of our own minds." You can almost see Spencer nodding in agreement. Yes, he would say, this is exactly what inconceivability amounts to—but that does not disqualify it as a way of accounting for truth or reasoning! Given the length of one's life, or the history of science as a tradition, those beliefs that *survive* the constant barrage of sense experience or experimental data are bound to correspond, at least in a rough sense, to the real world. If the contents of our minds are, in our own lives and in the accumulated experience of our ancestors, subjected to constant inputs from the outside world, then what follows is a kind of evolutionary process in which survival signals success. Here, years before the publication of the theory of natural selection or the full elaboration of his own evolutionary views, Spencer provided an account of mental evolution that is usually taken to have emerged much later. And, much like Darwin, he worked it out in the context of scientific, rather than natural, competition.[32]

Spencer's essay "The Universal Postulate" is replete with the language of instinct, which he seems to use in the hope of enrolling readers through mutual recognition of our shared traits. "Instinctively we recognise the truth above demonstrated," he insisted about halfway through. The hope was that skepticism could be deflected by reminding readers of their common mental traits—about which Spencer, after all, was casting himself as an expert. By this same line of reasoning, Spencer tried to redefine even abstract, scientific ideas in terms of beliefs everyone had in common. For example: "Logic is simply a systematization of the process by which we indirectly obtain this warrant for beliefs that do not directly possess it." Philosophy solved a problem we all face: testing our private beliefs to see if they hold up. In this spirit, Spencer later returned to his immediate, *physical* context, inviting his reader to "contemplate an object—this book, for instance." Building up from the reader's consciousness of the paper, the words on the page, and the ideas expressed through those words, Spencer established a developmental chain linking the basic sense perceptions of his readers and the very philosophy they were encountering. "It was hinted at the outset," concluded Spencer, ". . . that the analysis of Philosophy must agree with the synthesis of Common Sense." And it does. "The instinct justifies the logic: the logic accounts for the instinct." The system was intentionally circular, a reinforcement mechanism. Anything less than a theory binding scientific method to ordinary thinking amounted, in Spencer's closing words, to "logical suicide." Any good scientific system had to account for science, too.[33]

Spencer set out the same line of argument in an anonymous review of books on method, including Whewell's. Printed under the title "The Genesis of Science," the piece tackled the biggest irony of Spencer's point about scientific and everyday thinking: their similarity violated common sense! Many readers, he admitted, have "a vague notion that scientific knowledge somehow differs in nature from ordinary knowledge." To argue for their similarity was a tall task, which Spencer approached by appealing to a single process underlying both science and common sense: "In both cases their mode of operation is fundamentally the same." The difference, according to Spencer, is that "science may be called *an extension of the senses by means of reasoning.*" In other words, while science was not fundamentally distinct from other ways of thinking, its power to predict based on sensory and nonsensory information—and the benefits that accrue from such predictions—meant it could extend our senses spatially and temporally. Just as Bain was doing at the same moment, Spencer was seeking a unified account, a way to bring our "basic" ways of thinking together with the exalted capacities of reasoning that were exemplified, he felt, in his own work. Whereas Bain revealed the "lowly" affective origins of cognition, Spencer was emphasizing the "higher" cognitive effects of something like affect. The two men were coming at the same thing from opposite directions.[34]

The point of convergence, and the goal of methodology more generally in these years, was the "genesis" to which Spencer's title referred. For him and for Bain, the nature of that genesis was mental, an underlying process that was shared across all kinds of thinking and thinkers. The question was how to get at it, how to dig down (or up) to find the point at which science merged with all the other modes of inquiry—and existence—that are out there. For his part, Spencer proposed a broad, synthetic project, one that would unify history and philosophy with psychology and anthropology, the study of "mature" science with thinking that was not (yet) scientific. The goal, according to Spencer, was a natural history of the mind—which is precisely what both he and Mill praised Bain for pursuing. Such natural histories needed to be reflexive, of course: thinking about thinking, reasoning about reasoning, meant bringing oneself into the study. Men of science would be their own objects, not only in psychology but in all the fields from which Spencer proposed to draw. What united them, or would, was their commitment to elucidating origins, to peeling back the layers of mental life in order to discover the genesis of the shared process underlying every method under study.

Even more than Bain, Spencer was after the *utility* of methods, not least because he saw the pursuit of methodology as a means of accounting for success and failure. His criticism of Comte's hierarchy of the sciences rested on this very idea. Though Spencer approved of Comte's recognition of the "division of labour" in the sciences, he disapproved of his particular division because it elided the evolutionary relationship between its parts. "The fact is," he wrote, "that the division of labour in science, like the division of labour in society, and like the 'physiological division of labour' in individual organisms, has been not only a specialization of functions, but a continuous helping of each division by all the others, and of all by each." Otherwise, Spencer argued, instability reigned. In any such system, every part must benefit; if they did not, it would devolve into competition rather than cooperation. This was true at both social and individual levels. The genesis of science was "a psychological process historically displayed," grounded in the basic mental functions with which all minds approach the world. This suggested that Comte's hierarchy of the sciences was too limited. A true account of scientific development was the same as one of mental evolution: it not only had to *include* psychology, it had to be *based* on it. Science, for Spencer, was just another feature of the adaptive mind.[35]

This line of reasoning, from a methodological dilemma to mental evolution, led Spencer to commence work on a psychology textbook soon after he completed "The Genesis of Science." He set himself the task of applying evolutionary ideas to the understanding of processes ranging from the most basic and general to the most complex and specific, all arranged in developmental relations to one another. Spencer began what would become his *Principles of Psychology* with an edited version of "The Universal Postulate," anchoring the book's opening section on "General Analysis." From this simple beginning, he worked his way up to the operations of the intellect, including those seen as central to scientific thinking. Order reflected the evolution of intelligent processes, "which, as originally performed, were not accompanied with a consciousness of the manner in which they were performed, or of their adaptation to the ends achieved, become eventually both conscious and systematic." Science, no matter how technical, was part of life in the ordinary sense. Was it not obvious, Spencer asked, "that, both in respect of origin and applicability, no method is possible but such as consists of an orderly and habitual use of the procedures which the intellect spontaneously pursues, but pursues fitfully, incompletely, and unconsciously?" The answer, at least to Spencer, could not have been clearer.[36]

If all methods emerged this way, then the study of their emergence suggested a new role for psychology in enabling scientific advances. As Spencer put it: "If in the instances given, the method of forming methods was that of observing the operations by which from time to time the mind spontaneously achieved its ends, and arranging these into a general scheme of action to be constantly followed in analogous cases; then, in whatever directions our modes of inquiry are at present unmethodized, our policy must be to trace the steps by which success is occasionally achieved in these directions; in the hope that by so doing, we may be enabled to frame systems of procedure which shall render future successes more or less sure." This was the method of hypothesis rendered as an account of the mind at work in its natural state. And the reverse was also true: by emphasizing spontaneity and by insisting that the nature of science was known to psychologists in a unique way, Spencer turned hypotheses into the "indispensable stepping-stones" of personal problem solving and scientific progress at once.[37]

The insight was not Spencer's alone, but he garnered the lion's share of attention. This was especially true in the United States, where the serial publication of his *System of Synthetic Philosophy* in the 1860s made evolution's application to the mind and society—what came to be called "social Darwinism"—a Spencerian movement, first and foremost. His American adherents were zealous in the extreme, willing to extend theory well beyond what Spencer was comfortable with. Spencerians came in many persuasions, from liberals protecting their political freedoms to conservatives shoring up the status quo. Men of science were mixed on Spencer's work, but many cast him as a defender of scientific reasoning against superstitious opposition even if they disagreed with his conclusions or opposed his methods. Darwin himself acknowledged Spencer's role as both a popularizer of evolution and an original thinker in his own right, especially when it came to applying theories about nature to problems in social life. Famously (or infamously), the tagline "survival of the fittest" was Spencer's, not Darwin's, though the latter worked it into the title for the fifth edition of the *Origin*. Somehow, the idea had staying power.[38]

Darwin's debts were more than terminological. Spencer's imprint can be identified in both *The Descent of Man* and *The Expression of the Emotions,* where Darwin drew on Spencer's thought to overcome his aversion to using natural selection to explain human phenomena. Even the *Origin,* composed prior to Spencer's rise, was not immune to his influence. Beyond its title change, Darwin also edited the lines from the *Origin* about psychology to give Spencer his due. What was once prospective became, in the final edition published in Darwin's lifetime, a tip of the hat to Spencer's progress: "Psychology will be securely based on the foundation already well laid by Mr. Herbert Spencer." Bain, too, worked evolutionary ideas—Darwin's and Spencer's—into his approach, most obviously in a new chapter in *The Emotions and the Will.* Associationism evolved, in other words, with complex thoughts now seeming to emerge from simpler ones—if they were useful. Science functioned the same way, according to the mental evolutionists: it was spontaneity organized, a delicate dance of individuals and groups. Darwin, Spencer, and Bain made science cognitive. In their hands, hypotheses came to seem like natural consequences of the adaptive mind, a tool for getting along in a world that was constantly evolving.[39]

UNCONSCIOUS SCIENCE

Bain's work on physiology and Spencer's focus on evolution toed battle lines in an age-old dispute over whether free will and determinism were at odds. If events in bodies *caused* those in minds, and if the former were subject to physical laws, then some saw our private thoughts as rule-bound, if not predetermined. If this was so, then even science was not safe from the effects of determinism. What had seemed like freedom—the use of hypotheses to test the limits of human understanding and control the natural world—might be nothing more than yet another pattern unfolding according to the laws of the universe. Not everyone arrived at this conclusion, but for many the idea that science was a natural process entailed at least an anxiety about just how far it could be pushed and what it might enable. The pursuit of certainty, the effort to ground reasoning in something stable as a means of progress, now seemed to put the agency of human cognition, of the imagination and creativity, at risk.

This was especially true as methodologists drilled down further and further into the mind, seeking the subterranean roots of scientific inquiry in our evolved capacities. Not all cognition, they argued, was conscious. And this included science. After all, if science mirrored evolution in the mind, it had to be as unconscious, as habitual, as other aspects of mental life. Bain and Spencer pointed the way by bringing work on the mind closer to studies of nervous anatomy and other bodily systems. But a broader network of practitioners was needed to take it all the way, to turn talk of "nervous energy" into a basis for rethinking science and its authority. "Currents," "electricity," "impulses," and other terms borrowed from the physical sciences were being pressed into service to explain how the physical body gave rise to nonphysical mental forces. The problem was identifying precisely where these two realms came together—how Spencer's "genesis" could be translated into a bodily vernacular.

Though psychologists were centrally concerned with these issues and their potential impact on scientific methods, they were not alone in reimagining how this embodied sort of reasoning might work, and what sorts of evidence you would need to examine its nature. One path toward the study of science incarnate pointed inward, toward nervous organs, while another pointed outward—to society, or culture, as a whole. The first was the purview of neurologists, especially the emerging evolutionary neurology of the Spencerian John Hughlings Jackson and his follower, David Ferrier. The second was the stuff of anthropology, another discipline taking shape in the same years, often under the sway of evolutionists like Spencer. Though neurology and anthropology seem quite distant today, the assumptions underlying their mid-Victorian rise were remarkably similar, especially when practitioners saw themselves as explorers of human phenomena at one level of remove from the autonomous agent who had long stood at the center of the human sciences. In other words, neurologists and anthropologists pursued a new science of human nature by zooming in or out, attending to smaller or larger scales of analysis. By locating cognition in the nervous system or in cultural networks, practitioners of both fields resituated mental abstractions—including their own—in the evolving, embodied world.

As psychologists pushed inward toward the brain, seeking hitching posts for emotions and other mental states, they encountered a group coming the other way—from the brain to the mind. Chief among the neurologists they met as they read their way into the brain was William Benjamin Carpenter, the most famous English physiologist around and an outspoken proponent of fusing physical and psychological work in a synthetic account of the human mind. Carpenter followed Thomas Laycock in extending concepts like the reflex arc up the neural chain, from the spinal cord to the human mind. As physiologists, Laycock and Carpenter's first task had been to insist that the body was subject to the same laws as the universe—that is, that it was a legitimate scientific object. Having established this in a series of books in the 1840s, Carpenter worked to integrate psychology across the mind-body divide. Explaining what many saw as higher-order mental processes like ideation and creativity in physical terms alone was still counterintuitive and somewhat sensational in this period. As a result, keywords such as "mental reflexes" and "ideo-motor action" had a polemical verve, striking mid-Victorian readers as unusual (if not oxymoronic) tools. Adding more and more psychological material to subsequent editions of his popular physiology textbooks, Carpenter approached Bain and Spencer from the inside out.[40]

At the center of Carpenter's analysis was his concept of "unconscious cerebration." This was a process that unfolded below the level of our attention, implying not only that our minds performed tasks of which we were unaware—this clearly occurred in dreams, for example—but also that even tasks we *thought* we controlled actually proceeded according to their own logic. When we find that our thinking has changed on a topic since we last took it up, or that a solution occurs to us in a strange moment, we experience "a development which cannot be reasonably explained in any other mode, than by attributing it to the intermediate activity of the Cerebrum, which has in this instance automatically evolved the result without any consciousness." This way of thinking about thinking was both strange and familiar. Strange, because we see ourselves as willing agents of our own intellectual lives, commanding a mind and a body to do our bidding. But as Carpenter's appeal to common experience suggests, the idea of unconscious reasoning was also familiar. Introspection served the rhetorical purpose of inviting the assent of his readers, but it was also the only source of evidence for the kind of phenomenon he was describing.[41]

Much like evolution was for Darwin, the unconscious was a limit case for the scientific methodology to which Carpenter aspired. Unconscious cerebration, like natural selection, was up against evidentiary limits, enmeshed in a similar set of argumentative ambiguities with what felt like incredibly high stakes. Both theories posited an inaccessible mechanism to explain the world as we experience it, requiring both authors to appeal to their own imaginations and those of their readers. In Carpenter's case, the inability to precisely define his object was almost a foregone conclusion. Appeals to the unconscious mind were both frequent and hotly contested, the kind of practice that invoked increasing suspicion even as the connections between neurology and psychology seemed to make such appeals more necessary. Turning all the different minds out there, from one's own to those of one's peers and one's patients, into "the mind" was—and is—a high-wire act, one of elision as much as synthesis. Carpenter was confronting what one historian of psychiatry has called "the specter of ignorance," a persistent failure to characterize the field's object of inquiry because it was, by its nature, so hard to characterize. In doing so, he practically projected that mysteriousness onto the mind. Carpenter's reliance on "unconscious cerebration" literally embodied the field's methodological frustration deep in our physical beings.[42]

This effort to ground the unconscious in "mental reflexes" had a politics, one that Alison Winter has called "a physiology of authority." Carpenter's theory gave a material basis to the conservative views of Whewell and others who policed the boundaries of science against what they viewed as incursions. Basing science on "unconscious cerebration" made scientific authority the kind of thing either you had or you did not. If science was embodied in this way, only those with the right endowments could contribute to it. Or at least, that was one way of seeing the matter. It was also possible, however, to see the unconscious dimension of science as liberating. Immediately before introducing "unconscious cerebration," Carpenter seemed to do just that. Combining a meditation on exceptional figures like Mozart with a recognition of the ubiquity of the processes that underlay his ability, he cited a neurologist's claim that such capacities were universal, if hard to access. Thus, neurology pointed in two directions at once: you might be born with a better brain, but you could also improve the one you have. The theory of "unconscious cerebration" was conservative *or* radical, part of nature *or* nurture, depending on the needs of those who invoked it. As with social Darwinism, the embodiment of certain higher functions could be harnessed to a variety of political ends. The politics of embodied cognition boiled down to two simple questions: Were bodies (and thus minds) more alike than different? and What explained that relative similarity?[43]

The most influential response to these questions was in the evolutionary neurology of John Hughlings Jackson. A follower of Bain and especially Spencer, Hughlings Jackson took Carpenter's mental evolutionism deeper into the brain through consideration of moments in which it slowed down or even reversed course. An expert on epilepsy, Hughlings Jackson saw nervous conditions as the breakdown of complex associations described by Bain and Spencer. Seizures and lesions were evidence of what he called "dissolution," a term he introduced with a doff of his cap: "Here when for the first time in this article I use the term Dissolution, I most gratefully acknowledge my vast debt to Herbert Spencer. What I have to say of the constitution of the nervous system appears to me to be little more than illustrating his doctrine on nervous Evolution by what I may metaphorically speak of as the experiments of disease." Whether or not Hughlings Jackson was correct to credit Spencer in this way (he may, for example, have owed as much to Carpenter as to Spencer with respect to the physiological dimensions of his work), he continued to do so throughout his long career.[44]

Two aspects of this acknowledgment merit scrutiny. The first is the importance of the term "dissolution" itself. Hughlings Jackson's evolutionary neurology marked a new stage in a trajectory that began with Bain, if not before: the materialization of mind in the imperfect body. Our minds, like our bodies, were beholden to the same processes at work in the natural world—and not just because our bodies are products of evolution. Individual minds, Hughlings Jackson insisted, were themselves evolutionary, in the same sense that a population was. We change over a lifetime, and even on a given day. And as we do, the push and pull of our nervous systems is as naturally selective as the tangled bank Darwin asked us to imagine at the end of the *Origin*. This view of the mind led to an unusual view of his clinical tasks: "We should try to know diseases as the hunter knows his lions and tigers, but I think we should endeavour to know them also as the zoologist knows his species." If our minds are environments, filled with ideas or diseases that are competing with one another, then neurologists and psychologists seem like natural historians—if not hunters. Their responsibility was to identify the "creatures" in such a landscape, cataloguing those that were unfit or whose behaviors worked against the collective interest. Evolution, on this view, gave new meaning to neurological method.[45]

Evolutionary theory also supported an experimental approach to the study of the nervous system. Like his hero Spencer, Hughlings Jackson always saw his work in methodological terms. Medicine was experimental, and not in the sense of the field's increasing reliance on laboratories for diagnostic authority. As his last line makes clear, disease *itself* was experimental. In an earlier paper, Hughlings Jackson suggested that "paralysis and convulsion may be looked upon as the results of experiments made by disease," in which nature performs "just what the physiologist does in experimenting on animals." Nature conducted experiments, exposing cracks in mind and body that medical ethics—even then—forbade the doctor from exposing artificially. Like Darwin and Spencer, Hughlings Jackson found his preferred method written onto the natural world. The mind seemed "experimental" in its basic, everyday processes—not least because the doctor was unable to experiment upon them. "Experiments of disease," in this sense, did what Hughlings Jackson could not: they perturbed health and identity from the inside out. As the physiologist Michael Foster put it around the same time: "the so-called 'experiments of nature' as seen at the bed-side, are extremely useful in correcting experimental inquiries." Nature became experimental, right down to the nervous system, because practitioners of science and medicine alike obsessed over the meaning of experiment in their own work.[46]

Small wonder, then, that Hughlings Jackson was an outspoken proponent of hypotheses in scientific and medical inquiry. The model of mind he proposed, with "spontaneous" emissions and "experimental" setbacks, seemed purpose-built to justify the use of hypotheses, much like it had for Herschel in the 1830s. Hypotheses were natural, now, in ways Darwin had only dreamed of. "We should follow the method of science," Hughlings Jackson told an audience in 1885, "and investigate by the use of hypotheses. This may seem a strange remark to those who erroneously suppose an hypothesis to be a conclusion in which we may rest. It is only used for the methodizing of work by observation and experiment." Hypotheses were tools, not conclusions. And the best tool for making sense of the mind was the hypothesis of evolution—which, looping back around, turned the mind into an adaptive toolkit. Once again, striving for a science of the mind turned minds into scientific tools. The hypothesis of mental evolution naturalized the use of hypotheses in the human sciences, transforming not only the theories but the methods of Bain and Spencer into physiological processes grounded in the human brain. And as our minds became hypothesis generators, the use of hypotheses gained new authority by virtue of having been "discovered" in nature. Scientists were just doing what came naturally.[47]

Neurology was not the only field in which evolutionary theories of mind were emerging in these years. Something similar was occurring in the scientific study of culture. Ethnologists and anthropologists, like neurologists, began adapting evolutionary theories for their own aims. Their object of analysis, "culture," became something dynamic and adaptive over the 1860s and 1870s, the result of integrating evolutionary accounts of mind and society into a single system. "Culture or Civilization," wrote the anthropologist Edward Burnett Tylor in 1871, "taken in its wide ethnographic sense, is that complex whole which includes knowledge, belief, art, morals, law, custom, and any other capabilities and habits acquired by man as a member of society." Early practitioners like Tylor and James Frazer saw themselves as the allies of evolutionary psychologists and neurologists. Their task? To make sense of complex phenomena, bolstering their own scientific status by grounding their chosen methods in a new account of the mind. Whereas Hughlings Jackson had zoomed in on progress and dissolution in the nervous system, Tylor and others zoomed out to culture at large, seeking (in Tylor's words) a "primitive culture" that could explain the historical development of the "civilized" mind—including, he thought, the deep history of science and the minds of men of science like himself.[48]

Tylor was typical of the British anthropology in this period, which Henrika Kuklick has called a "self-examination framed in ethnographic terms." This could be benevolent, as when ethnographers argued for tolerance or relativism, but the same impulse led in sinister directions too—including the intimate relationship between anthropological inquiry and imperial ambition. The idea that other people were "behind" on a civilizational scale served to unify humanity in one moment and justify colonialism in the next. By insisting on "psychic unity," in other words, early anthropologists not only justified their own comparative methodology (including the ways they substituted present-day people for historical human populations) but also bolstered claims of imperialists who presented themselves as helping non-Europeans ascend the difficult ladder of cultural and mental evolution. The early history of anthropology was the history of simultaneous inclusion and exclusion, of a rhetoric of commonality paired, all too often, with practices of difference-making. Scientific study was emancipatory and punishing at once.[49]

Differences, including those that were used to justify colonial practices, could be drawn by men who had never set foot in the parts of the world where they were "discovered" and ultimately enforced. But calling these men "armchair anthropologists" was not enough to dismiss their work as invalid, nor is doing so now enough to explain their practices. Just because ethnographers like Tylor and Frazer were "armchair" scholars did not mean that they had unsophisticated views on methodological issues or that those views did not matter. Writing from the comfort of their own homes, they turned the difference between their labors and those of the "practical men" in the colonies on whose observations they relied into an asset. To Tylor and others, the distance they maintained by staying at home was a means of staying objective. This can be hard to square with what we know of the centrality of work in the field to anthropologists today, but that centrality was only coming into being at the end of the nineteenth century. The armchair was a powerful position, not just because it invoked the kinds of domestic or fraternal spaces in which science was traditionally practiced (though it did invoke those). It also implied a detached, even impartial observer, a theorist who could rise above the gathering of data and its imperial entanglements to deliver pronouncements about human unity.[50]

The role of the armchair in this period might also explain some anthropologists' adoption of evolutionary accounts of the mind and the idea of "psychic unity" they used them to support. Why? Because the theory made sense of the increasing diversity suggested by the data gathered from around the world. Unity, in the hands of Tylor and others, became less a *conclusion* of the study of other cultures and more a working hypothesis, a starting point for the examination of features so often obscured by radical differences between cultures. Assuming that certain mental traits were the universal marks of human cognition, or drawing on new ideas in psychology to make the claim more substantial, helped anthropologists characterize culture as just a part of the natural order. George Stocking has argued that this adoption of evolution did just this, justifying "the comparative speculations of 'armchair' anthropologists as legitimate contributions to a scientific understanding of man's place in nature." Comparison was synthetic by nature, which turned the armchair anthropologist into a kind of laboratory scientist, embodying the ideal of "mechanical objectivity" then in vogue with what verged on a God's-eye perspective.[51]

If adopting evolution as a framework for studying human societies helped turn anthropology into a natural science of sorts, it brought along new demands as well. The most pressing was the need to identify an evolutionary *mechanism*. Darwin's trouble with pangenesis show how hard it was to identify such a mechanism for species; doing so for societies proved even more challenging. But anthropologists saw such a mechanism as extremely valuable, not least because it might both explain perceived European superiority and offer a roadmap for imperial rule. Both aims hinged on identifying a source of change. According to Tylor, three possibilities existed: "independent invention, inheritance from ancestors in a distant region, transmission from one race to another; but between these three ways the choice is commonly a difficult one." In his first book, Tylor had favored the latter two; but when his blockbuster book *Primitive Culture* appeared in 1871, he swung decidedly toward "independent invention" as the explanation of cultural change. If societies, like minds, recapitulated the same steps at different rates, then a shared underlying process made sense. "Pebbles held in the hand to hammer with," Tylor argued, "and cutting instruments of stone shaped or left smooth at one end to be held in the hand, may be seen in museums, hinting that the important art of fixing instruments in handles was the result of invention, not of instinct." A practice invented in multiple places was, for Tylor, proof of a common cognitive pattern—and thus, cultural evolution.[52]

Tylor's emphasis on invention reveals another dimension of evolutionary anthropology, one that confirms Kuklick's claim about the solipsism of Victorian ethnography in general. The study of what one historian has called "savage numbers" pointed to a methodological reflexivity that is now taken to be standard in anthropology but was then only emerging. Tylor's chapter on "The Art of Counting," which begins with a discussion of the debate between Mill and Whewell over whether the concept of number is innate, linked anthropology to the philosophy of science. Spencer, in the first volume of his *Principles of Sociology* (published in 1876), devoted a lengthy chapter to what he called "primitive ideas." Echoing Tylor's claims, Spencer argued for a hierarchical development in scientific reasoning stretching back in time and outward in space from its European epicenter. When we look at other cultures, Spencer argued, "We must not look for an hypothesis properly so called: an hypothesis is an implement of inquiry not to be framed by the primitive mind. We must look for some experience in which this duality is forcibly thrust on the attention. As a consciously-held hypothesis is based on some obtrusive instance of a relation, which other instances are suspected to be like; so the particular primitive notion which is to serve as an unconscious hypothesis, setting up organization in this aggregate of primitive notions, must be one conspicuously exemplifying their common trait." Primitive groups, for Spencer, were limited to their senses. These people were pure empiricists. Out of their approach, or something like it, the hypothetical method employed by Spencer and others had arisen through a gradual transformation of "unconscious hypotheses" into conscious ones. The difference, for Spencer and for Tylor, was one of degrees: at some point, science had evolved.[53]

Mental evolution, in the hands of anthropologists, transformed existing groups of people into the ancestors of their own scientific method. The theories of Tylor, Spencer, and others turned their subject matter into a kind of prehistory of science, making methods of inquiry yet another adaptive response benefiting those who developed it. In 1881 Tylor concluded his introductory textbook *Anthropology* by writing the future of the field into his analysis of ancient thought: "Had the experience of ancient men been larger, they would have seen their way to faster steps in culture. But we civilized moderns have just that wider knowledge which the rude ancients wanted. Acquainted with events and their consequences far and wide over the world, we are able to direct our own course with more confidence toward improvement. In a word, mankind is passing from the age of unconscious to that of conscious progress." As it had for Spencer a few years earlier, scientific progress amounted to a transformation from unconscious to conscious procedure for Tylor. What was once a latent feature of the mind was now manifest in the scientific method Tylor himself used. The task of the anthropologist was to wield this adaptation to its greatest advantage: to intervene directly in the evolutionary process.[54]

But how? The answer was on everyone's lips in these decades: education. The classroom, in both its literal and its metaphorical senses, was the ideal site for applying evolutionary logic to the work of politics and progress. Here, as elsewhere, evolutionary theorists like Tylor saw capacities as both universal *and* racial. On the one hand, they argued that "the lowest man, or savage" was close to a nonhuman animal, far from European attainments. On the other hand, "the savage is possessed of human reason and speech, while his brain-power, though it has not of itself raised him to civilization, enables him to receive more or less of the education which transforms him into a civilized man." This amounted to redrawing the lines. For Tylor, the divide between human and animal was brighter than internal divisions in either group. The difference, he argued, was in your capacity to change through education. Thus, while all humans (including Europeans) carried animal natures within them, adaptability distinguished them. "Upon this lower framework of animal life," Tylor concluded, "is raised the wondrous edifice of human language, science, art, and law." What made humans special—and what made some more special than others? Answering that question was now the task of human sciences like psychology and anthropology.[55]

By naturalizing human achievements, evolutionary theories of mental processes ended up articulating a new foundation for scientific study. This had significant consequences for the fields involved and beyond them. Within anthropology, for example, studies of "primitive mind" and "primitive culture" ended up producing a new object: "primitive science," both as a topic of analysis for ethnographers and as an imagined historical antecedent to their own work. What began, for Tylor, as "savage or barbaric science" grounded in mythmaking became, for Franz Boas and his followers in the twentieth century, evidence of a fundamental identity between themselves and the people they studied. "It would be in vain to try to understand the development of modern science without an intelligent understanding of modern philosophy," Boas argued at the turn of the century, "and so it is in vain to try to understand primitive science without an intelligent knowledge of primitive mythology." More than the differences between the two, what mattered to Boas was the interlocking aspects of scientific and religious or philosophical thought in any place or period. Seeing science as a basic process, identifiable in its efforts to understand and control the environment, was a major consequence of the Victorian study of mental evolution, even if some—indeed, many—of its practitioners would have denied the commonalities Boas later identified.[56]

But the "primitive" roots of scientific method mattered beyond anthropology, too. In fact, questions of the universality of cognitive processes and the differences between styles of reasoning were intimately involved with the exercise of imperial power in the decades around 1900.

Tylor foresaw the extension of these ideas and ended *Primitive Culture* with the hope that anthropology might play a role in that vision: "Active at once in aiding progress and in removing hindrance, the science of culture is essentially a reformer's science." What he meant was that anthropology should be a part of state administration, both at home and abroad. This is what he was trying to signal by using the word "reform"—the potential to act upon our ideas, to make practice out of theory. The shape taken by such practices ran the political gamut, from emancipatory claims of human unity to imperial projects of difference-making. As with social Darwinism and the many applications of evolutionary theory to social phenomena, this had been the case all along. Accounts of balance, used to shore up the existing order or to insist on a return to something older, had always had this power: their appeal was precisely their flexibility, even if that flexibility sometimes got out of control.[57]

THE SCIENTIST WITHIN

Evolutionary studies of "primitive minds" cut both ways. Some anthropologists found things that supported their sense of superiority; others found things that brought them closer to the people they studied. Historians have shown how anthropology's imperial encounters were self-reflexive, consciously and unconsciously. Kuklick's title (*The Savage Within*) signals how capacities and proclivities of "savage" subjects were absorbed into British ethnographers themselves by way of the evolutionary logics they applied. Whether such "savage" qualities lay just beneath the surface or deeper down, studying other minds had the power to draw them out. Fears and hopes about certain traits—violence, say, or innocence—had relocated nature into the individual mind and the collective past, bringing an assemblage of racist ideas, colonial ambitions, and universalizing models along. Psychology, too, was pressed (often eagerly) into service in a wide range of social projects. The evils of such applications are notorious; less well known is how anthropology and psychology accidentally undermined the very politics they were meant to prop up, revealing unity where division was meant to be sown. Such studies often surprised or frustrated the scholars conducting them, as "the savage within" suggested similarities across group lines even as the same idea reinforced invidious distinctions.[58]

One such similarity was the capacity for scientific thinking, which was being redefined in light of evolutionary work in psychology, anthropology, and neurology. In the hands of Bain, Spencer, Hughlings Jackson, and Tylor—among others—inner mental lives became not just savage, but scientific. Alongside the articulations of "savages within" emerging from certain ethnographic encounters, another image arose that entailed a different set of consequences: "the scientist within." This was the culmination of decades of methodological embodiment, through which the capacity for scientific reasoning was rooted deeper and deeper in both mind and body. From science's affective roots and unconscious cerebration (even unconscious *hypotheses*) to the anthropology of "primitive science," there was an embryonic, if somewhat unstable, sense that scientific method was latent in the human mind. A basic approach to problem solving, guessing at potential answers and then testing to see if they worked, seemed to be "out there" in the world beyond the lab—and even, as we shall see, beyond the human mind, in instinctive and learned behaviors exhibited by certain nonhuman animals.

The "savage" and the "scientist" were not as far apart as some might have hoped. Given present assumptions about the nature of these two terms—the one a racist holdover, the other a modern honorific—their proximity, even interdependence, in the hands of these human scientists can be somewhat hard to square. But square it we must, for the connections these anthropologists made between their own minds and those they observed in others remind us that methodology was reflexive and projective at once. Human scientists recast their own scientific ambitions by making evolutionary theory the key to a new science of the mind. Ideas about "savage numbers" and "unconscious hypotheses" were not (only) ways of distinguishing between good and bad, between rough forms of problem solving practiced all over the world and the exact methods to which these figures aspired. Methodology encompassed both in their evolutionary age. Every approach had its function; all were connected, both neurologically and historically. This interweaving of object and subject, of self-conscious inquirers and the people they sought to study, is captured in the erstwhile title of George Stocking's unpublished project: *Scholars and Savages: From Evolutionism to Functionalism in British Anthropology.* What might seem an opposition could also serve as a mirror: the one was defined in terms of the other, and indeed depended on it for a full understanding of the human mind.[59]

Put simply: being a good scientist meant being "savage," too. Intuition, spontaneity, and the ability to throw out solutions to problems—these were the stuff of "savage intellect," but they were also essential to abstract scientific theorizing. The differences anthropologists articulated were often racist, insisted upon with the power of one-sided political authority. There is no way around the role such ideas played in the erection of boundaries and the justification of self-regard. Yet as with much evolutionary thinking, putative historical links led to uncomfortable, but often revealing recognition of relationships among those human "types" even when scientists tried to prize them apart. Marks of genius were, for men like Spencer and Tylor, not just *present* in the colonial subjects they considered—they were the *keys* to defining "primitive thought." What they found when they looked beyond their immediate context, even when synthesizing the observations of others, was a crucial connection between nature and method that seemed, if anything, stronger in those they observed than in their own minds.[60]

Naturalizing the mind in this period was often done under the familiar rubric of "common sense." Part appeal to populist ideals, part empty gesture on behalf of the status quo, public invocations of "common sense" were a lot like the evolutionary rhetoric of the human sciences in these years. Both could be applied to various political projects. In the case of thinking through the capacity for scientific reasoning, this was precisely how the term functioned. When Thomas Huxley said science was "nothing but trained and organized common sense," he was granting his own method the imprimatur of a long history. When George Henry Lewes laid down as a rule for philosophy that "no problem [is] to be mooted unless it be presented in terms of experience, and be capable of empirical investigation," he too claimed the everyday world of problem solving as not just a precursor, but a guide, to inquiry into the nature of things. For Huxley, Lewes, and many other evolutionists in the mid- to late nineteenth century, "common sense" held human minds together and formed a foundation for their further study.[61]

This work on science's natural foundations extended into some surprising areas. The logician William Stanley Jevons, for example, was famous for celebrating the role of geniuses in the progress of science. His *Principles of Science,* published in 1874, returned again and again to Isaac Newton as an exemplar of scientific reasoning in general and—*pace* Newton's claims—of the hypothetical method in particular. But the role of hypotheses in scientific progress was not a function of the geniuses who used them best; it was, instead, a corollary of the *limits* on human cognition being established around the same time by evolutionary psychologists. "The powers of the human mind," Jevons wrote, "are so limited that multiplicity of detail is alone sufficient to prevent its progress in many directions." We are so easy to overwhelm, he argued, that the form of induction attributed to Bacon was not just inaccurate—it was impossible. For Jevons, wielding hypotheses is what saved us; they were an example of the "economy of mental power" on which our "elevated intellectual position" as animals capable of science was utterly dependent. In this way, naturalizing hypotheses meant acknowledging our limits, the overcoming of which was the essence of science and the hypothetical method on which it increasingly depended.[62]

Of course, there was nothing "natural" about naturalization—its proponents had to make it so. Defining "common sense" was as thorny as identifying and justifying the foundations of logic—and at a certain point you just had to get on with whatever you were doing. Huxley himself recognized as much, admitting that his own assumptions about how the world worked were "neither self-evident, nor are they, strictly speaking, demonstrable." They just *were*—and assuming that you held the same ones as Huxley did, you were inclined to assent to a basic view of the world and move on to the really controversial things he had to say. Like Lewes, Huxley turned "common sense" into a kind of methodology, arguing that we can root scientific inquiry in simple, shared ways of thinking that we see around us in our everyday lives. By grounding scientific method in ordinary thinking, Huxley *naturalized* it: doing science meant doing what came naturally. Because it was his approach to science in general (and the study of evolution in particular), he naturalized naturalism. Soon, "natural science" was simply *how we think*.[63]

It is hard to see such transformations at work, either in the past or in the world around us. Huxley's naturalism was the winning bet in Victorian debates over how to pursue scientific theories, but it is precisely that victory that makes it hard to see. This is because the arguments Huxley made for it are still with us today, structuring how we think not just about science but about everyday thinking, too. Both evolved, he argued, from a "trial and error" process, and both bore the traces of a shared history in their connections with problem solving more generally. In other words, it was not enough for scientific naturalists like Huxley to "make their view of science seem obvious and inevitable," as Matthew Stanley has convincingly shown they did. The interplay of spontaneity and rigor, of hypotheses and tests, was not just a characteristic of scientific method—it was part of "common sense," too. By making the way he thought about the natural world *seem* natural, Huxley was able to close the loop by using naturalism as its own justification. Like "common sense," the idea of a single "scientific method" was a tricky tool: it was easy to conjure with but hard to contain once you did. Reckoning with how science and common sense were related, if not conflated, means watching psychology bolster both.[64]

As scientific psychology took shape in the late nineteenth century, the complex meanings of "experiment" invoked by its champions would ramify, blurring boundaries between the people *doing* the experiments and those being experimented upon. This was especially true in the part of the field devoted to using tools from evolutionary theory to probe the boundary between animal and human reasoning. Drawing on Darwin but also on Bain, Spencer, and others who filled in the idea of mental evolution, a new generation of scholars distinguished themselves by ratcheting up the reflexivity of the mind sciences. As psychologists founded departments and journals, and as they claimed expertise on patterns of thinking overall, they turned their theories of embodied cognition and adaptive tools to the task of disciplinary formation.

Their conflation of evolution and experiment carried on what Darwin had begun: reimagining hypothesis testing as nature's method, now observable in the habits of human and nonhuman animals. This continuation was most eagerly pursued not in Europe but in the United States, where evolutionary psychology and its reflexive application took hold most firmly. To see how—and why—we must cross the Atlantic.

5

A LIVING SCIENCE

"My Psychology still lags behind," William James admitted in 1888. "It is one thing to write chapters, another to write a book, on a subject of which the first principles are as yet undetermined." He had been learning this lesson the hard way for a decade. What James called "My Psychology" was supposed to be an introductory textbook in Henry Holt's "American Science Series," for which he had signed a contract way back in 1878. At the time, he thought the project might take him two years to complete. Even that optimistic estimate left Holt "a little staggered," since it usually took his authors about six months for books in the series. Ten years on, James was so far from finished that he despaired of completing the manuscript at all. Neither the profusion of competing textbooks nor James's own production of numerous, related articles did anything to quiet his biggest concern: that, in the case of psychology, it was still not clear what held the field together. Beyond a basic sense that psychology was the scientific study of the mind, the identity of the various projects pursued under its name was far from obvious. James began to wonder: was there really a "there" there? Doubts on the matter delayed him.[1]

While James might have foreseen certain causes of the delay—notoriously poor health, for example, or his obligations to a new job and a growing family—the field's lack of coherence was not something he had imagined in 1878. Even though psychology was only just emerging as a self-consciously scientific field, there were established texts that at least suggested a kind of identity for the science of mental life. Bain's two books from the 1850s were widely used, and Spencer's *Principles of Psychology,* originally released in 1855, had recently been rewritten for the "System of Synthetic Philosophy" series he started publishing in 1870. While a great deal divided Bain and Spencer, as we have seen, they agreed on fundamentals. To them, any proper psychology began with the physiology of nervous organs, and especially the brain. Evolutionary theory also had a role to play, as Spencer's adaptation of Bain made clear, and a new generation of psychologists soon picked up where they had left off, eager to distinguish themselves from their philosophical colleagues with the latest scientific tools. Psychology was in the air, in other words, and its core texts and basic assumptions appeared to be coalescing.[2]

Given the general optimism about psychology's prospects, it came as something of a shock to James—and still seems to surprise us, as historians—that after a decade of drafting his textbook for Holt, he still felt that the field's principles were "as yet undetermined." Indeed, his doubts remained even as he sent in the finished manuscript two years later. In a scribbled note in the envelope to Holt, James called it "a loathsome, distended, tumefied, bloated, dropsical mass, testifying to nothing but two facts: *1st*, that there is no such thing as a *science* of psychology, and *2nd*, that W.J. is an incapable." Though he expressed pride around the same time, he did so in terms that fail to capture what was actually new or exciting about his book—at least for its fans, then or since. The book is at once a textbook and an anti-textbook. While the late nineteenth century saw the genre of the science textbook solidifying into something like its modern form, James's book was an outlier: superficially a taxonomic review of the mind and its processes, a closer read reveals a text (and an author) divided on basic points, stuffed with new theories and received wisdom, equally snarky and straightforward. *The*

Principles of Psychology is a jumble, both loveably and maddeningly complex. Some, many even, have adored it for that complexity, which has made it required reading far beyond psychology. But it can be tough going.[3]

Like its author, the book is ambivalent about the very thing it symbolizes to readers then and now: the promise of scientific psychology at the turn of the twentieth century. Understanding that ambivalence requires getting a sense of the materials James had to work with. When he looked outward at the psychology practiced in his day, or inward into his own mind—what did he see? And why did he or his critics imagine that such a book should or even *could* be coherent, given the recent and scattered state of the field? Answering these questions means attending to both the state of the discipline in the 1870s and 1880s and the way James portrayed it. Spliced together from pieces he published over a decade, in various fields and a wide range of venues, the *Principles* captured the bits and bobs of psychology as it was. James's writing process was more an organic outgrowth than an organized assembly line, a fact that was plain to those who knew him and a target for those who did not. Demands for coherence featured prominently in early reviews of the book, as readers (some writing textbooks of their own) clearly yearned for the same synthetic account that James had imagined from the start.[4]

Part of the problem was that psychology was moving in these decades, in the United States as elsewhere. All the work that James surveyed *seemed* to be coalescing into an approach soon to be dubbed "the new psychology," but reality was messier. This (supposedly) singular movement was associated with a specific set of tools: the experimental apparatus of German laboratories, where so many Americans sought training in these years. Historians agree, deeming "the new psychology" a consequence of the ongoing "laboratory revolution" rising up around it. This account was cemented early (including in James's own writings), but real life inside and outside the laboratory was not so straightforward. There was no "new psychology," experimental or otherwise. Instead, there were (as Lorraine Daston argues) many "new psychologies plural," each with champions defining its terms and building its instruments. Competition and conflict characterized these early years, with practitioners positioning their preferred approaches as *the* way forward for psychology as a whole. The battles of psychology's "brass age" were fought everywhere, with weapons and stakes particular to each place but with plenty of cross-talk and transnational reinforcements as well.[5]

Even central concepts like "experiment" were themselves objects of heated debate. The celebrated laboratory instruments of the period were not simply *applied* to the mind; they were used to construct it, making some phenomena visible and calling others into being. Psychologists altered the very object they were trying to study by adapting new tools to the purpose of making minds legible. The rise of new techniques for measuring and quantifying the stuff of thought did more than access the mind; they remade it. In the process, the mind was *instrumentalized,* turned into an experimental tool like those being used to study it. And this instrumentalization of mind gave new meaning to experimental method. "Experimental psychology," in these years, meant experimentally analyzing the mind—but it also meant, for some, the psychology *of* experiment, the study of what it is we do when we think scientifically. This is not just apparent in hindsight. James himself was aware (even wary) of the mind's malleability, the constant give-and-take between evolving approaches to studying the mind and what that object was supposed to be in the first place. There was no single psychology, no one mind, under the banner of "the new psychology." Instead, a motley crew of psychologies jostled for position, the effect of which was a mind that seemed to be many things at once, a Swiss Army knife of possibilities.[6]

What Julie Reuben calls "the making of the modern university" took hold in the same era, involving both new admissions procedures and the empowerment of certain wings of the faculty. One feature of this shift was the replacement of traditional, capstone courses on Common Sense philosophy and character formation (often taught by the college president) with "disciplined" courses in departments of philosophy and, increasingly, psychology. This change had particular salience for the rise of the human sciences, not least because it helped lay the groundwork for seeing psychology as fundamental to liberal education. This helped propel "modernization" at American colleges, but these changes were not unvarnished goods. The kinds of boundary-work involved in the rise of psychology and the prioritization of the sciences also foreshadowed these universities' obsession with professionalization and credentialing—what James called "the Ph.D. Octopus." But the boundaries he and others decried were still porous when he began teaching at Harvard, and James took advantage. He learned a lot about the nature of scientific reasoning and the status of psychology—not just through courses, but beyond the walls of the university, as a part of discussion groups and through extensive correspondence. The informal world he found may have been fragile, even fading—but it gave method new meaning before it fell.[7]

THE NEW EDUCATION

The whig interpretation of science was born in the Age of Reform, and its American adaptation arose in its own era of transformation: the aftermath of the Civil War. Intellectual historians have long pointed to these years, also known as the Gilded Age and the Progressive era, as a hinge between a Puritan past defined by European debts and a future of American preeminence on a global stage. While accounts of "the American mind" or "American thought" have been rejected for all the diversity they leave out, the idea that a self-consciousness emerged in these years, and with it an emphasis on scientific attainment, is hard to refute. Not everything was new, of course. What has been called "the launching of modern American science" began before the war, with the founding of the Smithsonian Institution and the American Association for the Advancement of Science (modeled on its British predecessor). But the postwar years saw a new emphasis on avowedly "American" ways of thinking and calls for a national scientific office to put the United States on par with its European peers. This intellectual, even epistemic nationalism, centered on defining a uniquely American mode of inquiry, spurred a new wave of methodological thinking that took on a life of its own over the ensuing decades.[8]

Changes were clearest at colleges and universities newly reformed (or founded) after the end of the war. Clustered in the industrial Northeast (at least at first), some went from being the finishing schools of an established elite to engines of class transformation, at least in terms of their aspirations. Reform hit places like Harvard and Yale at different rates, while the founding of Johns Hopkins (1876), Clark (1887), and a revamped University of Chicago (1890) put pressure on the older schools. Underlying these shifts in procedure was a shift in focus, from the midcentury ideal of character formation to an emphasis on the sort of research for which most universities are known today. Reform-minded university leaders, tapping into sensibilities like those that had reshaped British science in the previous generation, began to imagine something once unimaginable in the United States: competing with Europe's schools for scientific status. Echoing and actively repurposing their British predecessors, reformers threw their weight behind this goal, hiring faculty they felt could carry them into a new, progressive era.[9]

A major factor in the ascendance of American universities in the latter half of the nineteenth century was the place they accorded to work in the sciences, and specifically to new experimental approaches. Just as German universities' embrace of the laboratory helped them leapfrog their French and British counterparts as destinations for international students around the same time, American administrators saw scientific training as a way to expand enrollments and telegraph their ambitions. In part, this was a lesson learned in Germany. American reformers flocked to laboratories in German-speaking Europe at midcentury, eager to imitate the successes they found there back home. But the fad for scientific growth cannot be chalked up to selective importation alone. Experimental science looked different at American universities, not least because its arrival entered a feedback loop with anxieties about national status and the desire to differentiate the United States from what had come "before." The role of science in university reform was not static or preordained, like some kind of fertilizer for intellectual change. The meanings that attached to "science" at American universities were different in telling ways from those the term (or its rough equivalents) had acquired elsewhere.[10]

The fate of Harvard is a case in point, both in terms of the particular transformations it underwent and the role that science played in those changes. Charles Eliot, the president who would shepherd Harvard through its postwar transformations and rise to national preeminence, was trained as a scientist—a chemist, in fact. At only thirty-five years old, Eliot had made the case for his administrative vision in a pair of articles published early in 1869, the same year he was appointed president. Called "The New Education," Eliot's proposal centered on updating semi-independent scientific "schools" by folding them into the university in a more formal way. The plan was to use that transformation to introduce new ideas such as electives (courses had been part of a set curriculum) and a balance of scientific and humanistic training. Science, Eliot felt, would bring Harvard new life. "A new system of education," he wrote, "crude, ill-organized, and in good degree experimental, has been brought into direct comparison and daily contact with a well-tried system in full possession of the field." Progress meant updating traditional methods, and Eliot saw the path forward in one method in particular: experiment. Why not make all ways of thinking experimental? The new education, like the new psychology, was premised on turning the life of the mind into a kind of experiment.[11]

Eliot was relentless in emphasizing the *uses* to which learning could be put: his attention to application was clearly part of what made him attractive to the businessmen who selected him for the presidency. But his utilitarian focus also shines a light on how he thought science worked. For Eliot, science supported a different view of the mind than the older one of "a globe, to be expanded symmetrically from a centre outward." The scientific mind was different: "A cutting-tool, a drill, or auger would be a juster symbol of the mind. The natural bent and peculiar quality of every boy's mind should be sacredly regarded in his education; the division of mental labor, which is essential in civilized communities in order that knowledge may grow and society improve, demands this regard to the peculiar constitution of each mind, as much as does the happiness of the individual most nearly concerned." Reimagining the mind as a tool encouraged a new vision of education. Each of us, sharpened or honed appropriately, was part of a division of cognitive labor, one piece in a larger social vision that came to resemble a toolkit. In his inaugural presidential address the following year, Eliot expanded: "As tools multiply, each is more ingeniously adapted to its own exclusive purpose. So with the men that make the State." Every person, every mind, was a tool. Together, they formed a toolkit.[12]

This vision was not Eliot's alone. Indeed, it played a broader role in justifying social and labor arrangements in the name of efficiency and underwriting increasingly technical directions in American higher education. But attending only to the direct consequences of the toolkit view of mind and society misses part of its power. Something more basic was at stake. If the mind is a tool, then it has a function. With this metaphor in place, the task of education became identifying students with functions. It was this essentially instrumental approach to minds and how to train them that had the biggest impact on Eliot's Harvard and beyond. Debates over the natural history of human thinking were only part of the story, as this utilitarian, means–end view took on new associations that linked institutional reform to individual development. In both, science took on the centralizing role formerly occupied by theology at many schools. Science and its benefits, both for society and for the individuals who engaged in it, were now central to understandings of liberal education and its relationship to progress. By helping to articulate this link, a generation of psychologists teaching or studying at Harvard turned their emerging forms of expertise into a node in the new scientific vision that Eliot put on the map in this period.[13]

But it was not all due to Eliot. A major factor in this transformation was a relic of the old Harvard, of what you might call (adapting Thomas Kuhn) its "pre-disciplinary matrix." Before Eliot's reforms standardized instruction and regularized disciplinary boundaries, lines between departments and ways of thinking were blurrier—at Harvard and elsewhere. This was not all good, of course. Standards were mixed under the old system; nepotism and entitlement were rampant (as, some insist, they remain). Disciplinary boundaries can help, especially when they enable rather than prevent coherent conversations across them. But the blurriness that preceded Eliot was a kind of cross-fertilization, occurring both inside and outside the university. The community around Harvard gathered an incredible range of scholars and intellectuals in informal associations and event series that paid less attention to affiliation and departmental politics than after reforms had taken hold. It was in this pre-disciplinary matrix, during the decade or so after the Civil War, that science's meaning was most hotly debated—not least because those debating it felt that the stakes, in terms of which disciplines were considered scientific and who qualified as what we now call a scientist, were incredibly high.[14]

The most famous fruit of Harvard's pre-disciplinary matrix was "The Metaphysical Club," a group convened intermittently in the early 1870s and credited with ideas that are still taught as a high point of American intellectual history. Chief among these was what we now call American pragmatism, a way of thinking about thinking rooted in methodological concerns, as we shall see. Though almost no records of the group's meetings survive (certainly not anything official), the Metaphysical Club's membership and matters of discussion have been pieced together from diaries, letters, and—less reliably—memoirs of the period. But histories of the group remain hazy. Louis Menand's *The Metaphysical Club* is halfway through before the Club is introduced, when Menand notes only one member ever mentioned it by name—in an unpublished piece written three decades later. Indeed, Menand's title is intentionally ironic, given the paucity of our sources and the fact that the Club's members actually opposed metaphysical speculation and seem to have named the group in jest. The diverse views and aims of those who attended its meetings point in many directions, from the most abstract logical theory to mundane aspects of the law and ordinary life. All that the members shared, as best we can tell, was a basic sense that philosophy needed to come down to earth.[15]

"Why," Menand asks toward the end of his book, "on a pragmatist account of ideas, did the idea of pragmatism arise?" His answer has less to do with its origins—a shared desired to get past the Civil War, he argues—and more to do with the group's *effects*. As Menand suggests, this is exactly how pragmatists themselves would want it, given the emphasis on practical effects that their name implies. But even that is not so simple. The name "pragmatism" was bestowed by Charles Sanders Peirce, adapted from Kant, and first articulated at a meeting when the Club's existence was already petering out. Neither the name nor the approach was widely known until it was publicized—made effective, you might say—by Peirce's friend and fellow Club member William James. But the core of pragmatism had crystallized early on, centered on what was later called its "method": a call to consider concrete results over abstract principles, to locate what a word or a concept meant in the effects it had on the real world. Before James made it famous at the end of the century, pragmatism's effects were felt in Club discussions and, soon after, in the work of its members—in law and politics, psychology and philosophy.[16]

The texts discussed at the Metaphysical Club will be familiar: Mill, Bain, and Darwin were all favorites; Spencer came in for critique and praise in equal measure. Experiment and evolution were highlighted by all, as were the psychology of scientific thinking and inquiries into the pervasive logic of adaptation. The scholars who huddled around Harvard in the 1870s made new things out of old materials. In their hands, Bain's embodied cognition and Spencer's evolutionary vision became inseparable: when we think, we are adapting to circumstances; and as we adapt, we evolve. To do so, our acts of thought need a practical bearing, something onto which an evolutionary process can grab hold and from which it can select. It was this idea, above all else, that formed the center of the group's discussions. Indeed, the importance pragmatists placed on practical results could be said to stem from their early, intensive focus on evolution—itself a way of describing life and its origins in the language of results, of fitness. Bain and Spencer, in other words, were given new practical meaning in a different environment, as their ideas were tested and adapted to new intellectual struggles.[17]

No member mattered more to this focus on evolution than the man Peirce called "our boxing-master," a passionate if tragic philosopher named Chauncey Wright. It was Wright who advocated the loudest for Darwin over Spencer, and it was Wright, more than anyone, whose status as an outsider helped secure the boundary-crossing, pre-disciplinary energy of the group, especially in its early years. An eager acolyte of Darwin's, Wright pushed his idol's line of thinking further into the mind than any had yet dared. Wright was also responsible, along with the philosopher John Fiske, for introducing "positivism" into the group's discussion—but also, just as importantly, for dissociating the positivist movement from its founder, Auguste Comte. At a general level, positivism was a lot like pragmatism. Both emphasized the scientific study of natural phenomena, and both set a great deal of store by the material world. But Comte took this one step too far for some readers, insisting that there was no other path to truth than the scientific one. During early discussions at Club meetings, Wright and Fiske introduced a version of this view, drawing on Mill's reading of Comte. In doing so, they presented positivism as more English than French, emphasizing its affinities with the (British) strands of pragmatism the group was exploring. Seeing positivism and pragmatism as both derived from British empiricism and evolutionary theory unites the two under what has been called "Big Tent Positivism." Under that tent, matters of mind and method dominated discussion for decades.[18]

Wright's union of these diverse strands was spurred in part by contact with Darwin himself. Having initiated correspondence during a back-and-forth with one of the *Origin*'s negative reviewers, Wright impressed Darwin with his grasp of the issues (and, most likely, his loyalty), leading the more famous man to open up about what he called "the will of man" and its role in evolution. Darwin's engagement produced a manic, rambling response from Wright, who pushed hard to apply natural selection to "species or structures or habits or customs," arguing that all revealed "the conditions of *choice*." Wright offered this application to the mental world as a solution for Darwin. Too many critics were caught up on the "natural" in "natural selection," or so Wright thought; the better task was to extend the meaning of "selection" into the realm of the mind. In a subsequent letter, Wright made the same point in an indirect way, suggesting there was a slippage between the mode of reasoning employed by naturalists and their ability to reason about the mind. "It must be a great advantage," he wrote, "particularly on such a subject as the mental powers and habits of animals to have gained the habits of independent observation which the position of the conscientious travelling naturalist tends to produce." The habits of naturalists, Wright implied, reinforced their naturalism *about* habits.[19]

What Wright had written, at Darwin's behest, was "more like an essay than the letter I intended," and he made good on his acknowledgment by publishing a version of his thoughts the following year. "Evolution of Self-Consciousness," which is what Wright called the expanded form of his letter, took up almost seventy densely reasoned pages of the *North American Review* in 1873. In both the letter and the article, he insisted that the "natural" and "mental" worlds were evolutionary and that they converged in acts of self-consciousness. The epitome of such acts was the experiment, which Wright insisted was nothing less than the materialization of the mind, a means of making inquiry into something stable and shareable. "Scientific research," he wrote, "implies the *potential* existence of the natures, classes, or kinds of effects which experiment brings to light through instances, and for which it also determines, in accordance with inductive methods, the previously unknown conditions of their appearance." Experiments, in other words, turn something imaginary into something real; they are how our minds gain a practical bearing in the material world. This is what makes experiment evolutionary: science, like nature, produces difference. What we see around us are the differences that worked.[20]

Wright's essay caught the central thread of discussions at the Metaphysical Club, which is no surprise given that he seems to have been their animating force. Drawing on Darwin so as to blur the boundary between the human and animal worlds was one of the group's persistent projects. Another was working out the implications of the theories of another mid-Victorian naturalist: Alexander Bain. It was Bain's theory of belief, in particular, that captured the attention of the young men in the Club, and especially that of Nicholas St. John Green, then a law professor at Harvard. Like Wright, Green was interested in basic philosophical issues like cause and effect, but only insofar as he could make them concrete and practical. This meant asking how notions of cause and effect were adjudicated in court, the site where matters of responsibility and harm were blended together with lay notions of cause, intention, and belief.

Other members of the group, including Wright and Oliver Wendell Holmes Jr. (who would go on to serve on the Supreme Court), were also concerned with how abstract concepts "cashed themselves out" in the real world. Indeed, the organic roots of abstraction and the physical fruits of philosophy were something of an obsession for the Club's members. But it was Green, more than the rest, who steered the conversation toward law, and specifically toward the places where psychology and the law intersected.[21]

This was the context in which Bain and his theory of belief were debated. For Bain, belief was "a concomitant of our activity"; it related, by necessity, to what we (might) *do*. But this was no mere relationship, according to Bain. Instead, he argued that "belief has no meaning, except in reference to our actions" and that "no mere conception that does not directly or indirectly implicate our voluntary exertions, can ever amount to the state in question." This is the definition that caught Green's eye—though according to Menand, Green got it secondhand from a legal treatise. Bain's own introduction of the idea followed a lengthy footnote on the law, in which he argued against a popular view that we are not legally responsible for our beliefs because they are out of our control. Bain saw this as an abdication of our freedom, and countered it by insisting on a relationship between volition and belief, making voluntary activity a part of belief's *definition*. A belief was, according to Peirce's summary of Green's borrowing from Bain, "that upon which a man is prepared to act." This complex chain of adaptation, from one formulation to another, seemed to prove the very concept. Each of Bain's champions *believed* in his theory enough to act on it, at least in conversation. They made the theory true by engaging with it.[22]

Wright, too, was enamored of Bain's theory, not least because it offered another way to fuse the mental world of private thoughts and the physical world of resources and constraints in which natural selection was supposed to operate. By making belief physical, both men opened up the mind to new ways of applying evolutionary theory—including to the scientific process. Belief was a part of science, after all; every scientist depended upon the work of others in order to push a field forward. In the pre-disciplinary matrix of the 1870s, one's ability to use results and theories from a range of different fields came along with the necessity of trusting, even of believing, whatever one was borrowing. The view that what we believe—or, as James would later argue, what is *true*—depended on how that idea functioned in the real world, was highly controversial. But it made sense for a group of young scholars, feverishly reading across lines of emerging disciplines, often incapable of judging the veracity of claims in their original areas of development or application. In such a context, the idea that truth or falsehood rested not on the relationship of an idea to some inaccessible reality but instead on its utility, made practical sense. Critics said this crassly reduced complex issues, and perhaps abandoned truth altogether. And to a certain extent, they were right. But champions argued that the approach put everyone on an even footing. Attention to consequences was something we could all put into practice.

Neither Green nor Wright lived long enough to turn their ideas to account. In an obituary for Green, Peirce quoted a friend as saying that Green's contribution was less a thesis on belief in particular, and more "an intellectual tendency" to see *all* meaning in practical terms. "The basis of his philosophy," this friend reflected, was "that every form of words that means any thing indicates some sensible fact on the existence of which its truth depends." Green had urged Club members, appropriately, to *apply* this basic view of things to "other branches of philosophy,—to Logic, to Psychology, &c." And this is precisely what they did. Holmes changed how he thought about the law, redefining it in terms of practical results that can be predicted from its application. Peirce saw something similar in logic, specifically the logic of science, and argued that our ideas and beliefs amount to how we put them to work. James turned this fusion of Bain's psychology and Darwin's theory into a new model of the mind. And it was James, in the end, who made the transformation of cognition into selection official. Arguing that our mental lives are not separate from the physical world of comparative advantage, he spent his career insisting that the intimate links between our minds and the material universe require us to rethink what matters to us and how best to pursue it. At its root, this view and those of Holmes and Peirce represent a kind of disciplinary effect of the ideas Wright and Green advocated for so strenuously in the meetings of the Metaphysical Club.[23]

According to Menand, what eventually killed the Club was—ironically—Eliot's reforms. Wright's unpopularity as a lecturer among undergraduates led to his classes being canceled, a stress that may have contributed to his premature death by heart attack. Green died after he left Harvard, where he thought the law school was becoming too formalistic. Peirce, too, struggled under Eliot's leadership. Despite deep connections to Harvard—his father, Benjamin, had been a prominent faculty member—Peirce was unable to secure a long-term academic position there (or anywhere else). This left Holmes and James, who would become two of the most distinguished figures in American intellectual history. But in the early 1870s Holmes was still a Boston lawyer, and James was teaching anatomy on an adjunct assignment. They were far from fame, in other words, and far from "pragmatism." Their sense of possibility, of open doors, mirrored the sense of playfulness and boundary-crossing at the Metaphysical Club. Out of this flexibility, and the group's *awareness* of it, arose new ways of thinking about thinking. Evolution and experiment, embodied cognition and a sense that science *mattered,* solved problems peculiar to the members of the Metaphysical Club. And as they did, these ideas became means for attaining and even imagining entirely new projects in turn.[24]

LESSONS IN LOGIC

The Metaphysical Club did not have a monopoly on scientific commentary in the 1870s—even in Cambridge. Eliot had his own agenda, though his emphasis on practical results was akin to the views being expressed by James and Holmes. But it was another Harvard associate, not a Club member, who made his name defining science in this period: Simon Newcomb. A friend of Wright's and, like him, a disciple of Mill and Darwin, Newcomb studied astronomy under Benjamin Peirce in the 1850s and, like Benjamin's son Charles, took a job working for the Nautical Almanac as an astronomer and "human computer." Born in

Canada and something of a self-taught man, Newcomb eventually became one of the most famous scientists in the United States. As early as the 1870s, his scientific results attracted renown and—eventually—job offers, which spurred him to adopt the role of a spokesperson for science throughout his career. As a result, he not only contributed to work in astronomy, mathematics, and political economy, but also offered what his biographer has called "a scientist's voice in American culture"—in other words, the perspective of a trained scientist on the public issues of his day.[25]

Like Babbage before him, Newcomb used this role to comment on the status of science as it was practiced in his own country. Though born in Canada, Newcomb's government positions (which had included a post at the Naval Observatory during the Civil War) led him to reflect on science and its method in the United States. Comparing American and European efforts, just as Babbage had done for England half a century earlier, Newcomb found his countrymen coming up short, especially in the matter of what he variously called "exact," "abstract," and "basic" science. In 1874 he published a widely read piece in the *North American Review* titled "Exact Science in America," which blamed the sorry state of national science on the fact that "the American public has no adequate appreciation of the superiority of original research to simple knowledge." Like Babbage, in other words, Newcomb located the problem in the lack of support (financial and otherwise) for the kinds of thinking that might not produce useful results right away. "Original research," in this sense, was not "the efforts to which we are impelled by our daily physical wants, but those to which we are impelled by the purely intellectual wants of our nature." The task, according to Newcomb, was to set up a "rigorous system of intellectual natural selection" to incentivize such work. The national mind, like its economy, would advance through an organized system of competitive progress.[26]

How would that work? Newcomb provided a surprising suggestion when, a few years later, he was asked to expand on his pessimistic essay for a centennial issue of the same journal. Its editor, Henry Adams, asked him specifically for "a diagnosis of the mental condition of our country which may offer some light as to the probable tendency of our future."

Newcomb's reply was published as "Abstract Science in America, 1776–1876," and it identified both the disease and the cure as matters of method. "What is required to insure us against disaster is not mere technical research," he argued, "but the instruction of our intelligent and influential public in such a discipline as that of Mill's logic, to be illustrated by the method and results of scientific research." What Americans needed, for the sake of their scientists, was now deeper than money or public will. In order to enable those necessary developments, Newcomb called for something surprising: a healthy dose of logic, specifically the logic of science propounded by a recently deceased Englishman. The future of science depended on looking to its past.[27]

Mill was a favorite of Newcomb's. He had bought Mill's *System of Logic* while a student, and Mill's views on how science ought to operate informed Newcomb's scientific career as well as his public campaigns on behalf of funding and abstract thinking. Methodology, to both men, meant you had to "direct the minds of the rising generation toward the methods of science"; the point was not just to make Americans scientific thinkers, but to make them better citizens as a result. Logic is not always what one thinks of when contemplating politics and personal affairs, but this is indeed what Newcomb—following Mill—saw as the subject's highest aim. For both, logic was supposed to be a living account of how minds work, whether in their most rarefied operations, such as "abstract science," or their most mundane. Indeed, the two spheres were not separate: the abstract and the everyday were linked in a balancing act revealed at the level of disciplines and at the level of private thoughts. For both men, logic contained lessons to live by, lessons that built on the history of science but also, more importantly, evinced a natural ability in everyone who used a scientific approach to solve the problems that mattered to them.[28]

Newcomb's essays made him a prominent spokesman for American science, but it was his lesser-known contemporary—incidentally, the son of his mentor—who sounded the strongest call for logic as a salve in the years after the Civil War. Born as close to the heart of Harvard as he could have been, Charles Peirce seemed destined for the kind of career Newcomb attained over the course of the Gilded Age. Peirce's father, Benjamin, played an active role in the training of both young men. As a prominent professor at Harvard's Lawrence Scientific School, director of the Coast Survey, and an early president of the American Association for the Advancement of Science, Peirce the elder was well positioned to launch them both into government and academic work alike. In the case of Newcomb, this support played out as planned; for Charles, however, an illustrious career was not in the offing. Indeed, holding down a steady job proved hard enough. Famously difficult as a colleague and even as a friend, Peirce's wide-ranging interests ran afoul of the lines being set up around disciplines and departments. Eliot's vision for Harvard, which Benjamin Peirce helped to realize, had no place for as strange a man as his son, whose views of logic and science, tested in the Metaphysical Club and ceaselessly promoted as the cure for all maladies, taxed the sympathy of even his closest friends and fellow members.[29]

The world was still Peirce's oyster in the 1860s, however, and high hopes for a career in academia kept him giving talks and publishing papers at a feverish pace well into the 1870s. His preferred subject was always logic, on which he gave his first public remarks in 1866. Offered as a part of the "scientific lectures" arranged at Boston's Lowell Institute, the idea was to meet the audience demand for scientific topics—which had been increasing since the Civil War—by framing science as both a pastime for busy people and a potential tool for social advancement. Peirce's added incentive was to present his topic in a way that made space for him, and his love of logic, at a university like the one he had left across the river in Cambridge. Both aims, it seemed to Peirce, could be satisfied if one took as broad a view as possible of the topic: that is, if science was made out to transcend disciplinary divisions, thereby calling for the overarching discourse of mind and method that Peirce had taken to calling "logic."[30]

Entitled "The Logic of Science; or, Induction and Hypothesis," Peirce's lectures drew on the work of one of his heroes: William Whewell, who had died that very year. This choice helps explain the enmity between Peirce and Newcomb: as we have seen, Whewell and Mill fought a decades-long battle that many—included Newcomb—took Mill to have won. Peirce's preference for "the loser" is telling. Springing, it seems, from what he saw as a fatal flaw in Mill's approach to logic, Peirce turned to Whewell to correct it. At times Peirce expressed his preference in terms of psychology: Mill (and Newcomb) saw logic as part of psychology, whereas Whewell seemed to insist that logic had nothing to do with the actual workings of the human mind. In 1865 Mill had insisted that if logic "is a science at all, it is a part, or a branch, of Psychology; differing from it, on the one hand as a part differs from the whole, and on the other, as an Art differs from a Science." While Peirce agreed that logic was a science, he was adamant that it had nothing to do with psychology—at least as Mill had defined it, in his father's associationist terms.[31]

That same year, just before he began preparing his Lowell lectures, Peirce staked out his own approach to the overlap of logic and psychology. Here things become a bit murkier. In the first draft of a project he was to pursue—and never finish—for the rest of his life, Peirce tried to make his view as clear as possible in the title: "An Unpsychological View of Logic." There he framed the history of logic as a clash between views "which do not and those which do give to logic a psychological or human character." Though he described the same dispute in different ways during his career, such as "thought in general" versus "thought," or "Formal" versus "Anthropological" logics, he always seemed to align himself against the idea that logic was part of psychology. Logic, he wrote at the time of that first draft, "has nothing at all to do with operations of the understanding, acts of the mind, or facts of the intellect." Mistaking one for the other was not just an error of taxonomy or a slip of reasoning—it risked sacrificing the potential Peirce saw in logic as a kind of independent arbiter for scientific disputes, a guide to *right* reasoning that was never beholden to the vagaries of how we embodied, imperfect human organisms actually try to solve problems.[32]

This was where Mill and his followers had gone wrong. According to Peirce, the problems began with their allegiance to the theory of associationism. What was a hypothesis, a useful idea in need of testing, had become constrictive, a mold cast by James Mill into which his son had forced all of logical reasoning. Pointing out this flaw, Peirce hoped, "will help to awaken that numerous class of general readers who have become impregnated with the ideas of Stuart Mill's logic into self-consciousness in reference to the intellectual habit which they have contracted." The method Mill propounded became a disease or an unwanted pregnancy, in Peirce's strange metaphor. The idea was to give the public—or at least the readers of *The Nation,* where the review appeared—a sense of agency over their patterns of thought. Peirce continued: "A philosophy or method of thinking which is held in control—the mind rising above it, and understanding its limitations—is a valuable instrument; but a method in which one is simply immersed, without seeing how things can be otherwise rationally regarded, is a sheer restriction of the mental powers." Associationism, like any other -ism, risked morphing from a question into an answer, from a guess into a certainty. The only way to prevent it from doing so was to insist that the hypothesis be subjected to test after test—to insist, in other words, that we learn to think logically about everything.[33]

To Peirce's mind, Whewell had avoided the psychological trap into which Mill had fallen. Whether or not this was true is a matter of dispute—indeed, as we shall see, whether or not Peirce himself avoided it depends on where you look in his writings. But as he drafted and lectured during the 1860s, Peirce became increasingly convinced that Whewell provided a model by which to work. In "Lectures on British Logicians" a few years later, Peirce praised Whewell for his careful attention to history and for his familiarity with scientific methods—both of which were virtues that, unsurprisingly, Peirce felt characterized his own approach. What he liked most about Whewell's work was the way these two features came together. Because it was "founded on the history of science in a truly scientific spirit and by a genuine inductive method," Whewell's logic was not only an excellent account of science but also an example of scientific reasoning. It was, in other words, "a Scientific Induction from the History of Science." Whewell not only *told* readers what induction meant—he also *showed* them induction, according to Peirce, by building his philosophy on the dataset of science as it had developed over time.[34]

Did Whewell's use of history make him less psychologistic than Mill? Yes and no. And this is where things get interesting. The two combatants disagreed about what both terms meant. For Whewell, science was primarily something that unfolded over time, whereas for Mill it was a feature of each individual mind. These two views were not mutually exclusive, but were instead separated for polemical purposes—first by Whewell and Mill and later by Peirce. All three saw that science was, or could be, both. Peirce, for instance, confessed that there was "doubtless something else which concerns the individual investigator and which does not appear in [Whewell's] publications which such a theory would be apt to overlook." One could trace ideas over the course of history, which Whewell did, following them from one generation to the next. If you did, you saw the development of science as a set of formal relationships between ideas in their perfect, or at least ideal, form. "But," Peirce asked, "what was the mental process, what was the change and what the law of the change in the individual mind by which an obscure idea becomes clear? This Whewell tells us nothing of and indeed seems to have no conception of the question." It was one thing to track ideas over time, another to show how they took hold.[35]

Answering this question—indeed, even *asking* it—came close to the psychologistic approach to which Peirce seemed so opposed. But a closer inspection reveals an openness, at least early on, to the sorts of issues psychology could explore. In both his lectures on logic and the famous essays he wrote to illustrate them a decade later, Peirce saw the value of studying individual, even ordinary, minds in order to provide a robust account of scientific thinking. In yet another abortive project (this one called, more honestly, "Toward a Logic Book"), he grounded logic in the personal sense of doubt we all share, all the time. "Doubt," he wrote in one of many drafts of an opening chapter, "is an uneasy and dissatisfied state from which we struggle to free ourselves and pass into the state of belief." If doubt is what Peirce called "genuine," then "real inquiry" would follow. Like his friends in the Metaphysical Club, to whom he read a version of this chapter in 1872, Peirce borrowed Bain's theory of belief for his own purposes. "The feeling of believing," he wrote later on, "is a more or less sure indication of there being established in our nature something which will determine our actions." Mind and matter were one, or at least came together, in our capacity for belief—a capacity that followed "naturally" from our efforts to allay the sensation of doubt. Reasoning, even science, was an act of self-soothing.[36]

Peirce's project, or a version of it, eventually made its way to a broader audience, but not in the form he imagined while drafting chapters for his magnum opus. What had once promised to be a grand book laying out a unified theory became more practical, but also more interesting: a series of popular essays for a new magazine called *The Popular Science Monthly*. Published in 1877 and 1878 under the title "Illustrations of the Logic of Science," the series seemed to make good on Newcomb's call for a logic "illustrated by the method and results of scientific research" for the general public. But Peirce's logic was not Mill's, as Newcomb hoped, or even Whewell's, but his own. It answered a question he felt his predecessors had failed not only to ask but to even comprehend: How do our ideas, from the most ordinary to the most rarefied, arise out of the interplay of doubt and belief that we all experience? This was a fundamental question of logic, and it was here that Peirce's sense of his own project linked up with what psychologists were asking at the same time. His answer, in other words, connected a psychological description of how each of us transforms doubt into belief in our daily lives to a methodological prescription for how to conduct scientific research in the pursuit of truth. This mental connection centered on a shared process that, Peirce believed, looked a lot like evolutionary change.[37]

The first essay in the series, called "The Fixation of Belief," began with a history of science. According to Peirce, "every work of science great enough to be remembered for a few generations affords some exemplification of the defective state of the art of reasoning of the time when it was written; and each chief step in science has been a lesson in logic." As an example, Peirce picked up "the Darwinian controversy," reinterpreting the debates surrounding natural selection in terms of the methodological, even mental, consequences of Darwin's attempt "to apply the statistical method to biology." The problem, for Peirce, was not whether God had a hand in creation or even whether the evidence Darwin used was adequate to his claims, but instead that methods developed to deal with one set of problems were being applied to another—with the result that "questions of fact and questions of logic are curiously interlaced."

In the case of Darwin and his theory, this interlacing presented the added complexity of shifting science's foundations from logical certainty to the adaptations of imperfect, evolved organisms.[38]

According to Peirce, all inquiry begins with an aversion. "Doubt is an uneasy and dissatisfied state from which we struggle to free ourselves and pass into the state of belief," he argued, "while the latter is a calm and satisfactory state which we do not wish to avoid, or to change to a belief in anything else." Doubt is something you feel, a problem Peirce framed as a universal experience. The questions we ask have their roots in affective states, just as Bain and Spencer had noted in their accounts of mental evolution and its origins in the feelings associated with thinking. Peirce built on their approaches by offering a way to "fix" our beliefs and avoid negative feelings. You can stick with your beliefs through what Peirce called "the method of tenacity"—but you might be wrong. Or one group can decide what everyone should believe and enforce it by "the method of authority"—but here again, though more people are involved, there is no assurance of truth. Better might be to work together, to test our beliefs against experience and share the results. But even here, with what he called "the *a priori* method," Peirce thinks we are beholden to whatever beliefs we happen to come in with, which will alter the terms of debate and might adapt to experience without revealing their fundamental flaws.

The solution was what Peirce called "the method of science." With science, our thinking has no effect on the truth—we come into contact with the world and adapt our ideas to *it*, not to those of other people. We do so individually or collectively, in private or in public. For anyone to choose this particular method was, Peirce stated grandly, "one of the ruling decisions of his life, to which, when once made, he is bound to adhere." It was a way to expand your "intellectual kit of tools," but it was also something more than that, as the closing lines of "The Fixation of Belief" make clear: "The genius of a man's logical method should be loved and reverenced as his bride, whom he has chosen from all the world. He need not contemn the others; on the contrary, he may honor them deeply, and in doing so he only honors her the more. But she is the one that he has chosen, and he knows that he was right in making that choice." Logic, for Peirce, was something like loyalty, or love. It was not only an account of our "intellectual kit of tools." It was a matter of right and wrong, not just in the sense of truth and falsehood, but also in a moral sense, of how to live one's life.[39]

The sequel to "The Fixation of Belief" was "How to Make Our Ideas Clear," by which Peirce meant opening up our private thoughts for public consumption. This was crucial for the practice of science, which began in individual minds but was always a community activity, not least because it extended so far in space, time, and complexity that no individual could hope to see the ultimate agreement of belief toward which scientists aimed. The way beliefs were made manifest was familiar: "by the different modes of action to which they give rise." Borrowing again from Bain, Peirce made the principle of practice key to everything: "There is no distinction of meaning so fine as to consist in anything but a possible difference in practice." Ideas are not only unclear, but are literally *nothing* without the "sensible effects" they (can be imagined to) have: "Consider what effects, which might conceivably have practical bearings, we conceive the object of our conception to have. Then, our conception of these effects is the whole of our conception of the object." This was the essence of pragmatism, as both Peirce and James recalled it later. Though it is a thorny way of putting things, the thrust is clear: in science, as in everything else, ideas exist both privately and publicly, subject to our senses but also to the social pressure of sharing them in groups.[40]

Subsequent essays in the series drilled down into the details, elaborating theories of chance and probability that would become more and more central to Peirce's approach to logic. But here, in the earlier and more accessible essays, we have his fundamental lessons of logic: Inquiry is born from personal experience, becoming science only when its products cash out in experience. This emphasis on the material and natural dimensions of thinking—on how science is *lived*—also extended inward, into the physiology of the brain. In the final essay in the series, for example, Peirce distinguished between the *feelings* of induction and hypothesizing, much as he had referenced "feelings of belief" earlier on. Whereas the feeling of induction "is the logical formula which expresses the physiological process of formation of a habit," that of hypothesis is "a single feeling of greater intensity." Their differences are affective, even emotional: "Hypothesis produces the *sensuous* element of thought, and induction the *habitual* element." (Deduction, Peirce added, is a matter of attention and volition, of "nervous discharge," but not of the emotions as he defines them.) Though sketchy on details, it is clear that in his "Illustrations" Peirce was keen to offer something approximating a physiology of science, an affective account of the *act of thinking,* rather than a simple set of formal rules.[41]

The tension between Peirce's earlier openness to psychology and his later aversion to it is clearest in what is perhaps his most famous idea: abduction. For Peirce, the "abductive method" avoided the rigidity of deduction and the endlessness of induction. In the series for *Popular Science,* he illustrated it with an example. Deduction applied a rule ("All the beans from this bag are white") to a case ("These beans are from this bag") to produce a result ("These beans are white"), and induction worked from case, via result, to rule. Abduction—which Peirce called a "hypothesis" at the time—did something in between. It began with a rule, like deduction, but then it made a leap: encountering some white beans, to use his example, abduction infers that the beans are from the bag in question. As Peirce later wrote, "the process of forming explanatory hypotheses" in this way was "the only logical operation which introduces any new idea." Science, for Peirce and his followers, only advances through abductive leaps; pragmatism, according to its founder, was nothing more than "the logic of abduction." What some now call "inference to the best explanation" was, in Peirce's early work, the common denominator of everyday thinking and scientific reasoning, of how ordinary problems are solved subconsciously and how the grandest theoretical advances are made. It was, in other words, where psychology and logic connected in the evolved (and evolving) human mind.[42]

Anyone who has taken a course in logic knows how unnatural the subject can seem, how hard it can be to wrap one's mind around its rules and notation. The difficulty so many encounter is what makes Peirce's insistence on its "sensuous" dimension (not to mention his logic-bride) a bit hard to understand. Maybe, you might think, it takes a particular sort of person to view logic in the ways he describes. Peirce himself acknowledged the gap between how we think and how we account for that thinking in logical terms. This may explain his ready, if irregular, insistence that the psychology of actual human thinkers played no part in logic as properly practiced. But there was also a side of Peirce that *wanted* logic to touch our psychology, even our physiology. His series for *Popular Science* was a case in point, insisting that logic was a universal capacity, an almost intuitive approach that education brought to the fore. This explains Peirce's argument that "man's mind has a natural adaptation to imagining correct theories of some kinds," not to mention his idea that true hypotheses are more "natural," more "readily embraced by the human mind" than false ones. We are built for this stuff, he seemed to say, and hypotheses were proof. If logic pointed the way for scientific inquiry, it pointed inward—to our minds, or our anatomy. The task was to discern what psychology could tell us about our chosen methods.[43]

SPONTANEOUS VARIATIONS

Just as the "Illustrations" series started, James took up the same issue. Offering Harvard's first undergraduate class in psychology, James soon signed a textbook contract with Holt. Though he had taught psychology to graduate students the previous year, "Natural History 2: Physiological Psychology" was James's attempt to wedge the subject into the college curriculum. He pitched the course in terms Harvard's scientifically minded young president would understand. "A real science of man," he insisted "is now being built up out of the theory of evolution and the facts of archaeology, the nervous system and the senses. It has already a vast material extent, the papers and magazines are full of essays and articles having more or less to do with it." Besides signaling that Harvard was falling behind work being done elsewhere (especially in Europe, where James had trained briefly), he was making a more interesting point: popular texts in psychology were exceeding the field's technical advances; students would get psychological ideas one way or another, whether they learned them at Harvard or in the newspapers.[44]

This was a strange argument. Courses were not supposed to mirror what "the papers and magazines" were publishing. What you taught in universities was supposed to be lasting, if not traditional, rather than the latest fashion. James was making an unusual pitch. But psychology was an unusual subject—and according to James, it would take an unusual instructor to capture it in all its complexity. Where could Eliot find someone capable? "Apart from all reference to myself, it is my firm belief that the College cannot possibly have psychology taught as a living science by anyone who has not a first-hand acquaintance with the facts of nervous physiology. On the other hand, no mere physiologist can adequately realize the subtlety and difficulty of the psychologic portions of his own subject until he has tried to teach, or at least to study, psychology in its entirety. A union of the two 'disciplines' in one man, seems then the most natural thing in the world, if not the most traditional." If Harvard wanted what was "traditional," then James could recommend a few Germans. But if they wanted "a living science," then they had to offset experiments with something more experiential. If they wanted to strike *that* balance, James knew just the man for the job.[45]

Eliot agreed, and the following year James got his first shot at putting "living science" in the Harvard curriculum. For a textbook, he chose the first volume of Spencer's *Principles,* in part for its insistent evolutionism and in part because it began where James himself had begun: anatomy and physiology. James had been reading Spencer for a decade, and though he disagreed with him, he saw Spencer's approach and prominence as a good starting point. But his opinion quickly changed. The experience of reading the new volume of Spencer's *Principles* turned him so violently against its author that, by the end of the spring, James was practically foaming at the mouth. "Of all the incoherent, rotten, quackish humbugs & pseudo-philosophers which the womb of all-inventive time has excreted," he exclaimed, "[Spencer] is the most infamous." The course got him so worked up that, when it finished, James composed and sent off what would become his first academic article: an all-out attack on Spencer's theory of mind entitled "Remarks on Spencer's Definition of Mind as Correspondence." It is not much of a stretch to say that his pedagogical disappointment in Spencer's *Principles* is what led James to sign on to write his own textbook for Holt the very next year.[46]

Spencer was not always James's nemesis. Early on, James had been a fan. Of his encounter with Spencer's *First Principles,* James recalled: "I read this book as a youth when it was still appearing in numbers, and was carried away with enthusiasm by the intellectual perspectives which it seemed to open." He admitted, in his notes, that "Spencer's account of the variability of feelings is admirable." And in an anonymous review of Spencer's *Data of Ethics* a few years later, James gave Spencer credit for how "the facts of evolution have crowded upon the thinking world"—in other words, for the soil in which James's own thinking was taking root. Even his "obituary" for Spencer, which mostly reads like an exercise in character assassination, gave him "the immense credit of having been the first to see in evolution an absolutely universal principle." As he put it years later, referencing a line from Byron: "'He left Spencer's name to other times, linked with one virtue and a thousand crimes.' The one virtue is his belief in the universality of evolution—the 1000 crimes are his 5000 pages of absolute incompetence to work it out in detail." This may not *sound* positive, but since James also extended evolutionary theory into new terrain, it stands to reason that he recognized this as a virtue. Great minds think alike, he might have said.[47]

James's obvious animus was doubtless tinged with jealousy. Spencer was a best-selling author and a minor celebrity; it was he, and not Darwin, who had made the theory of evolution (including its application to human issues) a talking point, especially in the United States. His American reputation was aided significantly by a set of influential Spencerians, all of whom made it their mission to preach the brilliance of their British prophet and, eventually, spread the gospel of social Darwinism. One of the most impactful was Edward L. Youmans, an author and publisher whose passion for evolutionary theory and knack for international copyright helped him serve as a major vector for scientific ideas in and out of the United States. Whether through the "International Scientific Series" or *The Popular Science Monthly,* both of which Youmans founded in the early 1870s, he helped turn evolution into a conversation around the country and Spencer into a household name. Spencer's much-publicized trip to the United States a decade later, during which he was greeted by adoring fans wherever he went, was capped off by a lavish banquet, organized by Youmans, at which Spencer was toasted by New York's rich and famous. This was the cachet that a mix of scientific vocabulary and social commentary could command—especially if its conclusions seemed to justify the economic status quo. Whether or not this was how Spencer saw his system, Youmans and others turned it into an American philosophy.[48]

When James called Spencer "the philosopher whom those who have no other philosopher can appreciate," then, we should see it for what it was: the dismissal of an outsider by a scholar who, though famous himself, no doubt chafed at the prominence of a man whose entire body of work James and others had subjected to decades of attempted refutation. It is no surprise that academic argument failed to undo what popular enthusiasm had made, but it irked James to no end. Spencer was, he told a friend, "as absolutely worthless in all *fundamental* matters of thought, as he is admirable, clever & ingenious in secondary matters." When James, around the time his own textbook appeared, wrote that he would love "to help stone Uncle Spencer, for all extant quacks he's the worst—yet not exactly a quack either for he *feels* honest," he signaled not only his animosity toward Spencer, but also his (admittedly complex) filial relationship to the man and his views on evolution. Spencer's popularity, after all, helped pave the way for later applications of evolutionary theory to mental and social life, no matter how distant they were from the system Spencer proposed. Not for nothing did James call him "Uncle."[49]

"Remarks on Spencer's Definition of Mind as Correspondence" was an attack on the analogy between minds and mirrors. James was most forceful in his famous conclusion: "I, for my part, cannot escape the consideration, forced upon me at every turn, that the knower is not simply a mirror floating with no foot-hold anywhere, and passively reflecting an order that he comes upon and finds simply existing. The knower is an actor, and co-efficient of the truth on one side, whilst on the other he registers the truth which he helps to create." According to James,

Spencer's view of the mind was passive and permissive, whereas James saw it as active—and unpredictable. All our thoughts and hypotheses "help to *make* the truth which they declare." Here again the ghost of Bain lingers, hitching the mind to the world. Where Spencer saw connections between mind and world as reactive, James thought we could make the first move. "In other words," he concluded, "there belongs to mind, from its birth upward, a spontaneity, a vote."[50]

But it was not just Spencer's *theory* that bothered James. It was the way Spencer articulated it, the way he put it to work—in other words, his "absolute incompetence to work it out in detail." Key to James's complaint was a single word, one that summed up Spencer's big ambitions and was increasingly anathema to how James viewed the study of the mind: system. What was wrong, as often as not, was not any particular view that Spencer held, so much as what he *did* with those views, forcing them into a system. In the face of this self-importance, James always claimed an anti-systematic stance. "Yes," he wrote to a critic of the *Principles,* "I *am* too unsystematic & loose! But in this case I permitted myself to remain so deliberately, on account of the strong aversion with which I am filled for the humbugging pretense of exactitude." Given his use of "humbug" again, it is clear which system-builder he had in mind. "Great thinkers," according to James, were not grand systematizers like Spencer, but instead those who "have vast premonitory *glimpses* of schemes of relation between terms, which hardly even as verbal images enter the mind, so rapid is the whole process." This was the mode James most admired, calling himself (in a letter to his wife) "a thing of glimpses, of discontinuity, of *aperçus.*"[51]

James expressed his misgivings about Spencer's system in various registers, including in his objections to even his most basic points. Take Spencer's assumption that we intuitively believe in laws of the universe—like the persistence of force—because mind and universe are part of the same system. The problem with such an assumption, James says, is that Spencer tries to "use it to deduce anything from it," including "his whole system." Although the idea was perhaps a worthy hypothesis, Spencer had served it up as *evidence* of the system that supposedly affirmed it was true. Later in the same notes, James came back to this point:

"Spencer's real motive: To make his system of phenomenalism absolute, & forestall criticisms. All I don't know is unknowable." What was wrong was not so much Spencer's idea of an "Unknowable" element in the universe, so much as the fact that (according to James) he *used* that idea as an illegitimate defense against the charge that his supposedly perfect system fails to explain key phenomena it should encompass. The issue was not theory, so much as method.[52]

If there was any truth to James's mean-spirited descriptions of Spencer—"equable and lukewarm in all his tastes and passions" and "a philosopher in spite of himself," for example—then it may be that what put James off most was not so much an idea as the temperament that Spencer expressed *through* his ideas. It seemed, James wrote privately, that Spencer's work "consecrates the dryness in him." James was attracted to the subjects with which Spencer was least comfortable: to the spiritual dimension of human nature, whether in the form of occult practices such as séances or in the more familiar guise of introspective attention to religious feeling. There was no space in the Spencerian system for these dimensions—or, rather, Spencer's approach was to translate them into terms that his system (and, one gathers, he himself) could comprehend. As the prevalence of other critics of the idea of "system" in this period makes clear, James was not alone in his dissatisfaction.[53]

Seeking a solution to Spencer's insufficiencies, James reached for a familiar source: the work of Charles Darwin. It is easy to forget that it was Spencer, not Darwin, who came first to mind on evolutionary matters in this period, especially for those accustomed to reading about science in popular periodicals. As James came of age, the ideas of both men were constantly debated and compared. A major difference between the two was that Spencer had consistently—and, to James, infuriatingly—pushed the boundaries of evolution and its applications in ways that Darwin was hesitant to do, if indeed he did so at all. It was Spencer whose books captured the attention of James's students, members of the Metaphysical Club, and audiences who knew nothing of such rarefied discussions. Indeed, it was partly *because* of his ubiquity that James sought alternatives to Spencer's approach, and it could be said that Darwin's followers set themselves the task of distinguishing his views from Spencer's for the same reason. Huxley and Tyndall, for example, saw the policing of the boundaries around science as part of their public duty—and, more and more, this meant promoting Darwin over Spencer in the popular press.[54]

James followed suit, holding up natural selection as a superior theory and Darwin as a superior theorizer. An emphasis on method was what set James's appropriation of Darwin apart from others. James's methodological preference came through in some of his earliest published pieces: a series of unsigned reviews in the 1860s that either focused on Darwin or else mentioned him by name. The first, written when James was only twenty-two, used a review of Huxley's view of evolution to explore "the play of the two great intellectual tendencies," which James labeled the synthetic (or imaginative) and analytic (or practical): "It is with the hope of one day reaching this sublime point of view that Science struggles ever forwards, spurred passionately on over the slow and difficult approaches by her synthetic followers, while her analytic ones moderate her speed and keep her from wandering away from the right path. To get her to the goal, the services of both are indispensable, for either class is infirm alone, and needs the other to make up for its shortcomings." Progress meant unity, and unity meant balancing tendencies. The evolution of life was mirrored in debates over evolution: new ideas would have to fight in order to survive. Though he emphasized that "the imaginative temperament, if left unchecked to deal with science, would run into endless excesses," James's sympathy for guesswork was clear.[55]

This preference was even clearer in a pair of reviews James wrote of Darwin's own *Variation of Animals and Plants under Domestication* for popular audiences a few years later. In one, he defended the book's "presumptuous" approach by arguing that "a bad hypothesis is far better to work with than none at all." Here, James came close to articulating a Darwinian defense of Darwin's own theory. By separating the production of variations from the selective process, natural selection turned anomalies into evidence: "The more idiosyncrasies are found," James noted, "the more the probabilities in its favor grow." Darwin's book, he concluded, "harrows and refreshes, as it were, the whole field of which it treats." In a subsequent review, James insisted on the value of "the instinctive guesses of men of genius," even if "the conclusions drawn from certain premises are assumed in their turn as true, in order to make those same premises more probable." Science, like nature, tumbles on, with novel adaptations put to work immediately to test their worth. What mattered, in Darwin's book or the book of nature, was not any specific result, which might pass away as soon as it appeared, but instead the basic method—what James called "the *kind* of speculation"—that underlay it all.[56]

Unsigned reviews were just the beginning. Soon James was framing an explicit analogy between life and science in detail. Unsurprisingly, he did so as part of a concerted effort to reject Spencer's approach to mind and method. His teaching notes record the beginnings of this use of Darwin to combat Spencer: "my belief that we can give no clear scientific description of the facts of Psychology considered as phenomena belonging to a purely natural plane without restoring to the inner at every step that active originality and spontaneous productivity which Spencer's law so entirely ignores." The problem with Spencer was that "in Psychology he repeats the defects of Darwin's predecessors in biology." Of course, Spencer *was* one of Darwin's predecessors, one of those "pre-Darwinians" for whom everything was adaptation. Darwin and his followers, including James, gave spontaneous variation a role Spencer would not countenance. Later, James's notes strayed into a fragmented list, with items like "Sugar" and "Wood-pile" seemingly signaling a turn to a to-do list. But along the way, he turned Darwin into a way of understanding the mind in general and science in particular: "Invention and discovery produced by Spont. Var. = Hypothesis." Here, Darwin's equation of nature with a man of science was inverted. Scientists and their method were now host to an evolutionary process all its own.[57]

Teaching Spencer sharpened James. His earliest essays, including "Remarks" but also "The Sentiment of Rationality" and "Brute and Human Intellect," all take direct aim at Spencer. But it was in his first signed piece for a general audience—"Great Men, Great Thoughts, and the Environment," published in *The Atlantic* in 1880—that James really hit his target. The article on "Great Men" picks up where his notes left off: "A remarkable parallel, which to my knowledge has never been noticed, obtains between the facts of social evolution and the mental growth of the race, on the one hand, and of zoölogical evolution, as expounded by Mr. Darwin, on the other." Though not exactly correct that the parallel was unknown, James gave new life to the analogy by emphasizing "spontaneous variations" in the face of Spencer's "environmentalism." The idea of separate spheres, of variations and selection playing out according to different logics, became the main means of distinguishing Darwinian interpretations from their alternatives. The two "cycles" were "relatively independent of one another," according to James; their connections only visible *"if we take the whole universe into account."* Their separation comes through with strange imagery: "The mold on the biscuits in the store-room of a man-of-war vegetates in absolute indifference to the nationality of the flag, the direction of the voyage, the weather, and the human dramas that may go on on board; and a mycologist may study it in complete abstraction from all these larger details. Only by so studying it, in fact, is there any chance of the mental concentration by which alone he may hope to learn something of its nature." To James, "the triumphant originality of Darwin [was] to see this, and to act accordingly."[58]

Causal separation proved central to James's later work. The *Principles,* comprised as it was of these early essays (including "Great Men"), leans on the distinction between cycles at every turn. Historians have noted how James's Darwinism was part of his lifelong goal of preserving free will in the face of the scientific materialism which both attracted and repulsed him. If our minds "have a vote," as James so often insisted, then our fates are not sealed by the time or place of our birth or even by the decisions we have made up to this point. He admitted that, from the view of "the whole universe," our spontaneous ideas might not seem spontaneous after all, but James saw such a view as infinitely far off—if not impossible to achieve. In the meantime, we "must simply accept geniuses as data, just as Darwin accepts his spontaneous variations." Accepting such data did not mean they were random, nor environmental. As James put it in the *Principles,* the world might "make a 'born' draughtsman or singer by tipping in a certain direction at an opportune moment the molecules of some human ovum; or she may bring forth a child ungifted and make him spend laborious but successful years at school." The bifurcation was, in a certain sense, a false one: it was always a balance of these forces. What mattered, to James, was that you recognize both sides.[59]

This account of the natural and social worlds worked for the mind as well. Balance, as ever, was key. We are, each of us, equal parts "brot [*sic*] in at door & born in house." Echoing Wright, who had written that "positivists, unlike poets, become—are not born—such thinkers," James put a twist on the idea by running these two types together. Each idea, poetic or scientific, emerges from "random images, fancies, accidental out-births of spontaneous variation in the functional activity of the excessively instable human brain, which the outer environment simply confirms or refutes, adopts or rejects, preserves or destroys." What is true in us all is especially true in the genius, in whom we find "a seething caldron of ideas, where everything is fizzling and bobbing about in a state of bewildering activity, where partnerships can be joined or loosened in an instant, treadmill routine is unknown, and the unexpected seems the only law." These ideas can be all sorts of things: "sallies of wit and humor; they will be flashes of poetry and eloquence; they will be constructions of dramatic fiction or of mechanical device, logical or philosophic abstractions, business projects, or scientific hypotheses, with trains of experimental consequences based thereon," and so on. What unites such mental elements is that "their genesis is sudden and, as it were, spontaneous."[60]

Spontaneity was central to natural selection, and James made this aspect of the theory a core part of a range of mental functions. Memory, for example, works best if "associations *arise independently of the will,* by the spontaneous process we know so well." The same was true of reasoning. Novel results depend on our ability to "spontaneously form lists of instances"—that is, "without any deliberation, spontaneously collecting analogous instances, uniting in a moment what in nature the whole breadth of space and time keeps separate." We are at our best in "a sort of spontaneous revery," when we admit—as he put it elsewhere—"the irremediably pluralistic evolution of things, achieving unity by experimental methods, and getting it in different shapes and degrees and in general only as a last resort." These were *not* the methods of those James called "prism, pendulum, and chronograph-philosophers," whose "method of patience, starving out, and harassing to death" James never tired of saying "could hardly have arisen in a country whose natives could be *bored.*" James wanted "generous divination," not "deadly tenacity." Spontaneity was the key to good science; without it, progress was doomed.[61]

"Whose *theories* in Psychology have any *definitive* value today?" James asked the man he hired to take over his laboratory at Harvard. "No one's! Their only use is to sharpen farther reflexion and observation. The man who throws out most new ideas and immediately seeks to subject them to experimental control is the most useful psychologist, in the present state of the science." James was a psychologist who preferred conversation to experiment. "When walking along the street, thinking of the blue sky or the fine spring weather, I may either smile at some preposterously grotesque whim which occurs to me, or I may suddenly catch an intuition of the solution of a long-unsolved problem, which at that moment was far from my thoughts." For him, spontaneity, even instability, was the key to thinking, scientific and otherwise. "The conception of the law," James argued, "is a spontaneous variation in the strictest sense of the term. It flashes out of one brain, and no other, because the instability of that brain is such as to tip and upset itself in just that particular direction." Both "triumphant hypotheses and the absurd conceits" were born of the unstable nature of the human brain. Everything beautiful had randomness in it.[62]

Behind James's arguments about how science worked best, we find what he called in another essay "The Sentiment of Rationality." Feeling, in some form, enlivens even the most calculating of activities. Science, like everything else, has an affective dimension. "The scientific hypothesis," James continued in his reverie, "arouses in me a fever of desire for verification." The spontaneity James defended in science—and preserved for his personal purposes—was also fundamental to the life worth living. Thinking, he had written in "Great Men, Great Thoughts," was the result of "that invisible and unimaginable play of the forces of growth within the nervous system." Ideas flash up out of nowhere, and then subside; hypotheses struggle for existence, first in our own minds as we play with them, and then out in the wider world of journals and debate. Science was a balancing act, a give-and-take that was always in motion—just like the natural world. In success and its failure, the triumphant feeling of being correct and the terrifying pain of being proved wrong, both were organic, pulsing activities. In a sense, science was alive.[63]

BRUTE AND HUMAN INTELLECT

Soon after signing his contract with Holt, James sent his editor some articles, "not to inflict on you the duty of reading them, but simply that you might know what I had been doing." But read them he did. As a loyal Spencerian, "Remarks" was what excited Holt (so much so that he read it twice). "Ciphered down to the lowest terms," Holt wrote, "your criticism of Spencer seems to be a complaint that he tells how mind acts without attempting to tell what mind is." This is a strange misprision of James's critique, in that article or the many other places he sharpened his knives on Spencer's work. James's response attempted to set Holt straight: "My quarrel with Spencer is not that he makes much of the environment, but that he makes nothing of the glaring and patent fact of subjective interests which cooperate with the environment in moulding intelligence." According to James, such interests "form a true spontaneity." Holt's reply extended to many pages, concluding that his failure to understand may have been "because I'm too stupid." And James may have agreed, as he begged off soon after. But not without a parting shot: "Objectively considered I can correspond to my environment either by assimilating roast beef and surviving, or by dying and feeding the worms. Which of these adjustments will you choose?" It was clear that James chose the roast beef. We are all animals, after all—eat, or be eaten.[64]

In a way, this was the theme of another essay that James sent Holt in that early batch. Called "Brute and Human Intellect," it appealed to readers' common experience from its first lines: "Every one who has owned a dog must, over and over again, have felt a strange sense of wonder that the animal, being as intelligent as he is, should not be vastly more so." At first glance, the article seems like one more attack on "Uncle Spencer," with James even admitting halfway through that "criticism of his theory will be the easiest manner in which fully to clear up what may still seem obscure in our own." But James also took it as an opportunity to nuance his use of "experience"—a favorite term of Spencer and his followers, but one of James's, too. "My experience," James wrote in a famous line, "is what I agree to attend to. Only those items which I notice shape my mind—without selective interest, experience is an utter chaos." To illustrate his argument about experience, James began "by taking the best stories I can find of animal sagacity" and articulating a definition common to human and nonhuman animals alike. It was possible, he argued, for all animals to exercise their agency in picking and choosing their experiences based on their interests. Their ability to do so, and the distinction between "brute" and "human" capacities in this regard, was proof that humans had heightened capabilities for attention and abstraction courtesy of evolution by natural selection.[65]

This ability extends all the way down into the brain. Our decision to fixate on one facet of the outside world over another has as its correlate a form of "physical selection" in the brain. It is a process in which "indeterminateness of connection between the different tracts, and tendency of action to focalize itself, so to speak, in small localities which vary infinitely at different times." The actual neurology of reasoning was, of course, a mystery. "Whatever the physical peculiarity in question may be," James concluded, "*it* is the cause why a man, whose brain has it, reasons so much, whilst his horse, whose brain lacks it, reasons so little." In nonhuman animals, "fixed habit is the essential and characteristic law of nervous action." The human (including the scientist) "owes his whole preëminence as a reasoner, his whole human quality, we may say, to the facility with which a given mode of thought in him may suddenly be broken up into elements, which re-combine anew." It is not enough, in other words, to have spontaneous hypotheses, or even to test them rigorously once they arrive. When James called humans "the *educable* animal," he meant the exercise of both of these capacities, and in particular our ability to rewrite what we have once learned—and to do so "suddenly," as though the random reshuffling of our ideas was a spontaneous discharge. The scientific method, rooted in adaptive play, was coming to seem like a matter of anatomy.[66]

If science is mental, and thus neural, and if our brains evolved over time and bore the traces of our relations with other animals, then how should we draw the line? "Brute and Human Intellect" provided an answer, but elsewhere James seemed to close the gap between human and nonhuman minds entirely. This was truest in "Are We Automata?," James's response to a debate over the so-called mechanistic philosophy of Huxley and other evolutionists that cast all minds, including human, as in some sense mechanical. James fought for an organic view. The mind is unique in its capacity to "choose out of the manifold experiences present to it at a given time some one for particular accentuation, and to ignore the rest." Or, as he put it in the *Principles:* the mind *"is always interested more in one part of its object than in another, and welcomes and rejects, or chooses, all the while it thinks."* All minds revealed what James called "selective industry," distinguishing them (and us) from nonliving matter. "Looking back," James wrote, "the mind is at every stage a theatre of simultaneous possibilities. Consciousness consists in the comparison of these with each other, the selection of some, and the suppression of the rest by the reinforcing and inhibiting agency of attention." We may differ in our abilities to distinguish, attend, or choose, but all minds present some version of these capacities, and in this regard even so-called lower animals could be, in some sense, scientific thinkers.[67]

But why stop there? These decades also saw the revival of Babbage's old ambitions to build a "calculating machine," one capable not just of programmed operations but of creative activity as well. The logician William Stanley Jevons, whom James credited for an emphasis on "quick invention of hypothesis" in scientific discovery, also drew his praise for "a logical machine out of which, when the keyboard is manipulated, all the conclusions consistent with given premises will at once appear." Although Jevons's "logical piano" would not replace all scientific thinking, it spurred some to imagine just that. Peirce, for one, extended the matter of "machine reasoning" into this domain: "Precisely how much of the business of thinking a machine could possibly be made to perform, and what part of it must be left for the living mind, is a question not without conceivable practical importance; the study of it can at any rate not fail to throw needed light on the nature of the reasoning process." Machines were not simple assistants—they could be used, like anything else in psychology, to probe the limits of the method being applied to them in turn. When Peirce built one with some graduate students in the early 1880s, his aim was to represent, not just replicate, the human mind. It was a far cry from Babbage's ambitions in the 1820s—but also not so far. One researcher's tools are another's objects of study; modeling a mechanical mind might achieve new ends (technological and otherwise) *and* provide a means of exploring old questions.[68]

This was precisely the task the next generation of psychologists set for themselves. Whether tinkering with brass instruments to measure the senses or building machines to mimic them, these new psychologists turned their tools on themselves, blurring subjects and objects together in a dizzy array of studies. Here again, Peirce had led the way, even if most were unaware of his founding role. Peirce had, after all, given the era its name, positioning himself to articulate a "method of methods" with which to attack it. But how to get there? The answer was to dig into as many fields as one could in order to master what they were about. "The data for the generalizations of logic," he wrote in an application to Johns Hopkins, "are the special methods of the different sciences," the ascertainment of which requires the "study of various sciences rather profoundly." Peirce's time at the Observatory and Coast Survey was not a distraction from his work, then, but a prerequisite to perform it adequately. Or at least, that was how he pitched what might be a surprising request to the Hopkins president: "Thus though I am a logician, I consider it necessary to have a laboratory." And although his job at Hopkins was terminated a few years later, Peirce's vision of a laboratory to study the roots of scientific reasoning was soon realized. It was just the kinds of mind under study that might have surprised him.[69]

6

ANIMAL INTELLIGENCE

Something was afoot in philosophy. Once thought to be an armchair discipline, by the 1870s the field was splintering, its practitioners crowded in on all sides by what must have struck them as a strange menagerie. William James evoked the decade's tumult with ironic panache: "Chaotic cohorts of outlandish associates, the polyp's tentacles, the throat of the pitcher-plant, the nest of the bower-bird, the illuminated hind-quarters of the baboon, and the manners and customs of the Dyaks and Andamanese, have swept like a deluge into the decent gardens in which, with her disciples, refined Philosophy was wont to pace, and have left but little of their human and academic scenery erect." The cause of the deluge was the vogue for evolutionary studies, which captivated one of the field's "disciples" more than the rest: psychology. Exchanging a few old assumptions about how to study minds for new ones, psychologists wrested certain topics from their philosophical mentors, turning them into empirical questions. In doing so, they introduced new, non-human minds into the scientific subject pool, pulling psychology and philosophy alike into unfamiliar territory.[1]

As psychologists made themselves at home among these "outlandish associates," how they understood their *own* minds changed, too. On the surface, nonhuman animals would have seemed like an easy contrast to the familiar objects of psychology: adult human minds, usually those of psychologists, examined introspectively. But the arrival of rats, cats, and the rest of the menagerie did not simply offset traditional studies; rather, the two blended together. Nonhuman animals were interpreted in terms from human psychology, while the scientists began to see their own thinking in the new light cast by animal studies. The emerging comparative psychology was comparative not just in the sense of discovering differences between kinds of minds. Instead, a feedback loop was inaugurated between the psychological models being developed to describe animal minds and psychologists' sense of their own mental labor. Like their predecessors, these theorists were after the roots of their own reasoning, using an evolutionary logic to drive science deeper into the shared past of what they came to call "animal intelligence." What exactly that intelligence was, whether it existed and in which forms, was the question of the day. Psychologists expanded outward to new minds and turned inward in new ways simultaneously, seeking the source of their scientific thinking in observable animal behaviors.[2]

The arrival of these new modes of analysis on the psychological scene did not sound the death knell of introspection—at least not at first. Reflecting on one's own mental states and then generalizing that reflection to broader claims persisted in both philosophy and psychology for a long time and is still an important aspect of both disciplines today. And even those who did turn to "other minds" in the nineteenth century did not necessarily do so in pursuit of a comparative approach or as a means of testing evolutionary claims. These issues were certainly relevant to such studies, but other topics mattered just as much to members of the expanding psychological field. If evolution is the thread tying Darwin and his followers to Dewey's five steps, it was also interwoven with other threads: reaction times, say, or attention, or habits, or intelligence testing. None of these topics was pursued in isolation, and each was—in principle—altered by work on the others. But it was possible and indeed common to work away at one in isolation, employing assumptions and tools that were increasingly distinct from those being used by one's colleagues. Thus, while the rise of comparative psychology is central to the history of the scientific method, it was only a piece of the expanding psychological pie—and a small piece at that.[3]

The early history of comparative psychology shows not so much an abandonment of the practice of introspection as its *externalization.* Over time, the things philosophers had examined in their own minds were sought in the minds of others—those of other humans, including one's colleagues, but also others in a wider sense that encompassed nonhuman animals as well. This transformation from internal to external observation was treated not only as justified but as inevitable. After all, early researchers asked, where else could they start except with the kinds of phenomena on which they were already experts? Work on what is now called "theory of mind" began with the theorization of one's own mind before proceeding to assign capacities of various sorts to the minds one could only study through external, observable signs. The result was a new criterion for mental states and inner life: no matter how private they were once thought to be, the minds of others increasingly had to reveal themselves through behavior in order to qualify as the proper objects of scientific study. Behavior, whether of a colleague or a capuchin monkey, could only be seen in light of one's own experience, including the motivations and affective states that accompanied it. Introspection remained a rubric for the "new psychology" in this way, furnishing its observations and experiments with both the questions to be asked and the means of answering them, even as practitioners claimed to leave the old methods behind them.[4]

The transformation of psychology into a laboratory science was thus far from simple. Old emphases on intuitions and other internal mental states persisted, and questions revolved around which of these were present in which minds—and how one could tell. Anxieties associated with searching for the roots of cognitive processes (including the scientific method) were tied up with an emerging critique of so-called anthropocentrism, or the tendency to see all animals in terms of human capacities and achievements. Though usually disparaged as sentimental or fallacious, it is important to note that anthropocentrism is only wrong under assumptions about what constitutes evidence or what the goal of animal studies might be. These assumptions were shifting in the last decades of the nineteenth century, riven as the period was by debates in and around the new field of scientific psychology. Old fights over naturalism and the proper subjects for scientific inquiry merged with new arguments about vivisection and the ethics of turning (living) animals into instruments for human gain, economic or otherwise. The use of animals by psychologists joined these two contested spheres together: the nature of scientific method was being debated alongside the ethics of using nonhuman animals as models for human intellect, including science itself. In this way, what might have seemed like insider battles over the standards of a budding discipline reflected political and ethical divisions, too.[5]

Various approaches to the evidence of experience emerged at the end of the nineteenth century, amid more general discussions about the rising authority of scientific work. The idea of studying other minds as a means of probing the limits of reasoning had been linked to anxieties about the methods of the "new psychology" from the beginning. Early founders of comparative psychology had turned philosophical questions into matters of methodology—questions that, in turn, fed back into how animal intelligence was understood. Figuring out what it was like to be a bat—or, more commonly, a rat or a cat—came to seem central to understanding the legitimacy of new forms of expertise that psychologists claimed, not least because that expertise rested on an account of evolutionary universalism they used to justify their approach. Efforts to build up their discipline, and disputes over how to do so, turned psychologists' experimental practices into a model for the sciences and a model of cognition in general. As it did, the forms of cognition that counted as scientific began to shift. From private mental states to public behaviors, and from consciousness to learning, minds revealed themselves to scientific observers in myriad ways. Quickly, those minds with the least to hide assumed priority of place.[6]

MIND-STORIES

"By what evidence do I know, or by what considerations am I led to believe," John Stuart Mill asked in 1865, ". . . that the walking and speaking figures which I see and hear, have sensations and thoughts, or in other words, possess Minds?" To Mill, the answer came from experience. "I conclude it from certain things, which my experience of my own states of feeling proves to me to be marks of it." We understand other minds through analogy from our own. To take the simplest case: other humans have bodies and exhibit behaviors a lot like mine, which makes it reasonable for me to infer that they think and feel the same way I do when certain things happen to them. This claim could even be expanded: "We know the existence of other beings by generalization from the knowledge of our own: the generalization merely postulates that what experience shows to be a mark of the existence of something within the sphere of consciousness, may be concluded to be a mark of the same thing beyond that sphere." Self-recognition was the key to all cognition. We imagine other minds in the same way we imagine anything that happens outside our heads: by analogy from our own experience. The minds of others were mirrors of our own.[7]

As he had with methodological arguments in the 1840s, Mill spurred new developments in the field of psychology when he formulated this analogical solution to "the problem of other minds" in the 1860s. Scholars bent on distinguishing themselves from "metaphysical" colleagues sought out new subjects for their inquiries, naturally settling on some that were close to hand: the minds of colleagues sitting across from them in the laboratory. The rise of new instrument-based studies of reaction times and sensory thresholds, first in Germany and even more enthusiastically in the United States, can be seen as the scientific pursuit of Mill's approach to the study of other minds. Introspection and observation merged, and so did the roles of scientists involved in the experiments. Sometimes the authors of experimental reports in the early psychology laboratories were those who observed their colleagues' reactions to stimuli; at other times the author was understood to be the person who did the reacting. In either case, the psychology of other adult humans was an initial step toward the objectification and, ultimately, the quantification of mental phenomena that has been the main thrust of experimental psychology ever since. Just as James and others began to gather material for textbooks with the explicit goal of stabilizing their new discipline, the studies of mind they were drawing on began to shift away from introspection and toward the examination of mental states "revealed" in others.[8]

In the years that followed, this foray into otherness extended across the species barrier into the minds of nonhuman animals. At the heart of this new "comparative psychology" was the mission to illustrate mental evolution by classifying animals according to the mental capacities they revealed in their behavior. Early on, practitioners pursued these revelations via naturalistic observations—sometimes their own, more often those reported by a varied network of scientific and non-scientific observers spread across the globe. Darwin's use of anecdotes and records from a range of informants set the mold for this kind of work, and his *Descent of Man* and *Expression of the Emotions in Man and Animals* perfectly exemplify the mid-Victorian effort to classify all animals (including humans) according to qualities of mind inferred from watching them behave. The reports that resulted from this approach were their own version of "animal intelligence": as dispatches from the front or news from far-flung places, their subjects and locations extending far beyond what any normal scientific observer might hope to catalog individually, data on the minds of others functioned a lot like intelligence in the sense of news, if not spy-craft.[9]

Such efforts built on a long tradition of anecdotes of remarkable animal achievements. While some scholars continued to use anecdotes as data within comparative psychology, a new generation came to regard them as suspect—if not embarrassing. Even Darwin was derided, his most famous anecdote revealing their continued use as data and their gradual decline. Seeking a "missing link" to explain natural selection, Darwin recounted an anecdote about bears: "In North America the black bear was seen by Hearne swimming for hours with widely open mouth, thus catching, like a whale, insects in the water. Even in so extreme a case as this, if the supply of insects were constant, and if better adapted competitors did not already exist in the country, I can see no difficulty in a race of bears being rendered, by natural selection, more and more aquatic in their structure and habits, with larger and larger mouths, till a creature was produced as monstrous as a whale." Though he removed (most of) this story from later editions of the *Origin,* Darwin never doubted its veracity. His books were stuffed with such stories, many beginning with lines like "I have heard it said. . . ." or "A friend reports. . . ." Darwin's dogs (and kids!) appear throughout his work, their behavior held up as evidence again and again.[10]

What embarrassed Darwin about the bear-whale was what he *did* with it. Even late in his career, he was worried about hypothesizing too boldly, sticking his neck out too far. A reviewer shamed Darwin's use of the story not because it was old or suspect but because it was worse than "any instance of hypothetical transmutation in Lamarck." Dialing back the idea that this revealed how evolution happened was not Darwin's idea, but the recommendation of Lyell—the mentor who had long counseled him on methodological matters. Though Darwin returned to the story off and on—suggesting seals instead, trying to back up the basic idea while hedging on details—he came to regard it as a "foolish & imaginary illustration," wishing the inference from observed habits to long-term processes had been made much more cautiously, or not at all. A founding figure in comparative psychology, one of the most careful students of animal habit and behavior, regretted *not* the kind of evidence used to study animal minds, but instead the nature of his mind, overlooking possible objections in the pursuit of supporting data. Complaining "how often that abominable animal has been made to worry me!!," Darwin was more vexed by his own habits as a naturalist than he was by the far-fetched animal behavior he invoked.[11]

In Darwin's wake, lines were drawn within and around comparative psychology as the field coalesced on the pages of new psychology journals and in special sections of scientific meetings. The central issues in the study of animal minds looked a lot like those racking human psychology: that is to say, they were methodological at root. Debates about the capacity of this or that species usually revolved, not around supporting or refuting accounts of animal behavior, but instead around the *scientific* behavior of gathering and analyzing such accounts. Work on animal intelligence, as it was called at the time, began to be countered by the charge that such studies begged the question: Was there anything there to study in the first place? After all, comparative psychology emerged in lockstep with the contest between Huxley and his critics over whether animals were even conscious, or whether they—and, potentially, *we*—were automata whose every act was determined by the machinery of the body and its interactions with sensory inputs. On one side of the battle over animal intelligence were those who saw themselves, at least in some form, in the animal minds that they studied; on the other were those who refused to do so, or at least expressed skepticism that any creature other than a human was capable of something worth calling "intelligence." The very object of such a science was at stake.[12]

The most vocal proponent of animal intelligence in this sense was George John Romanes, a physiologist turned psychologist and the man Darwin had appointed as his successor in matters of evolutionary theory. Born in Canada and raised in London, Romanes wrote to Darwin soon after receiving his degree from Cambridge, in principle to bring some essays he had written for *Nature* to his idol's attention. The letter led to a meeting and began a close relationship between the two men. Romanes ended up serving as a research assistant in the last years of Darwin's life, helping to shepherd his unpublished manuscripts to light after he died. Personal affection spurred intellectual loyalty in the younger man, who dedicated a great deal of time after Darwin's death in 1882 to defending natural selection from its detractors and expanding its scope to enclose and explain the many studies of animal minds and behavior that had begun to emerge in the wake of Darwin's own work in this area. Huxley may have been Darwin's bulldog during the fight over the *Origin,* but Romanes filled the role when it came to *Descent* and *Expression,* pushing natural selection as a sufficient explanation of mental states and complex behaviors across the animal world—from the way that mussels held onto rocks to the division between spiritual and scientific impulses that had characterized Romanes's own early education.[13]

Romanes cut his teeth on a series of disputes in the pages of *Nature,* where he positioned himself somewhere between the credulous purveyors of animal anecdotes and the skeptical men who wrote in to lampoon them. He accomplished this by turning the skeptics' arguments back on them: where certain contributors seemed to think that the onus fell on those, like Romanes, who saw human capacities in animal behavior, Romanes argued that human-like mental states were the better explanation of reported animal behaviors and that those who denied such states would have to supply evidence against them. When Romanes reported in print, "I have just received a letter from the Vicar of Carn, which relates an instance of mental reflection on the part of a poodle dog that has the merit of admitting neither of mal-observation nor unconscious exaggeration," he was acknowledging the weakness of many "animal stories" while affirming his status as a judge of the plausibility of "vivid and complex" animal reasoning. His book *Animal Intelligence,* published in 1881, was a compendium of such accounts, meant to serve as a kind of foundation for the theory of mental evolution in animals and humans that Romanes would go on to build over the next few years. And there, as in his *Nature* articles, he was careful to position himself as a neutral compiler and analyst of varied reports of animal mind and behavior, turning what was otherwise a scattered genre of observation and fable into the data on which the science of comparative psychology was to be based.[14]

Animal Intelligence was followed by two volumes: *Mental Evolution in Animals* and *Mental Evolution in Man.* As promised, Romanes used these sequels to reflect on the varied evidence in his earlier book, building an argument for a fundamental identity between the minds of human and nonhuman animals. Rather than denying the obvious differences between humans and dogs, say, Romanes insisted that these were differences of degree, not kind. Both he and his critics recognized that such arguments depended on how one defined terms like "intelligence," "evolution," and, most of all, "mind." Romanes tried to make his views on definition clear from the start of *Mental Evolution in Animals:* "The distinctive element of mind is consciousness, the test of consciousness is the presence of choice, and the evidence of choice is the antecedent uncertainty of adjustive action between two or more alternatives." Minds choose, Romanes said, and they do so in ways that exceed the demands of their bodies or the dictates of instinct. "The power of learning by individual experience," he wrote later on in the book, "is therefore the criterion of Mind." Evidence of other minds, then, manifested in the choices animals made and, just as importantly, in their ability to learn from those choices when confronted with similar ones later on. The roots of reasoning lay in what could be recognized as choice.[15]

Learning, for Romanes, was both the central criterion of mind and how animals escaped the confines of their bodies—and thus, the evolutionary process that led them to learn in the first place. The mind, if it is going to count as a mind, exceeds the body in which it sits: "From the time that a stone was first used by a monkey to crack a nut, by a bird to break a shell, or even by a spider to balance its web, the necessary connexion between the advance of mental discrimination and muscular coordination was severed. With the use of tools there was given to Mind the means of progressing independently of further progress in muscular coordination." It is here, Romanes argued, that natural selection blurred into mental evolution. While "the nervous system is not *prophetic,*" its pleasures and pains parallel what is good or bad in an evolutionary sense. Pain is how our bodies warn us about existential threats. "Survival of the fittest," Romanes wrote, "is thus provided with a ready-formed condition or tendency of psycho-physiology on which to work." We may not see the fitness benefit of a head cold, but that was only because the benefit made sense in a world long since left behind. Our minds indexed our evolution.[16]

At the end of the book, Romanes returned to tool-use to show that animals demonstrated not only intelligence, but reason. The analogical case for other minds—"the only instrument of analysis that we possess"—reappears at the end, this time as a "stage" of mental evolution shared by human and nonhuman animals. Analogies, Romanes argued, are "the stage in which objects, qualities, and relations are deliberately compared with the intention of perceiving likenesses and unlikenesses," and "the action which follows is therefore undertaken with a knowledge, or perception, of the relation between the means employed and the ends attained." Reason, for Romanes, was precisely this ability to link means and ends—whatever the means and ends might be.

While his opponents refused to attribute reason to animals on the basis of the nature of the mental states or observable behaviors involved, Romanes was more open: "*Wherever* there is a process of inference from [ideas] which results in establishing a proportional conclusion among them, there we have something more than the mere association of ideas; and this something is Reason." Irrespective of content, reasoning was about use—a bird clutching a rock revealed its reason in the same way a scientist did with a microscope. Reason was about choosing tools.[17]

If Romanes represented the "pro" side in debates over animal intelligence, the "con" was represented by his contemporary Conwy Lloyd Morgan. Like Romanes, Morgan made his name by writing for *Nature*—in his case, often by responding skeptically to Romanes's claims about animal intelligence in its pages. A rapid-fire exchange between the two, beginning with a critical review by Morgan of Romanes's *Mental Evolution in Animals,* set the terms of their subsequent engagement and clarified the stakes of their disagreement over the nature of other minds. The two agreed that we encounter other minds, including those of animals, neither as subjects nor as objects, but rather as "ejects," a term both men attributed to the recently deceased philosopher William Kingdon Clifford. To Clifford, ejects were "things thrown out of my consciousness." As Morgan summarized the idea in a review of Romanes: "My neighbour's mind is not and never can be an object; it is an eject, an image of my own mind thrown out from myself." Both Romanes and Morgan quoted these lines of Clifford's, and both saw "ejection" as a necessary consequence of the desire to know what others are thinking that is frustrated by the limits imposed by the mind's hiddenness. We throw our own minds into the void.[18]

For Morgan, this hiddenness was not a barrier to psychology's achieving scientific status; indeed, it was a prerequisite. "A science of mind," he argued, "only becomes possible when I am able to compare my own conclusions with those which my neighbours have reached in a similar manner." Introspection would not suffice. But the problem emerged when it came to the scientific study of nonhuman minds—in other words, comparative psychology. For Morgan, "the plain tale of behavior, as we observe and describe it, yields only, as I have put it, body-story and not mind-story." And mind-story was what psychologists wanted. Even our closest relatives are so distant that any subjective data with which we might try to verify ejective inferences about their minds would be too shaky to be useful. The problem was methodological. "These are the facts," Morgan argued, "which have to be taken into consideration when we seek an answer to the question 'Is there a science of comparative psychology.'" And the facts did not bode well. "Notwithstanding that it has won for itself a more or less recognized place among the sciences," Morgan concluded, "I venture to submit that our answer to this question should be an emphatic negative." Romanes, by implication, had dedicated himself to a field that could not exist.[19]

In a lengthy response, Romanes defended comparative psychology in methodological terms. While it was true that the capacity for language made human psychology easier, the simplicity of the minds of animals alleviated some of the difficulty of being unable to speak with them. "In both cases alike," Romanes insisted, "our ejective inferences can only be founded on the observable activities of organisms"—whether the activity was speaking or purring. Given that both seemed to admit the necessity of ejection, Morgan's response was surprising: "There is but one method in human psychology—that of introspection. By this method I obtain certain results. These results I communicate to my neighbour, and he by introspection verifies them for himself. This I call 'submitting the results to the test of subjective verification.' In this way and in no other can a science of human psychology be constituted." For him, the debate went beyond methodology—it depended "upon what we mean by 'a science.'" For his part, Romanes thought that "the possibility of a science is furnished wherever there is material to investigate." Morgan disagreed, arguing in a piece drawn from the exchange that the problem was "unconscious bias; the tendency to see what one expects to see, and to fill in missing links which one is certain were there—only we stupidly failed to observe them." Owing to how hard it was to separate facts and inferences, Morgan proposed "the most sparing use of the psychical element consistent with an adequate study of habits and activities." What science needed was humility.[20]

Given the methodological nature of this debate, there were two things Morgan might have meant by "the psychical element." On the one hand, it seems obvious that he was aiming to limit the *inferred* mental states of the animals about which psychologists like Romanes were theorizing. If you limit yourself to describing behavior, Morgan argued, you avoid introducing inaccurate assumptions into the science of animal minds. This means understanding Morgan in a direct, material sense: maybe nothing was going on inside an animal's head. On the other hand, perhaps Morgan was advocating "sparing use" of the scientist's *own* "psychical element." Here, ejecting too much of ourselves risked flooding other minds. Mind-stories were just that: stories, which threaten to displace the realities they seek to describe. Morgan was worried about both of these issues; his warning was material and methodological at once, part cautionary tale and part scientific claim. Ejection brought these together: a method *of* mind and a claim *about* mind. The dispute between Morgan and Romanes was about the natural limits of nonhuman cognition and the methodological limits that should be placed on human cognition as a result.[21]

Like Romanes, Morgan was a dedicated evolutionist who used the theory of natural selection as a source for philosophical reflection and methodological self-awareness. In *The Springs of Conduct,* a popular book subtitled *An Essay on Evolution,* Morgan intertwined science and nature into a single evolutionary narrative. Adapting Spencer and Darwin alike, Morgan was after lessons from the evolution of life for both science and ethics. "The traits which characterize the progress of science," he wrote, "are the traits which characterize the progress of organic life, the progress of society, the progress of the universe—the traits of evolution." Like Romanes, he saw the evolution of the mind in the same terms as that of the species—and extended the insight to the history of science. "Both in the individual and in the race the discovery of law is itself subject to law"; scientific development, in the mind of the child or the history of a field, unfolded according to the same sorts of laws that its practitioners outlined in their work. This was nowhere truer than in psychology, in which the object and subject were the same: minds studying minds, discovering laws that explained the very process of discovery. Somehow, as it had for so many of Morgan's contemporaries, his attention to mind and method led the two to blur together, each shifting the terms of the other. Applying evolution made science evolve.[22]

Within a decade, Morgan had written up his anxieties and the results they had spurred in a textbook on comparative psychology. The book, which soon became a standard, was partly a response to a different survey: William James's *Principles of Psychology.* In a review of James's book titled "Psychology in America," Morgan emphasized how James blurred subject and object, especially in his chapter "The Stream of Thought." The metaphor of the stream, which James made famous but which was drawn from a variety of sources, provided a vocabulary to explain both oneself and others: all are streams, burbling along within their banks but also, occasionally, merging and separating again. Psychology, on this reading of James, is a kind of river navigation or the map of an estuary, the gaps between individuals disappearing as the tides rise. The stream also helped Morgan get past his former stridency on the matter of using animals, since all could be framed in the same basic language without positing too complicated an account of cognition. His own pet dog, whom we will meet shortly, may have been on Morgan's mind when he wrote (with a wink): "I am far from wishing to dogmatize on the matter." He was, at least in principle, open to the evidence drawn from other minds, so long as drawing it did not mean attributing too much to those minds or to the sources they were drawn from.[23]

The question was: What had changed? While Morgan may have acclimated himself to the scientific study of the nonhuman mind, he also introduced a new rule to clarify what precisely such a science could say—and what it could not. What would come to be called "Morgan's Canon" was a prescriptive condensation of the skepticism he had expressed in *Nature.* Having introduced the idea at a conference in 1892, the rule assumed its canonical form in his textbook:

> *In no case is an animal activity to be interpreted as the outcome of the exercise of a higher psychical faculty, if it can be fairly interpreted as the outcome of the exercise of one which stands lower in the psychological scale.*

A version of Occam's razor sharpened through psychological debates, "Morgan's Canon" had an immediate effect on the field. Its success among a younger generation of practitioners effectively settled the dispute over how to study other minds in Morgan's favor, even if he rolled back his opposition to comparative psychology's scientific status in the meantime. Romanes's reliance on anecdotes was increasingly dismissed as anthropomorphic, not least because he imported the interpretations of his informants along with their observations. In this regard, Morgan won.[24]

At first, applying the canon seemed to buttress Morgan's hunch that animals, though often intelligent, could not reason—that is, there was no evidence they contemplated various means to a predefined end. Morgan illustrated his interpretation with a series of "experiments" he conducted at home in his garden with his fox terrier. In one test, the dog, named Tony, had to learn to carry an imbalanced stick in his mouth. Morgan described Tony's behavior as "solving in a practical way a problem in mechanics." When Tony learned to carry the stick, Morgan insisted that what might have looked like reasoning had in fact been "trial and error," an adaptation that was not (here Morgan drew on evolutionary theorists) limited to creatures with minds, much less reason. The same thing happened when Tony learned to open the garden gate: if you just saw the end-result, you might think he had reasoned it out; but in fact, Morgan argued that a "relation between means and end did not appear to take form in his mind." Watching Tony fail to escape the garden over and over, Morgan formulated his canon against the anthropomorphic tendency to find in the minds of animals the reasoning processes we pride ourselves on.[25]

The lesson for Morgan was, again, less about animal behavior and more about *scientific* behavior. "In zoological psychology," he concluded, "we have got beyond the anecdotal stage, we have reached the stage of experimental investigation." The point for Morgan was the same as for both Darwin and Romanes: what you did with the animals at hand. And while Darwin could be read as "observing" through the eyes of others, accumulating through letters and reports and eye-witness accounts,

Morgan wanted something else. Whether or not he achieved it, his canon became a weapon in others' pursuit of the same ideal. The irony of that eventual success was what it said about the scientists themselves. As Romanes himself had pointed out, not even humans cleared the bar where it was usually set: "A *general* idea of causality," for example, "demands higher powers of abstract thought than are possessed by any animals, or even by the great majority of men." So, what could you do? You lowered the bar or you raised it; either you granted reason to nonhuman animals or denied it to humans, too. And the strength of Morgan's Canon, in the hands of those who came after him, was such that, within a decade, it was being applied not just to dogs and cats but to humans, too—with unflattering results.[26]

ANIMAL STUPIDITY

Animal psychology flourished in the 1890s, spurred by the disputes between Romanes and Morgan and by the latter's perceived victory. But it was far from the only game in town. Even while the old method of introspection seemed to be in decline, the very people arguing against it were still noting its importance. In "On Some Omissions of Introspective Psychology," James located the field's problems much deeper than its reliance on introspection—which, in any event, we could not help but use. We reflect on ourselves as a means of imagining others' minds—there was no alternative. "Our mental life, like a bird's life, seems to be made of an alternation of flights and perchings," James wrote. Thinking was in the flights—what he called the "transitive" aspects of thought—as much as the perchings, or its "substantive" components. And the failure to account for either was not just a flaw in introspection, but in the psychologist's point of view: "The standpoint of the psychologist is external to that of the consciousness he is studying. Both itself and its own object are objects for him. They form a couple which he sees in relation, and compares together, and it follows from this that he alone can verify the cognitive character of any mental act, through his own assumed *true* knowledge of its object." Introspection—when properly practiced—was the savior, rather than the traitor, of the study of consciousness and its relationship to reality.[27]

It is interesting to note that, while James defended an appropriate use of introspection, he laid the groundwork for the objective study of minds other than our own at the same time. If the psychologist's task was to distinguish the act of studying from the ideas being studied (not to mention the reality to which those ideas point), doing so required observing other minds in the laboratory. One reading of James's methodological caution aligns him with Morgan's early, skeptical account of the scientific status of comparative psychology: only the scientist's own mind can ever be an *object* in the direct sense James was sketching. But as Morgan came to see around the same time, it was possible *in principle* for a psychologist to study other minds so long as they revealed their inner workings in testimony (in the case of humans) or behavior (in the case of all animals). What mattered was how these external signs were interpreted and how those methods were monitored by psychologists for accidental biases and assignments of faculties that were not *necessary* to explain the activity in question. Psychologists converged on this point: the science of the mind made mental processes observable, turning subjects into objects.[28]

Objectification was achieved most forcefully by a new generation of psychologists, eager to capitalize on the field's new scientific status. Their leader, in terms of his impact on followers and detractors alike, was a graduate student named Edward Thorndike. An acolyte of James and an enthusiastic adherent to Morgan's Canon, Thorndike did more than anyone in his generation to put psychology on the scientific map, using his eventual position at Teachers College in New York as an entry into American classrooms and, as a result, the minds of millions of Americans. His biographer called Thorndike "the sane positivist," an allusion to a colleague who saw his "sane positivism" as a necessary corrective for widespread "speculative" impulses in the young field. Cautious deference to empirical data was a personal as well as a professional trait for the New Englander. Thorndike was not given to introspection, even in personal correspondence, as the summary of a day's activities written at the time makes clear: "Up at 7:30. Experimented 8:30–9:30. Histology lecture 9:30–10:30. Experimented 10:30–12:30. Seminar 12:30–1:30. Experimented 1:30–3:30. Lecture 3:30–4:30. Made apparatus 4:30–5:15. At 6:30 managed to sacrifice myself long enough to start this pleasant epistle. . . ." An autobiographical essay Thorndike wrote years later betrays a similar seriousness, casting his own life in the terms he used to describe his animal subjects. "The most general fact about my entire career as a psychologist," he recalled, was "its responsiveness to outer pressure or opportunity, rather than to inner needs." This way of writing makes reading his biography a chore, but also makes a great deal of sense when connected to Thorndike's work and its wider impact.[29]

That work, begun under James and completed at Columbia under the psychologist James McKeen Cattell, was focused on the learning process in nonhuman animals. Having read the *Principles* as an undergraduate (recalling it as "the only book outside the field of literature that I voluntarily bought during the four years of college"), Thorndike enrolled for graduate study with James and others at Harvard. Though he began with a project on children, he shifted focus to nonhuman animals under pressure from the administration. Chickens were Thorndike's first subject, which he kept in his apartment until his landlady protested and they were moved to the basement of James's house. Logistical difficulties like these, and a fellowship that enabled him to focus on research rather than teaching, lured Thorndike to Columbia, where he completed his degree under Cattell in 1898. Space in Schermerhorn Hall—gained after some more difficulties with landlords and competing claims—helped Thorndike expand his original focus on chickens to include cats and dogs, forming a mental menagerie that he kept throughout the fall and spring of his final year. The work was laborious (as his account of daily activities suggests), and the finished product is a talisman of both experimental industry and the harsh self-denial that came through in both his methodological pronouncements and autobiographical reminiscences. The study of animal minds in this mode was a trying business.[30]

Thorndike published his dissertation as a monograph supplement to the *Psychological Review* under the title "Animal Intelligence: An Experimental Study of the Associative Processes in Animals." The title was a red herring, since Thorndike made it clear that he did not think a nonhuman animal (at least the sort he studied) demonstrated anything that deserved to be called intelligence. He was using the phrase to insert himself into the ongoing conversation about the limits of reason and what (if anything) made humans special from a psychological perspective. Thorndike tackled this question by presenting his animal subjects with a challenge: escape from this cage, and I will grant you mental agency. His box-cage experiments were explicitly framed as "tests" of a set of claims about animal reasoning that Romanes had included—in the form of anecdotes received in correspondence—in *Animal Intelligence.* Thorndike's dissertation, in title and tone, right down to the design of the experiments and his interpretation of their results, was a response to Romanes and, to a lesser extent, Morgan. "The cat," Thorndike summarized, "does not look over the situation, much less *think* it over, and then decide what to do. It bursts out at once into the activities which instinct and experience have settled on as suitable reactions to the situation '*confinement when hungry with food outside.*'" Animal intelligence, much less animal reason, failed to appear in Thorndike's New York laboratory.[31]

To see why, it is worth following Thorndike as he set tasks for his animals and observed how well they fared. By day and night, as his diary suggests, Thorndike built cages for the testing of various escape mechanisms and, thus, various possible impulses a given animal might bring to the solution of the conundrum of confinement. Thorndike's method, in every case, was "to put animals when hungry in enclosures from which they could escape by some simple act, such as pulling at a loop of cord, pressing a lever, or stepping on a platform." He would then record the details of their escape attempts; if they were successful, the same task would be repeated in order to track a decrease in the amount of time required. In all the cases in which subjects did not fail to find their ways out of the cages, he recorded a regular—if erratic—arc in the time it took them to complete the required tasks on subsequent trials. The result was a series of tables and time curves that represented "the learning process" in graphical form. These visual representations of the mind quickly became standard in experimental psychology, as illustrations and as talismans of what kind of exactitude was possible if the invisible could be made visible.[32]

The result of all this confining, timing, and graphing was Thorndike's polemical claim that a gap existed between human and nonhuman mental processes. Surprisingly, Thorndike arrived at the conclusion without conducting similar tests on humans. Despite repeated diatribes against anecdotes and other casual references to experience, the humans to whom Thorndike compared his cats were a kind of abstraction themselves: an "average American child of ten years," for example, or a "good billiard-player." Thorndike assumed humans had "a thesaurus of valuable associations" at their command, which would "make the process of acquisition in many cases quite a different one from the trial and error method of the animals, and in general much shorten it." The ten-year-old's associations "outnumber those of any dog" and the billiard-player "probably has more association in connection with this single pastime than a dog with his whole life's business." Thorndike could be certain that, after a hundred pages of diagrams and numbers, most of his readers would be inclined to believe him—drawing on memories of their childhoods or experience with billiards as necessary. Ejection, as ever, was the key to comparison.[33]

Yet even here, at the height of human exceptionalism, Thorndike admitted that these were differences of degree, not kind. "Small as it is, however, the number of associations which an animal may acquire is probably much larger than popularly supposed." Animals' relative dearth of associations was a matter of "motive," not aptitude. This was about how they tended to interact with the world around them. Unlike a human, a cat "does not form any association until he has to, until the direct benefit is apparent, and, for his ordinary life, comparatively few are needed." The equipment was there to do things differently, however. "For there is probably nothing in their brain structure which limits the number of connections that can be formed, or would cause such connections, as they grew numerous, to become confused." Elsewhere, Thorndike saw "homologous" associations in human and nonhuman minds: learning a sport was like how cats learned to escape, for example. Despite his scorn for Romanes, then, Thorndike's "Animal Intelligence" shared some basic conclusions about mental capacities with Romanes's *Animal Intelligence.* The difference between them, according to Thorndike, was one of method, not any specific claim or theory. No matter what was observed, the conditions of observation—in a laboratory or in your house, firsthand or gathered from elsewhere—made the difference. For Thorndike, our capacity to use controlled experiments was what made us human.[34]

For Thorndike and his followers, the superiority of experiment was predicated on two aspects of his early work. The first was familiar to friends and enemies alike. Science, in the sense toward which psychologists strove in this period, had to be *comparative.* This meant using a range of animals and standardizing the experimental setup in order to facilitate comparisons both between nonhuman animals and between nonhuman animals and humans. Experiments, in this regard, eliminated or greatly reduced both the differences between types of animals and the peculiar tendencies of the scientists themselves—their so-called personal equations. Comparison between organisms was supplemented by a subtler shift in practice, emphasizing comparisons of the same animal's trials, time after time. Elucidating "trial-and-error" learning was only possible if the same cat or dog attempted the same escape again, and again, and again. Thorndike's tables and graphs capture both dimensions of comparison: "Cat 1" and "Dog 8" appear alongside one another, and the repeated experiments with each illustrate change over time. The conceit of both is obvious today, so it is easy to privilege the first, evolutionary sense of comparison over the second, experimental sense. But both were central to Thorndike's vision of science, and to those that came after him. Comparison had both spatial and temporal dimensions.[35]

A second aspect of Thorndike's scientific approach made the first one possible. In order to compare across species or within individuals, you needed to be able to abstract from any given trial to a more general picture of the mind under study. Abstraction was accomplished partly through the rigorous quantification to which Thorndike subjected his results (and himself), but it also came prior to it. After all, to measure and count in the laboratory, one had to decide *what* to count, choosing an object or set of objects that were not specific to subject, species, or situation. When Thorndike labeled something "an impulse" or "an act," when he picked out "behavior" as a proxy for "consciousness," he was engaged in just this sort of abstraction. Identifying his cats by numbers, not names, and describing them in tables more than words, enabled Thorndike to toggle easily from any instance of a behavior to a description of behavior *as such*. If an impulse could shed its cat-ness, then the study of animal intelligence could reveal the roots of reasoning. "The main purpose of the study of the animal mind," Thorndike wrote in his introduction, "is to learn the development of mental life down through the phylum, to trace in particular the origin of human faculty." While he is remembered, along with Morgan, for cautioning against any easy comparisons, Thorndike retained the field's founding focus on animal minds as reflections of scientists' own thinking, inside and outside the laboratory.[36]

The combination of comparison and abstraction, achieving a kind of generalization, was not just a methodological feature of Thorndike's work. It was a feature of the minds he studied, as well as of his own mind as he did so. Just as "the activity of an animal when first put into a new box is not directed by any appreciation of *that* box's character, but by certain general impulses to acts," so Thorndike's observations extracted a "general tendency"—he generalized about generalizing. The task of the comparative psychologist was to see past any individual capacities to the "general slope" of learning over time or across species. What is more, the slope represented not only the mind but the brain: "gradual" slopes revealed "the wearing smooth of a path in the brain, not the decisions of a rational consciousness." Though he did not pursue the question in his own dissertation, Thorndike made it clear that one could even smooth a slope that aggregated many different individuals, thus fusing the developmental logic of one process of learning into an account of species capacity in general. The links between the "average slope" this practice would produce and the (inherited) neural processes on which learning depended were not just underscored by timing and graphing—they were *represented* by those practices. Time curves thus traced mental, even neural activity; they generalized a skill—generalization—that Thorndike sought to cultivate.

Thorndike's experimental method also indexed mental progress. When he said that any association in humans "is built over and permeated and transformed by inference and judgment and comparison," and that human reasoning "takes frequently the form of long trains of thought ending in no pleasure-giving act," he was describing *himself*. Science, as he defined it, was the crowning achievement of the species. Its detached, sober comparisons were the sort of thinking to which the mental processes he studied were tending. Where cats could contemplate things only in a very literal way, humans could abstract from their immediate needs to other, invisible or unobservable phenomena. The generalizing he traced in animals confronted with new boxes was a nascent form of this basic capacity, linking comparative psychologists to the animals they studied. From anecdotes about specific animals to comparative studies of many, the field was following the arc of mental evolution, from particular to general phenomena. Science was evolving, and Thorndike's work both accounted for and embodied that arc.

Evolution, for Thorndike, had produced experiment—specifically, the experimental method he advocated in his dissertation and for the rest of his career. But beneath this view of scientific progress, he had also shifted the conversation about cognition. Though it had long been held "that very minute bits of consciousness come first and gradually get built up into the complex web," Thorndike saw mental evolution in a different light: "This our view abolishes and declares that the progress is not from little and simple to big and complicated, but from direct connections to indirect connections in which a stock of isolated elements plays a part, is from 'pure experience' or undifferentiated feelings, to discrimination, on the one hand, to generalizations, abstractions, on the other." Abstraction, generalization—these were the signs of an advanced mind. They were no "bigger" than the simpler links that preceded them. What set them apart was precisely that they were set apart. His invocation of both "isolated elements" and "indirect connections" naturalized the detached approach to the study of the mind he preferred. Here again, evolution and experiment blurred into a unique account of mind and method.[37]

Unsurprisingly, Thorndike's arguments drew a lot of criticism. Just as he had, Thorndike's antagonists focused on methodological matters in their critiques of his work and in defense of their own. Animal intelligence remained a proxy for scientific intelligence: nonhuman associations were a basic form of what psychologists themselves were doing. One of the most common lines of critique was of Thorndike's application of Occam's razor. Was it those who saw human reasoning in animal minds, or those who refused animals the higher faculties, who had more to prove? "The burden of proof," an early reviewer wrote, "lies with those who deny them, and this remark applies to feelings as well as intellectual processes, though to a less degree." After all, the "failure" of an animal to imitate another may result from conscious or unconscious inhibition, rather than inability. It was the same for him, the reviewer insisted: "When I read a chapter on psychology written in the fascinating style of James, one exemplifying the profundity of a Ladd or a Hall, the bold constructive character of a Baldwin, or a vigorous plea on behalf of modern psychology by Cattell, . . . there is an impulse to imitate, but I have not as yet taken the first step." Just as Thorndike defined his own approach as evidence of human superiority, certain reviewers drew the opposite conclusion. In accounts of the mundane practices of psychology, of daily scientific life, was evidence of science's animal nature.[38]

The same was true for another major line of criticism: that Thorndike's cold, artificial apparatus unfairly stacked the deck against animal intelligence. Morgan, whom Thorndike had damned with faint praise throughout the dissertation, put the point most directly: "The sturdy and unconvinceable advocate of reasoning (properly so-called) in animals may say that to place a starving kitten in the cramped confinement of one of Mr. Thorndike's box-cages, would be more likely to make a cat swear than to lead it to act rationally. And he may further urge that where the string passes out of sight and the bolt is hidden from view, the opportunities of understanding the situation are excluded. All the kitten could think would be: here's something loose and unnecessary to the normal constitution of a box; I'll try that on chance." Unnatural conditions were unlikely, according to Morgan and others, to reveal natural aptitude. Controls like these may have been necessary for abstraction, comparison, and generalization, as Thorndike insisted, but something was lost in the process: the very object of investigation. If Morgan was right, then Thorndike's dissertation did the opposite of begging the question implied in its title: rather than produce the phenomenon of animal intelligence, it made animal stupidity the only option.[39]

Morgan's imaginary account of feline cognition mirrored one of Thorndike's. "Does the kitten," Thorndike asked, "feel '*sound of call, memory-image of milk in a saucer in the kitchen, thought of running into the house, a feeling, finally, of "I will run in"*'?" The difference between the two ways of imagining nonhuman thinking could not have been starker. Thorndike's stream of consciousness parodied the theory that cats associated feelings, memories, and ideas much as humans did; Morgan's tone of exasperation expressed his own lack of patience with Thorndike's unwillingness to see things from a cat's perspective. Both were reacting, in their own ways, to a common tendency to overestimate the inner lives of nonhuman and human organisms. Both, too, drew on similar, abstracted images of human behavior. Morgan noted Thorndike's invocation of "the impulses which make a tennis player run to and fro when playing," pointing out that he had used the same image to make the point in his own work. Imagined inner lives, anecdotal human performances, line-drawings of the experimental apparatus and calls for controlled conditions: all practitioners shared cognitive and literary devices like these, even as they disagreed about how to interpret them or practice their discipline.[40]

Such disputes over means and ends were not the culmination of animal psychology—they were its beginning. Students working in laboratories at these same institutions quickly adapted to the stakes of such debates, forging techniques and interpretations from either side of the divide. Abstraction and comparison were here to say, but the aim of "naturalizing" experiment framed a new set of studies in the decades that followed. "Animal intelligence" remained a contested idea, but comparative psychology expanded despite—or perhaps, because of—the inconsistency of its ostensible objects. Indeed, animals may have been central to American psychology in these years precisely because of these disagreements over the content of animal minds and the best methods for drawing them out for analysis. A generation of psychologists had turned their field into a site for analyzing the nature and scope of scientific thinking. As a result, the methodological crises of the study of other minds improved, rather than undermined, the field's claim to be addressing real issues about mind and behavior. Cats clawing at cages could be both a proving ground for psychology's scientific status and a metaphor for the process of scientific reasoning.

RUNNING THE MAZE

Thorndike inspired champions and critics in equal measure. Outright rejection of his work was less common, or at least less influential, than reaction and modification. Followers fixed up their own boxes and extended his studies to new groups or new problems; detractors did the same with an eye to Morgan's critique and a desire to produce situations that were more natural, or at least more familiar, to those animals under study. Observations in the field, many in anecdotal form, continued to stream in from various sources—often with the express purpose of refuting this or that experimental result—but the antagonism Thorndike had set up between such accounts and his own methods of designing specific encounters began to disappear. On the model of some of Morgan's early work (and Thorndike's appropriation of it), stray recollections became a source of hypotheses to be tested in controlled settings, which might produce ideas that could be used to filter news from far-flung places (including other laboratories). In this way, a vast network of comparative psychologies—professional and amateur, scientific and nonscientific, human and nonhuman—began to emerge at the turn of the twentieth century.[41]

The field also began to reflect self-consciously on its methods and on the identities and relationships of those who used them. After all, observations and experiments on animal behavior had been conducted, published, and debated for a few decades at this point. By the turn of the twentieth century, psychologists like Thorndike were converging on topics that until then had been explored under the rubric of natural history. And of course, not every observer of animal behavior identified as a psychologist, even in the age of Morgan's Canon. Charles Otis Whitman, a zoologist who help establish ethology (the scientific study of behavior), came to his work on instincts from the study of leech embryology. Like Thorndike, he saw behavioral analysis as a way to renew his chosen field—which, for Whitman, was biology rather than psychology. Instinct was like any other feature an animal might present: it had evolved and could thus provide clues about the "phyletic descent" of the species. In the same series in which Thorndike announced his work with animals, Whitman argued that the only way to find such clues was for scientists, "remembering that this may be deceptive, [to] observe and experiment under conditions that insure *free behavior.*" Citing Morgan, Romanes, James, and Darwin, Whitman made the case for naturalistic observations of the behavior of animals just as Thorndike pushed psychologists to do the same.[42]

Whitman's subsequent work makes clear he was as skeptical of anecdotes about animals as Thorndike. However, what we see in Whitman, and in reactions to Thorndike, is a willingness to press the boundary between naturalistic and experimental work in order to test anecdotes but not stamp out intelligent behavior. This balancing act was the task a new group of psychologists set for themselves over the next few years, keen to identify animal intelligence where Thorndike seemed to reject it out of hand. Starting with a psychology instructor at Clark University named Linus Kline, we see the kind of integration Whitman was aiming at. Kline sought to explicitly unite what he called "the *natural method*" and "the *experimental method*" of animal studies by comparing mammals, birds, insects, and protozoa. Testing the limits of instinct, intelligence, and imitation, Kline insisted that each depended for its demonstration upon the balance of natural and experimental conditions. Thorndike never gave his animals a chance. Had he wanted to observe imitation, he would have needed "the opposite to those [conditions] created in his experiments, viz., freedom, security from harm, satiety, in a word—well being. Nothing so shrinks and inhibits completely the fullness and variety of an organism's activities than prison life and fear." Kline's notes point in many directions: naturalistic observations and morphological studies are side-by-side, all in the service of his vision of evolutionary-experimental studies.[43]

Kline's article looked a lot like Thorndike's (which he cited frequently). In practice, the two committed to extensive note-taking and, at least in principle, to walling off interpretation from the observations they made of the animals in their charge. In this regard, both adhered to standards for laboratory notebooks that emphasized attention over interpretation. The differences are in the details of their respective approaches. By granting his chickens a measure of freedom and tracking them over many days, Kline made his tasks harder in the interest of cataloguing a wider array of behaviors—and, by inference, of mental states. His chickens stretched and peered, "uttered notes of surprise" and "had a sumptuous breakfast." Thorndike's laser focus on hunger and escape left no time or attention for other activities. What is more, Kline's location—in the yard, at least for the chickens—meant the weather became relevant, as did the social interactions between birds. This difference in setting presaged critiques of experimental study from a century later, when "natural" social conditions were restored to the study of laboratory rats and canonical conclusions from earlier studies came under question. Where Thorndike sought control and the artificial conditions enabling it, Kline sought naturalism as a corrective to the privileging of experimental artifice. Each man believed he had the better view of minds at work.[44]

Kline turned his methodological ideas into a pedagogical program. In "Suggestions toward a Laboratory Course in Comparative Psychology," he set down a rationale for using animals as teaching tools, articulating an otherwise implicit factor for the field's growth. The article began with two telling epigraphs. One was Morgan's canonical prohibition against the invocation of higher faculties; the other was a counterargument by the psychologist Wesley Mills: "But why should we bind ourselves by a hard and fast rule . . . ? Is it not the truth at which we wish to get? For myself, I am becoming more and more skeptical as to the validity of simple explanations for the manifestation of animal life whether physical or psychical." Adopted from Morgan, Mills, and others,

Kline's suggestions amounted to borrowing tools from classic studies, rendered as clearly as possible for reproduction in classroom settings. Each tool was presented with an explanation of the mental states it was designed to access—and, in some cases, those it seemed to reveal even when the original experimenter had some other process in mind. The result was a practical manifesto for naturalistic experiments in comparative psychology, with additional attention paid to creatures—like slugs—that were left out of studies explicitly aimed at elucidating the roots of human reasoning.[45]

In the end, Kline's bizarre menagerie was not as influential as his commitment to studying it in seminaturalistic settings—and to sharing the results of his studies in the dry, abstracted form that Thorndike adopted as well. This writerly decision set them apart from some of their predecessors. Even Morgan, whose canon has such a prominent place in histories of the field, allowed himself a measure of homey familiarity in his articles and books. While Kline agreed with Morgan in his critique of Thorndike's artificiality, his aesthetic self-presentation was more like his fellow American's. Turning the mind into a scientific object meant shearing it of name and location, with such "irrelevant" details confined to notes about method. Not all psychologists followed the trend, of course, but rhetorical abstraction and the datafication of mental lives played a huge role in the field's bid for scientific status. Whether with Kline's naturalistic methods or Thorndike's avowedly artificial ones, practitioners cast themselves in less familiar, less domestic terms. This mirrored science's changing location, to be sure—out of the British garden or country house, into university laboratories. But it also made the shifting sense of science visible: not just divorced from everyday life, but disembodied, abstract, and (ideally) repeatable anywhere. New methodological standards in the field helped, in other words, to make method itself something that could be decoupled from it, and even from the mind.

For those aiming to elicit "natural" demonstrations of intelligent activity in this way, the ideal model organisms were not protozoa or slugs (despite the fact that their small size made them convenient and comparably cheap), nor were they dogs or cats (despite the fact that their ubiquity and familiarity made them easy to work with). The ideal creature to conjure with was the one Kline treated last in his methods piece: the white rat. Albino variants of the brown rat *(Rattus norvegicus)* had long been used in European laboratories, but were only established in the United States—simultaneously at Clark and the University of Chicago—at the end of the century. Kline picked up the practice from a Clark biologist named Colin Stewart, who prized rats for being "small, cheap, easily fed and cared for; and best of all, when placed in revolving cages they spend most of their time, when not eating or sleeping, in running." This quality made sense for Stewart, who was studying diet and exercise; for Kline, what mattered was something simpler: "that the rat is a 'gnawer'!" Setting rats the task of gnawing their way *into* a box (rather than *out* of one), Kline insisted on "the importance of working with animals in as natural moods as conditions permitted," if the goal was to catch their mental lives. Again, this did not translate over to the *way* these animals were presented—but the differences mattered to Kline.[46]

It was not the box that made the rat famous—it was the maze, first used on rats at the turn of the twentieth century. The iconic "rat-in-a-maze" setup owed its development to the same set of methodological preoccupations, and seems to have sprung from a conversation between Kline and Edmund Sanford, both of whom worked with the influential psychologist G. Stanley Hall at Clark. Kline later recalled telling Sanford about rats burrowing beneath a cabin owned by his family, which led Sanford to propose a mock-up of the famous Hampton Court Maze in England as a "'home-finding' apparatus" for lab rats. Having never heard of the maze, Kline looked it up and immediately saw the use it might serve—though he failed to pursue it immediately. The task of constructing the first maze thus fell to a graduate student in the laboratory, Willard Small, who seems to have been similarly inspired by Sanford. (Indeed, Kline and Small recall such similar episodes that it seems certain their memories of priority blurred together in the years between these developments and writing about them thirty years later.) What was unambiguous in Small's mind—though again, one must be careful about such recollections—was that his project "was in no way influenced by

Thorndike, as I did not know of his work when I made my approach to the problem." Still, by the time Small's study made it into print, Thorndike's was well known, and it was to the younger man's advantage to address it at the start of his article. "The chief advantage I derived from it," Small later recalled, "was to make me more certain that I was on the right track in seeking to conform experimental procedure to the native tendencies of my animals and to sharpen criticism of my own methods and results."[47]

Small published two articles on rat experiments, both in the relatively new *American Journal of Psychology* that was run out of his own department at Clark. His first piece began with a confession: "The white rat exists, so far as I am able to learn, only in captivity, and so, though especially suited for laboratory study, may be expected to present some slight variation from its wild congeners." The issue was related to the one he aimed to correct in Thorndike's work: that extrapolating from a laboratory experiment to lived experience was a risky move. If the mind was the object of study, tests had to be designed for specific animal capacities. "The chief difficulty of such experimentation," Small continued, "lies in controlling the conditions of the problem without interfering with the natural instincts and proclivities of the animal, and thus distracting or deflecting its attention." He even quoted from an author of popular natural history books to prove his point: "An animal should be made to do difficult things only in the line of its inherent abilities." Despite its artificial existence, the abilities of the white rat were assumed to be those of its wild relatives. Gnawing, burrowing, or at least running were in order.[48]

The maze, which Small introduced in his second article, continued his exploration of human and animal method simultaneously. "Primarily a study in method, an attempt to observe this animal under approximately experimental conditions," Small's maze work blurred his interests in addressing comparative psychology's methodology with the learning abilities of its animal subjects. Stretching six feet by eight, with walls and paths measuring four inches, the maze was designed to test the abilities of hungry rats to navigate toward a food goal located at the center. Intriguingly, Small asked his readers to vouch for the demands the maze would make on his rats. "A glance at the maze," he wrote, "will be sufficient to convince one of the difficulty of the problem." He used the same trick to bolster his argument at various times in the article. "A glance at the diagram," he wrote later, "will show that these mistakes [in navigation] may be rectified without a return—indeed they are not, properly speaking, mistakes at all, but rather failures to select the shorter path." In this way, readers' decisions when solving the maze in their heads reflect the choices that rats themselves make as they seek the food at the end of the tunnel. The links—instinctive, neural, or otherwise—between mice and men are implied by comparative psychology and reinforced by Small's presentation.[49]

Small drew a few lessons from his work with the mazes, some more explicit than others. For one, he questioned the tendency to assign a greater share of animal learning to "accident" when this was probably no truer than it was for humans. Yes, accidents occur, but "the term *accidental* must be used with reservations." This is because rats, like humans, "seemed to profit at once by experience [and] the conduct of the rat, his hesitation more than his avoidance of the error, indicates, rather, recognition and selection." Like the language of "error," the language of "trial" applied both to his own practices and to those of his animals. At times, rats are "subjects" of trials conducted by Small; at others, the rats seem to be living their (mental) lives for their own purposes as scientists do the same. "The fact that the trial and error method plays so large a part in human mentality, and especially that it predominates with children, still further supports the view taken in the beginning of this paper that animal intelligence works almost exclusively by this method." Reason, taken in its broadest sense to include "practical adaptation to varying conditions by direction association," was on display in each attempt that Small observed, and seemed to explain his own approach to the building and testing of his maze apparatuses as a graduate student as well.[50]

The reflective links wrought by the maze were made explicit in Small's second article, in a lengthy quotation from the philosopher David Hume. Noting how experience affects animal behavior—when a greyhound, for example, lets younger members of the pack tire an animal out before finishing it off—Hume concluded that learning does not always involve reasoning, in the sense comparative psychologists demanded. But this conclusion came with a twist: "Animals, therefore, are not guided in these inferences by reasoning, neither are children, neither are the generality of mankind in their ordinary actions and conclusions, neither are the philosophers themselves. Animals undoubtedly owe a large part of their knowledge to what we call instinct. *But the experimental reasoning itself, which we possess in common with beasts is nothing but a species of instinct or mechanical power that acts in us unknown to ourselves.*" Hume's "philosophers," like his dogs or Small's rats, acted according to mechanical impulses in ways that some scholars found it hard to stomach. "His expression 'experimental reasoning' seems to me singularly happy and accurate," Small opined, "guarded, as it is, by the suggestion of its subconscious character." Hume provided a defense of "chastened anthropomorphism," which Small used to combat "scientific pedantry" among comparative psychologists.[51]

The "imperfect instruments of experiment and the law of parsimony" were partly to blame for this pedantry, but the real problem was "the faulty and imperfect analysis of human experience." Though Small singled out Romanes for criticism on this score, he seemed also to have Thorndike in mind. After all, Thorndike's allusions to human capacities were as abstract and anecdotal as those he criticized among earlier animal psychologists. Despite disagreeing, Small and Thorndike converged on one point: whatever animal study revealed in the laboratory, its true meaning was in reflecting early or basic aspects of human cognition. This interest, as we have seen, led to a slippage between the perspective of subject and object, which Small made explicit but was also present in Thorndike's study. Though he did not go so far as calling his cats "experimental," Thorndike shared Small's slippery use of words like "trial" and "method." His cats demonstrated "methods of escape," a "method of formation," the "animal method of acquisition," and, most to our point, "the trial and error method." And Small's rats revealed "the method of animal intelligence in actual operation," which he then presented using "the method of graphic representation in curves." In both cases, and many others, method was the word that tied human and animal together.[52]

Methodology's collapse of the human–animal boundary was clearest in studies of a new animal, to which Thorndike turned after his dissertation: primates. Confronting the limits of his tests on cats and dogs, he pointed to "an investigation of the mental life of primates" as the obvious next step for the field—one he himself planned to take. Thorndike apologized that "the monkey which I procured for just this purpose failed in two months to become tame enough to be thus experimented on," but promised to continue "as soon as he can get subjects fit for experiments." Their attraction was obvious: primates were the closest relatives of humans and could thus help to fill out the analogies between human and nonhuman minds on which the field was based. Their "incessant curiosity" and "love of mental life" made them ideal subjects for such studies—and, potentially, links between the psychologist and those he studied. In "The Mental Life of the Monkeys" (1901), a monograph-length follow-up to his dissertation work, Thorndike made good on his promise, reporting experiments with three New World monkeys he kept caged in his house for the duration of the study. Once again, experiments involved escaping from puzzle-boxes adapted to the increased size and dexterity of his subjects. And once again, "we find no evidence of reasoning in the behavior" in the learning curves that he recorded.[53]

However, Thorndike's work with monkeys seems to have pushed him further down the path toward recognizing that scientific thinking was related in some fundamental way to what he observed in nonhuman animals. "In their method of learning," he concluded, "the monkeys do not advance far beyond the generalized mammalian type, but in the proficiency in that method they do." Rather than leave things there, as he had in his earlier work, Thorndike now suggested a parallel—or an evolutionary link—between this achievement and human reasoning, including his own: "For it seems to me highly probable that the so-called 'higher' intellectual processes of human beings are but secondary results of the general function of having free ideas and that this general function is the result of the formation after the fashion of the animals of a very great number of associations." This recognition, Thorndike suggested in his final lines, pointed to "a general psychological law," one which tied basic trial-and-error learning to the very activities in which he himself, as a scientist, was engaged in describing (and ultimately teaching).[54]

Younger scholars picked up where Thorndike left off. The final citation in "The Mental Life of Monkeys," for example, was to a junior colleague named Robert Yerkes, who arrived for graduate study at Harvard the year Thorndike left. Yerkes is famous for his work with primates, including studies conducted at Harvard and Yale as well as an experimental station that became, after his death, the Yerkes National Primate Research Center at Emory University. But like other comparative psychologists, Yerkes had many interests: his first publication was called *The Dancing Mouse,* and he made his name among the wider public by developing intelligence tests used on the army during World War I. His studies of primates were the most momentous, but for a different reason than the one that had excited Thorndike. Primates closed the gap between the menagerie crowding in on comparative psychology for over a decade and what remained the ultimate topic of all its work: human consciousness. Insofar as Yerkes's primates displayed human-like capacities for learning and memory as well as affect and creativity, laboratory studies of nonhuman primates might function as a "missing link" in the evolutionary and experimental studies of the mind through which a new discourse of method was being formed.[55]

Yerkes was working toward a disciplinary transformation decades in the making. What had been the science of mind was becoming, by and large, the science of *learning.* Timing how long it took animals to complete set tasks became a default mode for psychologists, not least because it replaced the obscure, hidden, and immeasurable phenomena of consciousness with something concrete, observable, and measurable. As mice learned to associate colors with food, or primates gradually adopted complex behaviors, the object of analysis in psychology became more and more synonymous with the field's most obvious area of application: education. While human learning was an object of attention in its own right, nonhuman primates were a convenient stand-in. Indeed, Yerkes concluded they were more than that: "other primates," he wrote years later, "may prove the most direct and economical route to profitable knowledge of ourselves, because in them, basic mechanisms are less obscured by cultural influences." As Donna Haraway has famously argued, Yerkes inaugurated a world in which "monkey, apes, and human beings seem to move about, reflecting each other in elaborate show, in a house of mirrors." Through all that reflecting, some of the mirrors warped, leading psychologists to see their own methods in the animals they studied.[56]

As the field grew, Yerkes remained focused on methodology. The name "comparative psychology," he insisted in 1908, "should designate a method of investigation rather than a division of the field of psychology." Elsewhere, he pushed back against a German movement to collapse comparative psychology into sensory physiology, complete with a new vocabulary grounded in objective descriptions of physical attributes. Such language, Yerkes wrote, "is highly artificial, unnatural and roundabout." What might pass for good science for physiologists just would not cut it for psychology, which "rests upon a system of inferences" of the sort that *comparative* scholars should be the first to acknowledge: "It is the business of comparative psychologists to investigate the reasons for our inferences concerning consciousness in other beings and to evaluate the bases of these inferences. The inferences exist and they are unescapable, hence it is not only legitimate but also desirable to bring into full consciousness the objective facts which condition them." Comparative psychology, according to Yerkes, was not just the study of animal minds. It was the study of human minds, too—specifically, the minds of scientists and the inferences they drew from the behaviors they observed.[57]

The proximity of human and nonhuman primates is also a reminder of a tension within the emerging field of comparative psychology and related efforts in physiology, neurology, and anatomy. If the objects and subjects of such research were similar—which they had to be, if analogies could be drawn between them—then the same similarity posed moral problems. When Yerkes called the chimpanzee "the servant of science," he pointed in the same direction as Morgan had when he suggested that Thorndike's animals were more "victims" than "subjects." Predicated on the existence of an "ejective sense" that enabled such comparisons in the first place, similarities between scholars and "servants" could potentially limit the range of (ethical) scientific practice. Yerkes's use of shock treatment on mice, not to mention later collaborations with neurologists attempting to locate complex behaviors in specific brain regions using surgery, put the utility of scientific research before the reality of animal pain. Vivisection debates at the turn of the twentieth century hinged precisely on this point: the highest gains of invasive research seemed to depend on using human-like animals—whose suffering seemed like ours, too. Public attention to psycho-physiology contributed, in part, to the closing of the laboratory to outsiders—and altered the stakes of methodological reflection in the process.[58]

STRUGGLES AND SCREAMING

The years between Thorndike's publication of his monkey work and its reappearance in *Animal Intelligence* were some of the most fruitful in the history of experimental psychology. Yerkes's emergence as an icon for naturalistic experimenters, and the cementing of that "ideal" method in a series of textbooks in the period, helped push the field into universities around the country. In classrooms and laboratories, animals of all sorts were used as proxies for their human observers and, gradually, as stand-ins for animals generally. Their behavior, abstracted from the conditions in which it was observed (whether artificial or natural), was simultaneously loosened from the minds to which it was supposed to be a key. Both the promises and the pitfalls of this approach to psychology were embodied not in Thorndike's curves or Small's maze, but in the method to which they gave rise. A decade after their back-and-forth over the nature of animal intelligence (or animal stupidity), their shared emphasis on measuring observable behaviors was turned into a new, self-contained approach to psychology dubbed "behaviorism." Versions of this view were a powerful force in the social sciences throughout the first half of the twentieth century, not least because leaders of the movement adopted Thorndike's polemical tone to make their case.[59]

Behaviorism was, in a sense, the apotheosis of method: to its adherents, all that existed was the behavior you could observe. Claims about mental states, hidden beliefs, or even minds were ruled out by the most zealous behaviorists. The approach was not without its critics, of course, some of whom came close to articulating the central thread of this analysis: that the process of abstracting, comparing, and generalizing mental phenomena was never purely about method. Rather, doing the work of psychology altered the phenomena under study, gradually changing the mind into an organ of behavior—before pushing it out entirely. One skeptical reviewer of Thorndike went so far as to call his claims to objectivity an exercise in self-justification: "Comparative psychologists would, indeed, deem themselves fortunate if they could assume with Thorndike that introspection adds nothing to our knowledge of behavior. . . . Indeed, the view of consciousness which is here presented seems to be designed more to justify a procedure than to explain the findings obtained by that procedure." Thorndike gave as good as he got. He called one critic's evidence "a good record of Professor Mills' attitude toward animal psychology, but . . . worthless as far as concerns the dog." Thorndike argued that the problem was Mills's "habit of occasionally mixing up opinion with observation." That this critique came in a discussion of *animal* habit underscored his point: the behavior of psychologists was as much an object of study for the field as that of the nonhuman animals they observed.[60]

If primates linked dogs to humans in the study of mental processes, a gap remained between even the most anthropoid of great apes and how scientists imagined their own minds, brains, and behavior. Filling that gap became a priority for many psychologists of Thorndike's generation, especially as criticism mounted of the conditions in which such animals were studied and the arguments that were extrapolated from them. One letter to the editor put it: "If Dr. Thorndike tried his intelligent 'Experiment No. 11' with a two-year-old cat, why didn't he try it with a two-year-old human? I guess he would have found an equal amount of ignorance of the mechanism of door fastenings, which comes only with teaching, and would have produced only struggles and screaming." Such criticisms dated a long way back and were reflected elsewhere in the field: in Romanes's doubts about human capacities, say, or Small's citation of Hume. And they pointed to fundamental issues, to whether and how scientists could and should understand nature. Was it a resource on which to draw—or a reminder of our own origins?[61]

Part of the reason for the frequency and power of this criticism was the ongoing debate over vivisection and the ethical problems associated with invasive work on animals. Anxieties about animal experimentation were always also anxieties about *human* experimentation. This ambiguity was precisely what enabled vivisection, if not all comparative psychology: the animals in question had to be close enough to humans to serve as models or approximations, yet far enough from them to license the procedures being used. The issues raised by tensions between human and nonhuman ethics had been dramatized a few years earlier in *The Island of Dr. Moreau.* An exploration of a researcher who overstepped his bounds and conducted the kinds of hideous experiments that gave antivivisectionists nightmares, the novel exposed a broader reading public to the issues at the heart of debates over the kind of work Thorndike was doing. With the continued rise of the eugenics movement and the waxing and waning of public trust in scientists in general over the next half-century, the anxious infighting of Thorndike and his critics takes on a foretelling, if not foreboding, tone. Debates over the human–animal boundary reveal what lay just beneath the surface of what can seem like dry matters of method.[62]

The ethical implications of psychological study were laid all the barer as evolutionary psychologists branched out from animal cognition into a new subject pool: children. Serving as a kind of "frontier instance," in the sense Herschel had picked up from Bacon, children marked a limit case for both the study of nonhuman animals and of adult humans. A child's mind could be a boundary object: by studying it, psychologists hoped they could articulate the line between nonhuman and human psychology, between great apes and adult humans. Boundary work meant more than classifying various kinds of minds, deciding whether each fell on this or that side of a line that meant something socially. It also involved constructing and trespassing boundaries around different approaches to science. As they had been for everyone from Herschel to Thorndike, matters of mind were also matters of method, as important in policing the behavior of scientists as they were when it came to what happened in the wider world. Methods developed to study one kind of mind could be adapted to another—from cats to rats, say, or now from rats to children. This was "comparative" psychology in its dominant sense, aimed at extrapolating across lines of species.

But comparison was not only about difference or sameness—it was also a way of preserving methods, making them useful in new contexts. Translating the maze into something humans could do, or turning children's puzzles into tests for dogs, were both ways of keeping practices fresh and moving the field forward. Here again, matters of mind and method were worked out simultaneously. The drive to maintain methods, to apply them across lines of species as a way of making good on all the work that went into developing them, helped blur the lines they were crossing. This evolutionary psychology was only partly about theoretical debates over the differences between kinds of animals. It was also about justifying one's own approach, ensuring what psychologists now call "paradigms" lasted in struggles for disciplinary survival. Today, the same thing happens all the time: methods for studying dogs are tried out on human infants, and vice versa. And of course, such translations of technique help answer questions about development and comparative cognition. But they also help sustain the field—just as they helped launch it, in the early days of animal and child psychology around 1900. Then and now, methods helped psychologists see beyond their case studies—and proved the adaptive fitness of various tools and techniques in the process.

The stakes of introducing children into psychology were high. Child development was a final link in the evolutionary chain from slugs to scientists, a way to renew old methods with a new (practical) upside. This was why Thorndike urged his colleagues "to study the passage of the child-mind from a life of immediately practical associations to the life of free ideas." Of course, this was not as easy as it sounded: entering the classroom required permission and the good will of the public. Suspicion of such claims, not to mention of experimenting on children, was not what it would become later in the century—but such doubts existed and would have to be overcome. If psychologists could not go to where children already were, they would have to bring them into the laboratory—either into existing facilities or else by building new laboratories that blended the demands of science and school. Though more costly than fanning out into the community, bringing the community into the laboratory dangled the advantage of greater control, the promise of precision in a space you could govern yourself. If you wanted a science of the mind, and if those minds were as unruly as children's could be, then you had to turn the school into a laboratory. And this is precisely what they did.[63]

7

LABORATORY SCHOOL

Rats were not psychology's only new models. In the same laboratories where mazes were run and boxes were escaped, psychologists used a new organism to reveal the developmental roots of their own reasoning. These subjects were closer to home: human children. Kids could be brought into the laboratory, but it was more common to try to find them in their natural habitats. At the end of the nineteenth century, changes in the education system made primary schools the ideal sites for this work. Increasingly compulsory (and crowded), classrooms seemed to provide ideal settings for observing the behavior of children as they acquired and manifested new capacities. Studies of children dovetailed with those on animals: what nonhuman minds revealed about the historical development of science, human children demonstrated on a day-by-day basis. As psychologists set tests for kids at a range of ages, mental evolution almost seemed to happen before their eyes. Combined, as they often were, these studies of rats and children rounded out an emerging account of the scientific process as both a natural phenomenon and an artificial tool, as something at once fundamentally human and more than human.[1]

What was soon called "child study" rested on a set of assumptions that practitioners shared with the founders of comparative psychology, not least because the groups overlapped significantly. Studies of children and animals alike were framed in terms of recapitulation: the changes witnessed in early childhood were thought to mirror mental evolution over generations. An infant's ability to recognize shapes or a toddler's ability to navigate a room were versions of similar capacities in great apes, rodents, and the rest of the psychological menagerie. The parallel between humans and animals also broadened the meaning of "comparative" psychology. Leaving laboratories and entering schools (or turning schools *into* laboratories) did not mean abandoning the techniques developed to study nonhumans—it simply meant translating them into forms that would work on human children. Psychologists saw children in much the same way as they saw nonhuman animals: as proxies for adults, including themselves, and as windows onto their own ways of thinking. Blurring the lines between human and nonhuman subjects helped cement the basic behaviors being studied—including problem solving and learning—as fundamental, deeply natural processes. Even science came in for this kind of naturalization.[2]

Sometimes, kids came first. Thorndike's dissertation work and the controversy it caused had played out in response to recently published work by another young psychologist. Ernest Lindley, a student of Stanley Hall's, used puzzles to study children the same way Thorndike and Small studied cats and rats. While administrators had squashed Thorndike's studies of children, Lindley was allowed to test students in the Worcester public schools for his dissertation research. Armed with simple puzzles designed for children, Lindley timed how long it took students to finish them in successive stages. Much as his colleagues were doing with animals, he worked toward a simple goal: time curves that could be used to demonstrate the learning process as well as to compare across individuals. Hundreds of third-, fifth-, eighth-, and ninth-graders completed the tests, and Lindley published their results (and his analysis of them) in Hall's *American Journal of Psychology* in 1897. The article, "A Study of Puzzles with Special Reference to the Psychology of Mental Adaptation," had an immediate impact on comparative psychology. Citations appeared in the work of Small, Thorndike, and others who were turning psychology into the experimental study of mental behaviors as the nineteenth century came to a close.[3]

In Lindley's eyes, children were not just animals—they were scientists, too. From their innate sense of play shared with nonhuman animals, children gradually developed new means of adapting to the world around them through playful maneuvers and subsequent adjustments based on the *effects* of their maneuvers. Watching children work, Lindley thought that "the rising curve of puzzle interest marks the prepubertal age as the time to hasten transition to the higher mental methods." It was clear to Lindley and his readers that the distinction between higher and lower methods mapped onto the gap between conceptual reason and what he called "sense-trial and error." But in reality, the gap between child and scientist was not that wide. Both seek "joy in the overcoming of difficulties," from simple gestures to abstract theories. "Many movements of the young," Lindley wrote, "thus represent a kind of experimentation." And the converse was also true: a scientist's procedure, when confronted with a difficult problem, "may descend almost or quite to the lowest 'levels.'" Unfamiliarity and complexity were not the only causes of "descent," Lindley argued, "but also fatigue, temporary loss of interest, a fleeting state of mental muddle may produce a relapse into the animal method." Affective states that would have been familiar to readers were, for Lindley, the links between children, scientists, and nature's method.[4]

The idea of studying scientists was not lost on Lindley. "Of great value for the psychology of scientific method," he wrote, "would be the detailed account of the procedure of the most successful experimenters." Though such sources were rare, those we had suggested that science depended a lot more on error, accidents, and "fumbling" than was commonly believed. In other words, scientists were human: they screwed up, sometimes intentionally, and learned from their mistakes. Combined with the evolutionary arc of methods over time, this suggested a new goal for psychology: "a genetic view of the natural forms of adaptation, of the natural logic which organisms employ in dealing with novel situations." Lindley was optimistic: "Of manifest importance to biology and psychology would be the natural history of such processes, from the lowest forms of conscious life to man, as well as from primitive man and the child to the adult scientest [*sic*]. Some studies of animal method, notably those of Romanes, Lubbock, Lloyd Morgan, Binet and Hodge have shown the richness of the field. Similar studies of children have as yet scarcely passed the anecdotal stage." From "animal method" to "child study," a generation of psychologists was forging a unified approach to methodology—including their own.[5]

Their work found a ready educational audience in the years around 1900. Expanding enrollments at all levels put a premium on effective teacher training and finding efficiencies in the classroom. Many psychologists in this period delivered lectures on the field's applications to teaching, paving the way for several inroads into what was soon called educational psychology. Schools were not the only site in which the sciences of mind were applied in this period, but the field's increasing focus on learning found an eager audience of teachers, superintendents, and politicians for the next few decades. These clients, in turn, shaped the priorities of psychological expertise and the directions of research as the field matured. One form this influence took was the distillation of psychological theories into practical manuals for teaching and learning. Almost everyone seemed to have advice to offer, in the form of lectures and, soon after, books. While it can be hard to quantify the impact of this pedagogical turn on practices in the classroom, the growth of schools of education and the role of childhood psychology in the training of teachers are easy to see. Psychologists' entry into the classroom expanded their reach into other areas of American life as well.[6]

This expansion changed how psychologists pursued their own projects—the issues they found important and how they went about studying them. One aspect of this change was a new focus on what came to be called "applied psychology," of which work in classrooms was only a part. On the street and in boardrooms, as part of governments and in advice manuals of all sorts, theories built in laboratories and over decades were quickly coalescing into a concertedly practical subfield that put pressure on theoreticians in turn. And not only that: even those whose work was far from "applied" focused, more and more, on learning and problem solving in their studies. The turn to such subjects was not reducible to the market for books or the lure of public attention. Rather, just as the animal psychologists were finding down the hall (or, at times, in the same laboratories), learning fit the emphasis on measurable, observable phenomena that proto-behaviorists like Thorndike championed at the turn of the twentieth century. Determining if an animal—or a child—had learned a task was easy, at least compared to studies of consciousness that prevailed among those who were more philosophically inclined. Gradually, simple problem solving stood in for more complex mental states—including scientific experimentation.[7]

CHILD STUDY

Much like animal psychology, the early development of child study was riven by debates over methods and evidence. Anxieties about the authority of anecdotes were widespread, as were concerns about something similar to anthropomorphism: namely, the attribution of mental states to children, even infants, because scientists were reminded of their own actions. Evolutionary assumptions were as central to the study of children as they were to work on nonhuman animals, though in slightly different ways. Whereas nonhuman animals were thought to reveal a longer evolutionary history, children modeled in miniature how adults learned to solve problems. Pessimistic observers of phenomena such as childhood insanity argued that children "like brutes, live in the present," but this did not prevent other researchers from treating children as lenses and mirrors. In the minds of these psychologists, childhood development revealed distant aspects of our shared animal past *and* reflected fundamental features of human cognition. Children, like animals, could stand in for almost anything—including scientists themselves.[8]

Just as he was for comparative psychologists, Darwin was a founding figure in the new field of child study. Spurred by a French article on language development that was translated in the journal *Mind,* Darwin responded with a set of observations recorded thirty years earlier, in 1840, about the developmental milestones of his firstborn son. Entitled "A Biographical Sketch of an Infant," the piece is an exhaustive record of minute behavioral changes. Activities like "sneezing, hickuping, yawning, stretching, and of course sucking and screaming" appeared within hours of his son William's birth, while others emerged more gradually. Though Darwin took these notes before he had fully articulated his theory of natural selection, he was—as we have seen—well on his way, and his interpretation of William's behaviors both at the time and looking back reflect growing enthusiasm for an evolutionary account of human origins and development. Allegiance to his theory led Darwin to read as "instinctive" William's earliest activities, which he thought gave way gradually to experience and, eventually, learning. Emotional development dominated the essay, presaging later work that would become Darwin's *Expression of the Emotions in Man and Animals* (and for which these much earlier notes furnished important material).[9]

Darwin's "Biographical Sketch of an Infant" occupied a developmental stage between the older, naturalistic observations in which he had been trained and newer, experimental studies that were rapidly becoming standard practice in the science of mind and behavior in these years. This meant that Darwin's followers could read his work as observational, experimental, or even both. Battles over the methodological direction of psychology thus played out in child study much as they did in the psychology of animals, not least because many of the combatants were the same. George John Romanes, for example, relied equally on observations of children and nonhuman primates in building his theory of mental evolution and laying claim to the Darwinian mantle. And James Mark Baldwin similarly made his name with *Mental Development in the Child and the Race,* which borrowed methods from comparative psychology for the study of human learning. Romanes and Baldwin saw the use of children for their work not only as convenient, but as necessary. Extrapolating from animal minds to their own meant finding a bridge between the two, a model organism close enough to the motivations and behaviors of nonhuman animals to enable leaps across the species barrier, but also similar enough to adult humans to enable them to complete the comparison.[10]

Children fit the bill. On the one hand, their behaviors from a young age manifested what the psychologists studying them saw as "savage" energy, uncontrolled and—some imagined—untainted by the manifestation of willpower that came only later in development. Exuberance, joy, and a whole host of other affective states were just what it meant to behave and think like a child. Science, if it could tap into those qualities, would be even closer to nature. On the other hand, the gradual development of the capacity for self-control provided an accelerated model of how human society may have emerged over the course of evolutionary history. Science was not all fun and games, after all. It was distinguished from other ways of thinking by its combination of playful experimentation with grueling rigor. Balancing energy and control, enthusiasm and attention to detail, science was (supposed to be) the best of both worlds, childlike and mature at once. In their exuberance *and* their growing ability to channel it, children stood in for precisely the virtues of equilibrium toward which theorists of method and scientists of mind had been driving for a few generations.

Psychologists' effort to ground this balance in their study of children comes through most clearly in their fascination with the topic of play. Linked by Darwin and others to the universal pursuit of pleasure, play seemed like the perfect activity to connect animals and humans. "I was at first surprised at humor being appreciated by an infant only a little above three months old," Darwin reflected, "but we should remember how very early puppies and kittens begin to play." The instinct for playmaking exposed, to Darwin, an "aesthetic feeling" in animals that presented it. Though he did not cite it, Darwin's take on playmaking reflected one of Alexander Bain's early essays, on toys. Bain thought that toys revealed "a passion for handling," and that playing with them kept the minds of children and adults alike limber. For Bain, associationism explained the pleasures of play, as the experience of novelty (with a new toy) was recalled each time it was confronted again. Because they were more open to this kind of experience, children enjoyed it more: "With us [adults], reason constrains the mind into certain limited channels, and though our faculties are stronger, and our Past more copious than the child's, yet the child probably riots among coincidences, and they already experience pleasures of the Past, more profusely than we do." As in his account of "trial and error," Bain invoked play as an explanation for the interest we take in certain things—the driving force in the learning process.[11]

Darwin layered an explicitly evolutionary interpretation over this view of play. Bain's "toy principle" was deeper than the impulse to play or laugh, bigger than the behavior of any particular child. Darwin argued that what was true of the individual child's play was true of the species. The appreciation of beauty and the exercise of the imagination were connected, for both the single child and the larger group, to natural or sexual selection (or both). After all, if play was as ubiquitous as it seemed, it had to entail some advantage for either survival or reproduction. Its advantage could be atavistic or current, a relic of the past or a tool of the present. Bain's "passion for handling" did not provide a strong enough rationale for the depth Darwin observed in the "aesthetic feeling," given how widespread behaviors associated with it were. And where Darwin hinted, others staked careers, transforming the evolution of play and its relationship to cognition into a central question for child study—and thus, for psychology—after Darwin.[12]

One influential account of play's utility came courtesy of a German psychologist named Karl Groos. Like Romanes and others in the period, Groos split his most famous work into a first book on animals and a second one on humans. Rather than focus on mental evolution in both groups, Groos zoomed in on one aspect of their relationship: play. He located the advantages of play (for both human and nonhuman animals) less in the past, as a record of milestones reached, and more in the future—as forward-looking. According to Groos, play let animals practice new activities—such as fighting—that were essential to survival later in life. When puppies play-fight within days of being born, they are acting instinctively *and* learning non-instinctive behaviors in the low-pressure environment of the family. The randomness and creativity we associate with play were, for Groos, neither the by-products of the imagination nor the chance occurrences of individual animal development. Rather, the accidental nature of playmaking was precisely the point: play emerged from what Groos called "experimentation" (*das Experimentieren*), a process that built spontaneity into development, giving rise to everything from play in the young to the more mature attainments of art and—tellingly—science.

The Play of Animals and *The Play of Man* appeared in English in 1898 and 1901, just a few years after their original publication in German. Both were translated by Elizabeth Baldwin, the wife of the psychologist James Baldwin—who provided introductions and editorial additions to each book. It was the link between animals and humans, between roughhousing and science, that attracted Baldwin to this theory of play. As Groos put it in the second volume: "Veritable thirst for knowledge, with its unappeasable questioning, gradually develops from this [experimentation], making without difficulty the transition from the realm of play to that of genuine scientific investigation." Orienting the play instinct toward the future strengthened links between imitativeness in nonhuman animals, games played by human children, and the creative dimensions of adult human reasoning. And this was what made Groos's theory so enticing for Baldwin, who was busy making his own case for the connection between behavior and heredity, in the form of what he called "organic selection" (later termed "the Baldwin effect"). Groos cited both Baldwin and Morgan favorably for their work in the area, which only heightened Baldwin's enthusiasm for what he referred to in his introduction as the "practice theory of play."[13]

Not everyone was so enamored. Stanley Hall, then the president of Clark University and eager to shore up his power in the field, called Gross's theory "partial, superficial, and perverse." Rather than pointing ahead, Hall thought, play unearthed the past. "True play never practises what is phyletically new," Hall insisted, adding: "In place of this mistaken and misleading view, I regard play as the motor habits and spirit of the past of the race, persisting in the present, as rudimentary functions sometimes of and always akin to rudimentary organs. The best index and guide to the stated activities of adults in past ages is found in the instinctive, untaught, and non-imitative plays of children which are the most spontaneous and exact expressions of their motor needs." Play was a relic. "Thus," Hall concluded, "we rehearse the activities of our ancestors, back we know not how far, and repeat their life work in summative and adumbrated ways." The child's mind, which Bain regarded as a mystery (because a child is not a reliable narrator of subjective experience), was turned into the key to adult cognition. A child's mental development was a window into the past. "It is reminiscent, albeit unconsciously, of our line of descent," Hall wrote, "and each is the key to the other."[14]

Child psychology was more than fun and games for Hall. Widely known (and just as widely reviled) for his relentless self-promotion and polemical tone, Hall was an evangelist for his personal take on evolutionary theory and scientific psychology. Like both predecessors and contemporaries, Hall's interest in evolutionary approaches to the mind was grounded in a search for balance. In his case, scientific psychology was an answer to questions Hall developed as an undergraduate interested in philosophy. Fondness for Romantic literature and, soon after that, the philosophy of John Stuart Mill left Hall looking to make room for intuition, inspiration, and instinct in an account of reasoning he and many others adopted from Kant in the middle of the nineteenth century. Eventually finding his way to Hegel, and thus to what felt like a natural way of balancing reason and understanding, Hall settled on psychology as a means of translating a calling for philosophy into the kind of concrete problem toward which he tended to gravitate. With this idea already in place, he decided to pursue an advanced degree in the science of the mind under a rising star in the study of physiology and psychology at Harvard: William James.[15]

The first recipient of a PhD in philosophy at Harvard (and of a PhD in psychology anywhere in the United States), Hall soon achieved what his mentor had not: a stint in Wilhelm Wundt's laboratory at Leipzig, that calling card of method for American psychologists in the late nineteenth century. Having founded an experimental laboratory at Johns Hopkins upon his return (the priority of which, relative to those at other universities, Hall would spar about for the rest of his career), he trained a cohort of "new psychologists" first as a professor in Baltimore and, after 1889, from the president's office at Clark University. The term Hall would eventually adopt for his approach to the mind sciences was "genetic psychology," by which he meant both its evolutionary framework and his insistent focus on development (from stage to stage in human life) as the key to uncovering those changes. While he began as a champion of the laboratory, Hall soon turned away from the methods of his training, preferring the questionnaire—and, gradually, statements so general that they were not conducive to empirical elucidation at all.[16]

Even before he graduated, Hall was dissatisfied with American higher education and eager to rethink it along (European) psychological lines. In a letter to *The Nation,* he pointed to the "application of scientific methods in psychology by Spencer, Lewes, Lotze, Wundt, and others," as opposed to what he deemed the American impulse "to tell *what* to think, [rather] than to teach *how* to think." How, rather than what, was the way forward. Hall's sense of inferiority was heightened in Germany. By 1878 he followed up his earlier critique with a celebration of Wundt, which *The Nation* printed anonymously. "The breadth of the field which Professor Wundt has opened to the student of philosophy," Hall wrote, "[and] his acquaintance with and vigorous criticism of Mill, Spencer, Bain, Darwin, etc., . . . indicate more, perhaps, than the writings of any of his contemporaries the direction which philosophical thought is likely to take during the next decade." He called it "The Philosophy of the Future," and he meant it. The next year Hall returned to his earlier topic, concluding that the United States was "yet too young" for philosophy. "The minds of business and working men," as Hall characterized his countrymen, ". . . have short, plain, and rigid methods of dealing with matters of pure reason or of faith."[17]

Although Hall believed that Americans fell short in philosophy, he had hope for science in the country. Evolutionary theories remained controversial in certain European nations, but there was no such problem in the United States (or so Hall thought). Openness to evolution, too, he attributed to youth. Sometimes it got out of hand. "In a country of such remarkably rapid development as our own," he wrote, "where the ploughboy is never allowed to forget that he may become a millionaire or even President if he wills it earnest enough, the catchwords of evolution often excite an enthusiasm which is inversely as the power to comprehend its scope and importance." Americans were like enthusiastic children, often abandoning "the patient mastery of scientific details."

Still, there was hope in their energy. The best that American science writers had to offer, according to Hall, were the essays Charles Peirce had just published in *Popular Science Monthly* on the logic of science—"Illustrations of the Logic of Science." Hall read Peirce's essays as promissory notes for an American synthesis, one in which science was the culmination of mental evolution—and one toward which Hall's own work would have a lot to contribute. What Hall saw in Peirce was a blueprint for a new way of thinking, practiced in an evolutionary key, to take over American education and bring the national mind to maturity. Science, specifically Hall's vision of science, was the key to doing that.[18]

Hall cast his own intellectual development in the same terms in which he framed the state of American education. He turned personal experience into a call to arms: "I believe that no one has much knowledge of the inner workings of his own soul until he has served an apprenticeship in the psycho-physical laboratory . . . [which] lays bare the geology of the soul . . ." Classroom conduct was a kind of excavation: "Into the well conducted seminary all the hereditary influences from all the council camp fires and stories of our forebears, a little of the esoteric spirit of all the secret organizations of savage life from the immemorial past have gone and in it they find one of their highest expressions in the modern life of culture." Classroom discussion, like the child's mind, was a window into our past, a vision of the history of struggle and progress that culminated in the practices of scientific experiments. "There is a vagueness and mysticism about youth that is inevitable at the time when sentiment is ripening into thought and reason," Hall concluded, and it was this energy that he sought to capture in his teaching.

Just as Peirce had done, Hall invested his chosen field with almost spiritual significance: "psychology raises the interest in life which all feel to a higher potence and intensifies the desire to see, know, touch it at every point, to enlarge our experience as far as possible toward becoming commensurate with that of the race." Where Peirce thought of his logical method as "his bride," Hall was "smitten with a pedagogical passion of helping young men." This blend of the personal and the professional, of the pedagogical and the psychological, was there from his earliest essays. Critical letters in the 1870s, celebrations of the "new psychology" in the 1880s, and more prominent pieces on the state of the field in the 1890s all made matters of mind and method not just momentous, but almost mystical. At times, his high-flung rhetoric comes off as self-aggrandizing (if not a bit empty), but at others it reveals a commitment to making a science of the embodied mind central not just to the study of philosophy, but also to the practice of teaching and the reality of living. The important thing to note is that, while his zeal is striking, his focus was not exceptional: psychologists were bent on using their tools to intervene in American classrooms throughout the late-nineteenth century.[19]

A glance at Hall's publication record reveals a careful balance between research on what he called "the contents of children's minds" and efforts to put such research into practice in the classroom and beyond. Gradually Hall's pedagogical and psychological work blurred together, such that by the 1890s titles like "Child Study as a Basis for Psychology and Psychological Teaching" and "The New Psychology as a Basis for Education" start to become a rule, not an exception. Some psychologists saw their duties as researchers and as educators coming apart in this period, but Hall sought something like the opposite: an inextricable link between his theories of mental evolution and the recommendations he made for teaching. This relationship between psychology and pedagogy had a political edge, both in the small-p sense (Hall was attempting to consolidate support for his vision of the field) and in the big-P sense (he was trying to make the political utility of his work more apparent). This eagerness to combine psychology and pedagogy in a new politics of method turned Hall into one of the most prominent scientists of his day.[20]

In part, Hall was able to put his child-centric view of science to work by controlling the organs of publication. He founded the *American Journal of Psychology* in 1887 and, in 1891, *Pedagogical Seminary.* A year later he was named the inaugural president of the American Psychological Association, for which the *AJP* was made the official journal. All of this happened around the time Hall became the president of Clark, a combination of events that enabled him to get his students' results—and his vision for the field—into the hands of most of the young field's practitioners in short order. This is part of the reason Small's experiments on rats and Lindley's studies of children, conducted with Hall's theories in hand and under his supervision, reached such a broad audience despite the authors' comparative youth and inexperience. Though his entrepreneurial facility existed in tension with his interpersonal difficulties—Hall made as many enemies as he did friends through his efforts to promote his ideas and dominate conversations in the field—his enthusiasm for application put Clark on the map. Hall's insistence that his students spell out the implications of their research meant that even when colleagues at other universities disagreed with their results, his students' projects were taken seriously and played an outsize role in the child study movement.[21]

Hall's peculiar ability to both influence and infuriate is nowhere clearer than in the response to his two-volume *Adolescence.* Published in 1904, the book—like Darwin's *Origin*—is presented as an abstract of a magnum opus that never appeared. Instead, *Adolescence* stands in for that grander work, a testament to the range of Hall's interests, his disciplinary ambitions spelled out in the subtitle he chose: *Psychology and Its Relations to Physiology, Anthropology, Sociology, Sex, Crime, Religion and Education.* The book is credited with turning adolescence into an object of scientific study—distinct from childhood on one side and adulthood on the other—as well as introducing the term into wider use. But Hall saw the book as even more than that. Adolescence, he implied in the preface, stood in for the human species as a whole, at least as he found it at the dawn of a new century. "While his bodily form is comparative stable," Hall wrote, "his soul is in a transition stage, and all that we call progress is more and more rapid." What was true of the species was true of the scientist: "The view here represents a nascent tendency and is in striking contrast to all those systems that presume to have attained even an approximate finality." Method, like the human species, was undergoing a transition; psychology was moving from one phase to another. What better way to reveal its nature, to inaugurate "the long-hoped-for and long-delayed science of man," than by studying the stage of mental life that best corresponded to that methodological moment: the adolescent?[22]

The book's first volume consisted mostly of biology. The second turned inward, to adolescent psychology and behavior. An "unconscious" and "spontaneous" love of nature, Hall argued, emerged in puberty and was essential to science's development. This "first sentimental response" is followed by "popular science," including its history, after which "applied" and "pure" science can be taught to students and, eventually, practiced by them in turn. Throughout the book, Hall insists that learning depends on a genetic understanding of the subject to be taught, both in the sense of its history as a field but also of the stages by which students become capable of thinking in its terms. And after all, these two meanings of "genetic" were one and the same for Hall: adolescent development mirrors the rise of modern science, and only by studying both can you take advantage of the natural energy and interests of the students in your classes. Sadly, Hall reflected, "science is often taught in a way to destroy the love of the very department of nature it should develop. The only corrective," he concluded, "is to introduce evolution as a conscious method, a goal to which everything focuses to a great unity." The theory's wide applications and deep explanatory power are perfectly suited to the needs of adolescent minds: it combines the unconscious with the conscious, the natural with the artificial, in just the balance developing minds are seeking. This had been Hall's own experience of adolescence, and he wrote it onto the species. The solution for science and for adolescent longing was the same: a new method, modeled on the process of evolution.[23]

Child study, in Hall's hands, was not just a new window into the adult mind, nor was it only a lens to focus our shared history. It was also crucial to Hall's broader disciplinary agenda: turning psychology into an obligatory passage point for public and political discussions. In Hall's hands, the mind of the child, and especially of the adolescent, became a means of remolding science as a (natural) political tool modeled on the practices Hall and his students were already pursuing. To make science seem like the obvious starting point for broader discussions, it had to exceed or avoid the issues and differences that divided those involved in such conversations. One way to turn it into something natural was, as we have seen, to find science's roots beyond humans—and thus beyond politics. Hall's students pursued this line of study, building on work in animal psychology to remake both psychology in particular and science in general. Hall, on the other hand, extended back in time, casting child development as the key to science's history. Both ways worked: whether it was animal or child, laboratory or classroom, present or past—Hall and his students helped turn science into an organic process on which a divided society could depend for new directions.

Of course, turning science—or anything else—into a natural phenomenon does not void it of moral significance, even if that is the goal. A glance at the moral weight we assign to human life, to learning milestones or childhood trauma, is a reminder of the unavoidable normativity of any such effort. These moral dimensions were not particular to child study, of course, but there was something about child psychology that brought them to the fore. If child development made the hidden values of science clear, it was just a first step to recognizing them in other places. Hall was revealing, in his work and that of his students, a latent effect of turning science into a natural evolutionary process: what seemed like *descriptions* of thinking very quickly become, in the hands of psychologists and in the classrooms they entered, *prescriptions* for thinking—scientific or otherwise. By introducing "moral kinds" into the language of science, by subjecting social phenomena to scientific observation, child study highlighted the continued political salience of matters of method.[24]

REFLEX ARCS

During his brief stint at Hopkins, Hall worked closely with Charles Peirce, alongside whom he taught a young graduate student named John Dewey. Given that he studied with William James at Harvard as well, Hall's career displays an extraordinary centrality to American intellectual history for someone who is so much less familiar than his more famous peers. Hall's contested place in the period was owed, at least in part, to his personality: awkwardness almost seemed like a conscious program at times. Hall routinely pushed away mentors, colleagues, and students who got too close and drew distinctions where others might have formed coalitions. Some historians have followed Hall's lead, accounting for his institution building as though it was separate from the main developments in science, philosophy, and the other fields with which he engaged from the 1870s through his death in 1924. Even Hall's famous invitation to Sigmund Freud, which led to the Austrian analyst's only visit to the United States and propelled him to wide recognition, has been framed as the passion project of an American impresario rather than community building among human scientists—not least because Hall presented himself in these terms at the time.[25]

As is so often the case, however, a closer inspection reveals something more complicated. Hall's eagerness to distance himself from his peers stemmed as much from similarities as from differences. This is perhaps clearest when it comes to the relationship between Hall's research, including his reflexive attention to method, and that of his erstwhile graduate student, John Dewey. Long before he became the elder statesman of the "high tide of American liberalism," Dewey forged a path from philosophical idealism to psychological naturalism. After graduating from Hopkins, Dewey pursued this line of research first at the University of Michigan, then the University of Chicago, and finally (after 1904) from a perch at Columbia University. Dewey's status as an American icon has been cemented in endowed chairs at each of the schools at which he taught, hundreds of books dedicated to his work, and the continued currency of his ideas among academics, educators, and activists. Though often hard to comprehend—Oliver Holmes famously likened Dewey's writing to how "God would have spoken had He been inarticulate but keenly desirous to tell you how it was"—Dewey's publications played a key role in cementing science's new cultural authority in the Progressive era.[26]

Looking back in 1930, Dewey gave a name to a transition he felt he had undergone in the 1890s: "From Absolutism to Experimentalism." Fascinated by German idealism, which he picked up first as a student at the University of Vermont and then at Hopkins (not from Hall or Peirce, but instead from the school's third instructor in philosophy, George Sylvester Morris), Dewey's early interest in psychology was spurred by

Hegel, not Mill. What most captivated the young New Englander was an image of what Hegel called the "Absolute Idea," a construct that seemed to unify what Kant had separated into things-in-themselves *(noumena)* and things-in-use *(phenomena)*. To Hegel, this unification was "the sole subject matter and content of philosophy." Dewey was far from alone in his fascination with grand ideas like "the Absolute": across Europe and, perhaps especially, in Britain and the United States, a popular vogue for idealism held a great deal of sway in the late-nineteenth century. While James was famously skeptical of the movement, Hall had begun as Dewey did: animated by Hegelian idealism. Many contemporaries would have shared "the sense of divisions and separations that were," Dewey later recalled, ". . . borne in upon me as a consequence of a heritage of New England culture." Idealism, or what Dewey called absolutism, was a common response to those conditions.[27]

"Absolutism" named Dewey's idealistic roots, but "experimentalism" was what he was groping for in the 1890s. Still, he struggled to say exactly what that goal entailed, even many years on. Experimentalism, according to Dewey, was "too much the self that I still am and is still too much in process of change to lend itself to record." Part of its meaning is captured in the different names Dewey gave to it over his career. The most famous is "pragmatism," though James was fonder of the name than Dewey, who preferred "instrumentalism." (Peirce, too, was unhappy with the label, ultimately opting for "pragmaticism" to differentiate his mature views from what he viewed as James's "kidnapped" version.) Other "-isms" Dewey applied to his work included "functionalism," "naturalism," and, of course, "experimentalism." Each encompassed a phase of Dewey's life and, he felt, of intellectual history; both his own development and broader philosophical stages were "unstable, chameleon-like, yielding one after another to many diverse and even incompatible influences." For Dewey, "absolutism" was solid and "experimentalism" was flexible. The latter term picked out the shifting nature of mind and method as he and others experienced them. Dewey's many efforts to name it were themselves experimental.[28]

The transition from absolutism to experimentalism, as Dewey recalled it, occurred in 1891. The precise moment of rupture was crystallized in an essay he published that year entitled "The Present Position of Logical Theory." Dewey did not use "logical" in the sense that his onetime teacher Peirce had—far from it. Dewey sought what Peirce now denied was possible: a logic rooted in psychology, a science that could "determine the exact and concrete relations of subject and object, individual and universal within consciousness." While in his Hegelian phase, in the 1880s, Dewey saw the individual and the universal as one and the same, tying himself in knots with claims like "individual consciousness is but the process of realisation of the universal consciousness through itself." By 1891, with his ideas "undergoing reorganization," Dewey tried to shed these idealistic overtones while holding onto a reflexive link between psychology and the practice of science more generally. Still defending Hegel as "the quintessence of the scientific spirit," Dewey's "The Present Position of Logical Theory" marked a subtle shift away from the "Absolute Idea" toward "the inner anatomy of the realm of scientific reality." Logic, in his view, was "the theory of scientific method," its subject matter no longer fundamental or transcendental, but basic: "the various typical methods and guiding principles which thought assumes in its effort to detect, master, and report fact." Science was gradually being grounded.[29]

Logic was a human science in two senses. First, its subject matter was human: the real, lived efforts of human thinkers. They were scientists, at first, but Dewey soon extended the circle to include all cognitive practices—to grapple with any thinking he could find. In this regard, logic and psychology blurred into one another in just the ways anti-psychologists like Peirce were decrying in those decades. Second, it was not only the subject matter of logic that was human: it was the subjects, too—the people putting its methods into practice. This may seem trivial, given that all science was "human" as such, but recall that Dewey was specifically interested in the reflexive potential of what he called "experimentalism." As a science in the experimental sense, logic reacted to its own results; as new insights about thinking emerged, they redefined what any practitioners were capable of and how they might best go about achieving more. When inquiry was the focus of inquiry, when reflexivity was the norm of a scientific field, a peculiar kind of stable instability began to set in. Every new claim or new way of thinking shifted the conditions of possibility for future claims. Dewey, more than most, saw this not as a risk or a downside, but as essential to living experimentally in the modern period.

Dewey tested this scientific reflexivity on one of the icons of the "new psychology": a phenomenon called the reflex arc. Embodying the ideal of measurement and observation that was taking hold across the human sciences at the end of the nineteenth century, the reflex arc stood in for decades of laboratory attention to the precise moment when the mind reacted to an external stimulus. What James derisively called "the *elements* of the mental life"—the mental states that associationists and others were removing "from the gross results in which they are embedded, and as far as possible reducing them to quantitative scales"—were points of pride for laboratory psychologists. This was, ironically, the world James had helped bring into being in the 1880s and in which Dewey had been trained. But like James, Dewey wanted to complicate this emphasis on the mechanical and the objective, the separable and abstract, that had consumed the field. So by attacking the very idea of the reflex arc, insisting that it was complex and organic rather than simple or at least simplifiable, Dewey was going after the heart of what made psychology seem like a (potential) science to so many around him.[30]

Doing so meant wading into a standing debate over what exactly reaction times were. On one side was Edward B. Titchener, the champion of what he called "structuralism." Titchener's goal was to describe what he saw as the universal features that underlay all cognition, or what he referred to as "structures." His preferred method was introspective, though not (he insisted) in the old sense of the term. Titchener's introspection was "objective" and "experimental," careful self-assessment that was centered on the "simple reaction-time," or the measurable interval between a sense impression and a movement it elicited. Against Titchener's structuralism, a set of avowed "functionalists" emphasized the "how" over the "what" of mental life. While Titchener saw the reaction-time as a universal unit, his antagonist James Mark Baldwin claimed he had broken it down into different "types." In other words: reflexes were not universal. The dispute was arcane, but at root it was methodological: "The reaction-time experiment," said Baldwin, "becomes of use mainly as a *method*." Their disagreement boiled down to "the method of science in general," what Baldwin called science's "machinery." According to Baldwin, Titchener had some outdated machinery; it had blinded him to the fact that science needed to evolve.[31]

Functionalism, according to its proponents, was the next stage of that evolution. Proving it so fell not to Baldwin but rather to a psychologist at the University of Chicago named James Rowland Angell. Along with his co-author, Addison W. Moore, Angell sought to "combine and reconcile some of the principle contentions of both sides of the 'type' discussion" in the form of what they called a "dynamo-genetic" view. They sought, in other words, to turn functions into a set of explicitly evolutionary capacities. "Taking the simple reaction as the type of voluntary action in general," they wrote, "and voluntary action as action under the direction of attention, it seemed that the key to any explanation adequate to all the facts, the individual peculiarities and the effects of practice, must be found in the functions of attention and habit in their relations to each other." The difference between stimulus and response was "not one of *content,* the stimulus being identified with the ear, the response with the hand, but one of *function,* and both offices belong equally to each organ." Attention had been redefined "as the adjustor, the mediator" of embodied cognition.[32]

The dispute over reactions was soon a clash between competing "schools": structuralists and functionalists. As so often happened, their differences boiled down to methods of study, not theories of mind. For Titchener, structuralism was about preserving introspection as a valid way of reading minds. In his laboratory at Cornell, and those on which it was based at Leipzig, this meant continuing to rely on interactions between experimental subjects who registered internal experiences using instruments in pursuit of objective measurement. While nothing was stopping the functionalists from introspecting, they tended to observe the behavior of others rather than their own minds. This distinction became clearest among animal psychologists, who were soon overwhelmingly functionalist in their approach. On Titchener's view, animals could never reveal the structures of the *human* mind—hence the continued reliance on introspection. For Morgan and others, nonhuman animals were as legitimate as colleagues for the analysis of functions in the laboratory. The rapid rise of comparative psychology around 1900 went hand in hand with the success of functionalist approaches to the mind, not least because Titchener's structuralism was predicated on a practice—introspection—that quickly became impracticable.[33]

Yet, while the debate grew into something of a scandal for the new field, few observers recognized how the acrimonious rhetoric of Titchener, Baldwin, Moore, and Angell papered over a fundamental agreement: that there *was* a reaction-time, a measurable and meaningful unit, and that psychology could be based on quantifying its existence in the appropriate way. Structuralists and functionalists both claimed the label "experimental" for their work; both groups emphasized calibration and control. One, the structuralists, directed these ideals inward toward the self; the other, the functionalists, looked outward, to other minds. While they disagreed about where to look, they agreed that the objects of one another's studies *did exist.* Reaction-times, as Angell and Moore insisted in their effort to combine the two approaches, had structural and functional aspects. Whether you came from one school or the other, you agreed that psychologists studied a set of interlocking phenomena that could be visualized in much the way that physiological phenomena had been studied by the scientists who helped train them. What looked, on one level, like an acrimonious dispute, was actually a fundamental agreement on another.

Or so it seemed. Angell and Moore's diplomatic efforts were not the last words in the fight over the "new psychology," nor was their essay the founding document in functionalism as it came to be understood over the next decade. That honor falls to two of Angell's mentors: James, under whom he did his doctoral research at Harvard, and Dewey, who had first pointed him toward James as an undergraduate and who was now his colleague at the University of Chicago. The impact of James's *Principles* and Dewey's early articles was vast and varied, but contemporaries tended to assign pride of place to their respective roles in starting and solidifying functionalism as the main school of American psychology by the end of the century. Angell and Moore, for example, credited Dewey and another colleague with their initial inspiration: "we are indebted," they confessed in a footnote, "to Professors Dewey and G.H. Mead, for suggestions without which the following interpretations would not have been reached." And, if their essay articulated a halfway point between structuralism and functionalism, it was Dewey himself who pushed the field into functionalism for good. In the same year Angell and Moore published their landmark essay, Dewey brought out a piece that soon became the school's founding document.[34]

"The Reflex Arc Concept in Psychology," published in 1896, remains Dewey's most famous essay as a psychologist. The piece was also foundational to his later work in other fields. In it, Dewey argued against his own title—and, in doing so, spoiled the alliance his students had tried to forge that same year. His argument was simple: "the reflex arc idea, as commonly employed, is defective in that it assumes sensory stimulus and motor response as distinct psychical existences, while in reality they are always inside a coordination." In other words, "stimulus" and "response" were not so much *separate*—as most psychologists had assumed, even in the recent studies of reaction-times by his friends and students—as *separable.* It was scientific convention that held the two apart. In the real world, stimulus and response were "functions" of a larger process: "It is the coordination which unifies that which the reflex arc concept gives us only in disjointed fragments. It is the circuit within which fall distinctions of stimulus and response as functional phases of its own mediation of completion. The point of this story is in its application." But that application, Dewey concluded, would have to wait for "a more favorable opportunity" to come around. For now, functionalism was largely reactionary, a set of claims about how other approaches to the mind fell short.[35]

Readers would not have to wait long, however. Dewey's critique of the reflex arc was just the beginning, and he and his students soon began to march on behalf of functionalism in all sorts of venues. Two years later, when he published his essay "Philosophical Conceptions and Practical Results," James brought their approach to a wider audience. Angell wrote immediately to say how much he liked the attention—and in doing so, pulled James into the functionalist fold: "I was greatly interested in your presentation of Peirce's notion of 'pragmatism,' for it fits very closely with the life-long bias of my own untutored thinking and it is in many respects surprisingly like what Dewey is driving at and upon this I was of course more or less brought up." Angell had in mind the work Dewey put into his "Reflex Arc" paper, as well as his own on the coordination of habit and attention. A few years earlier, this argument would have surprised James. He had found Dewey's *Psychology*—a textbook published in 1887, during the high tide of his Hegelianism—"a great disappointment," for example. Their early exchanges had consisted largely of Dewey's (failed) attempts to convert James to Hegel. Given James's well-known scorn for Hegelian notions, Dewey had to learn to suppress it before James took him seriously.[36]

And that is precisely what happened. By the time of the "Reflex Arc" article, James was primed to see the connections Angell implied in his letter. And when, a few years later, Dewey and his students and colleagues at Chicago published a collection of essays under the title *Studies in Logical Theory,* James understood it to be a sign of the times. Endorsing the book in the *Psychological Review,* James wrote: "Chicago has a School of Thought!—a school of thought which, it is safe to predict, will figure in literature as the School of Chicago for twenty-five years to come." Though he never named it as such, James's summary made clear that he saw Dewey's work as a *functionalist* critique of the cleavage between thinking and action, which is just what Angell implied when he wrote him a few years earlier. James's writings had moved in the same direction, both in what he called his "radical empiricism"—a way of breaking down the barriers between mental and physical phenomena—and in ongoing research into occult phenomena under the umbrella of "spiritualism." James was the consummate boundary-crosser, searching for the limits of the self across lines laid by the guardians of discipline and propriety alike.[37]

To James, functionalism was a matter of method. He shared with Dewey an evolutionary view of the mind that informed a new way of thinking about science. After all, if humans are just one more organism doing its best, then what kind of warrant did their claims really have? It was in responding to this dilemma, James thought, that his view came closest to Dewey. Both, he argued, were functionalists, and for them "a fact and a theory have not different natures, as is usually supposed, the one being objective, the other subjective. They are both made of the same material, experience-material namely, and their difference relates to their way of functioning solely. What is fact for one epoch, or for one inquirer, is theory for another epoch or another inquirer. It is 'fact' when it functions steadily; it is 'theory' when we hesitate." James and Dewey both used psychology to redefine fundamental scientific categories. Most famous (or infamous) was a new definition of truth, most closely associated with James but which he credited first to Peirce and then, in his review, to Dewey: "'Truth' is thus in process of formation like all other things." It was the task of the psychologist and the philosopher alike to catch truth in the act of becoming. Doing so required a different sort of "School."[38]

WORKING HYPOTHESES

Dewey knew just how radical it was to critique the reflex arc in the 1890s. After all, psychology was already a mess—especially in the United States. Competing approaches were pulling it in opposite directions, with practitioners increasingly torn between "pure" and "applied" methods in their shared field. Theoretical and methodological chaos called out for something to stitch it all together. Dewey recognized this need at the start of his "Reflex Arc" paper—only to carry on with his radical critique. "That the greater demand for a unifying principle and controlling working hypothesis in psychology should come at just the time when all generalizations and classifications are most questioned and questionable is natural enough," he wrote, adding: "The idea of the reflex arc has upon the whole come nearer to meeting this demand for a general working hypothesis than any other single concept." So why deconstruct a concept with so much riding on it? Because the reflex arc failed to live up to the hype: "in the idea of the sensori-motor circuit," Dewey argued, "conceptions of the nature of sensation and of action derived from the nominally displaced psychology are still in control." The problem was not that proponents of the reflex arc had gone too far in basing the "new psychology" on it; the problem was that they, and the concept, had not gone far enough. The field desperately needed a new foundation.[39]

This search for a new working hypothesis began in an unlikely way: a chance encounter with a "newspaper man." While on the faculty in Ann Arbor, Dewey met a radical editor named Franklin Ford who had recently moved back to his home state of Michigan from New York City. To Dewey, Ford represented the prospect of turning philosophical and psychological theories to account in the real world. While Dewey was drawn to idealistic philosophy out of "some sort of instinct," Ford had been "led by his newspaper experience to study, as a practical question the social bearings of intelligence & its distribution [and] had found idealism." In other words, the two men had found their way to one another, and to the nexus of theory and practice, by moving in opposite directions. Ford held Dewey in thrall. Dewey's letters in these years were full of allusions to Ford: to his personality, his energy, and above all his orientation toward practical social problems. Dewey was so taken by his new friend that, in a letter to James, he vented an almost millenarian optimism: "I believe that a tremendous movement is impending when the intellectual forces which have been gathering since the Renascence [*sic*] & Reformation shall demand complete free movement, and, by getting their physical leverage in the telegraph & printing press, shall through free inquiry in a centralized way, demand the authority of all other so-called authorities." This focus on authority would have been familiar to James, but Dewey had found a figure in whom he felt it was personified: Franklin Ford.[40]

Dewey's enthusiasm culminated in a strange project. Called *Thought News,* it was an abortive attempt to turn Ford's focus on philosophy's practical side into something tangible—into, as Dewey put it to James, "inquiry as a *business.*" *Thought News* was supposed to be just what it sounded like: a periodical that turned the thinking process into a commercial product. According to a write-up in Michigan's *University Record,* the goal was to use "philosophical ideas in interpreting typical phases of current life." Despite these ambitions, however, *Thought News* never materialized. It was satirized in the press almost as soon as it was announced, and no issues ever appeared. Today, we have very little evidence of the project or its demise, but what we do have suggests that, either before or during its early collapse, something soured between Dewey and Ford. Dewey called Ford a "scoundrel," and seems to have lost track of him soon after their plan fell through. Decades later, he would remember *Thought News* as "an over-enthusiastic project," a symptom of youthful yearning within the constraints of the academic life he was living. To the relief of James and others, Dewey soon turned away from the organic idealism that he had desperately attempted to wed to the commercial interests Ford represented. As Dewey would have put it, his thinking evolved.[41]

But like his youthful Hegelianism, Dewey's dalliance with Ford left more than a mark on his mind going forward. Even if his aim to turn philosophy into a business (of a sort) was never fully realized, Dewey whet his appetite for the kind of practical, public-oriented projects that would become a defining feature of both his philosophy and his career. What he saw in Ford, beneath the grandeur of expectations and the need that drove him to them, was the prospect of making our ideas matter—not just in the sense of secure employment, or even changing the minds of students, but in a broader political sense. Ford may have been a phantom, but his commitment to turning ideas into "intelligence" by putting them to work struck a chord with Dewey and some of his colleagues in Ann Arbor—so much so that after *Thought News* evaporated, he and other members of the philosophy department were on the lookout for another opportunity to have an impact on the wider world. To them, this meant more than spreading the results of their research beyond the academy or testing hypotheses in the real world. It meant *making* those theories out there, adapting to the circumstances of application. This was what Dewey meant by a working hypothesis: he wanted to find one "out there," at work.[42]

And this is precisely what Dewey did—not in Ann Arbor, but in Chicago, where he soon joined a new university endowed by John D. Rockefeller and built on land donated by Marshall Field. Riding the wave of university reform that spawned Johns Hopkins and swept up Harvard and other older colleges, the University of Chicago was soon producing more PhDs in the sciences than any other American university. Thanks to the ambition of William Rainey Harper, its first president, the school quickly assembled a word-class faculty, largely by poaching from peer institutions. Harper's primary target was another new university: Clark, where Stanley Hall lost a huge proportion of the professors he had painstakingly assembled in what was soon called "Harper's Raid." Just as crucial as this recruitment binge to the shape of the new university was the specificity of its setting on Chicago's South Side at the tumultuous close of the nineteenth century. Torn apart by labor unrest and enlivened by a social reform movement, both of which were responding to rapid changes in industrial society, the city was primed for just the kind of engaged scholarship the abortive *Thought News* project was meant to embody. From Eugene Debs and the Pullman Strike to Jane Addams and the settlement house movement, Chicago and its new university were an ideal setting for rethinking the relationship between science and society.[43]

This is where Dewey set up shop—or rather, school. For it was the chance to work with kids that drew Dewey to Chicago. He had tried his hand at child study already, first with Hall at Hopkins and then, off and on, during his time at Michigan. But it was the prospect of expanding this area of focus, in addition to the desire to reach a wider audience (as exemplified in the Ford episode), that had his attention at the start of the 1890s. Just as Hall learned in his move from Hopkins to Clark, the ability to control the administration of one's own research and to trespass at will across the boundaries between various sub-fields was more than a luxury for researchers of Hall or Dewey's temperament. It was a necessity, if one was to control the application of one's own work to the areas in which it might have an effect. In a popular essay published just before he left Michigan, Dewey argued that the enforced separation of theory from practice was causing a "chaos in moral training" that he noticed in the undergraduates he taught: "Here as elsewhere our greatest need is to make our theories submit to the test of practice, to experimental verification, and, at the same time, make our practice scientific—make it the embodiment of the most reasonable ideas we can reach. The ultimate test of the efficacy of any movement or method is the equal and continuous hold which it keeps upon both sides of this truth." The prospect of such a testing ground led Dewey to leave Michigan. When Harper's offer came, he later recalled, "one of the factors leading to its acceptance was the inclusion of Pedagogy in the department with Philosophy and Psychology." The "test of practice" would be education.[44]

Dewey arrived at the University of Chicago in the fall of 1894, leaving his family back in Ann Arbor so he could get the lay of the land. Significantly, part of settling in meant arranging for his children to start school in Chicago, something that proved more difficult than he had imagined. Seeking a solution to this most practical of dilemmas, a new idea occurred to Dewey. "There is an image of a school growing up in my mind," he wrote to his wife Alice, ". . . the material & methods for such a school all exist now lying round loose in scattered form." The image was a "constructive" one, in which students would learn by completing manual tasks that would ground "a social training on the one side and a scientific on the other." The "material & methods" to which Dewey referred were, in one sense, embodied in an ongoing education reform movement now centered in the nearby Cook County Normal School, headed by Francis Wayland Parker, who had already made his name as a reformer. In another sense, however, what Dewey meant by "material & methods" were the techniques and results of child study in particular and the "new psychology" in general. In his hands, pedagogy would become a human science.[45]

Part of merging pedagogy together with psychology and philosophy was the emphasis that all three placed on practice, or what Dewey more often called "experimentation." Pitching his idea to Harper, Dewey proposed it as a "complete experimental school" for a reason: the term signaled a scientific mission, one suited to the university's goal of preeminence in an era already dominated by the rhetoric of the laboratory and experimental progress. In addition to securing an education for his children, Dewey was carving out prestigious space for himself in the university and in the broader academic community. Having a school at hand would help in his mission of developing working hypotheses for the field of philosophy. It might not have been quite what he imagined with Ford for *Thought News,* but the rhetoric of experiment clearly signaled something crucial. Later in the letter to his wife, Dewey made the stakes of the matter clear: "The school is the one form of social life which is abstracted & under control—which is directly experimental, and if philosophy is ever to be an experimental science, the construction of a school is its starting point." Philosophy, psychology, and pedagogy were converging on experimentalism, just like Dewey was. Experiments became how ideas were set to work in the world.[46]

Indeed, experimentalism was more than that: it was *changing* the world, turning every aspect of it into a laboratory. This transformation went in both directions. As Rebecca Lemov has shown, psychologists in the twentieth century succeeded in conquering new terrain for their work: beyond schools and colleges, they gathered data in stores and from polls, calling subjects and going out to find them where they were. The world was a laboratory in the familiar sense, then, with scientists fanning out to collect evidence and turn it into publications. But the other direction held as well: ordinary Americans began to see aspects of their daily lives in terms set by, or tied to, the research conducted by those psychologists. The rise of "experimentalism" as an ideal happened all over the place: in government and literature, in parenting and art. While the expansive sense of science in Dewey's work did not prevail in all these applications, the meaning of experiment was not uniformly reduced to white coats and sterile laboratories. Something of the buoyant, boisterous ideal of experimental living that Dewey identified with children lived on in modernist art and poetry, even if the scientific ideals to which such ventures pointed would be replaced by a more static image of experimental practice. Dewey's pedagogical work was part of a much broader, deeply cultural movement.[47]

Late in 1895 Dewey signed Clara Mitchell on as the school's head teacher. Mitchell was an instructor at the Cook County Normal School, which was soon to be folded into the university (as were other local institutions, part of Harper's consolidation of intellectual and political power in the city more generally). The school opened in January, with a dozen students (including, of course, Dewey's children) and Mitchell as the head teacher. Dewey first described the school as a "laboratory" in the context of recruiting Mitchell, telegraphing his interest in developing methods of effective teaching. Later, he gave the credit for calling it "The Laboratory School" to Ella Flagg Young, another reformer whom Dewey recruited to the cause. Young was a superintendent in the Chicago school system in the 1890s and, in her fifties, became a star student in Dewey's early seminars at the university. After resigning from the schools in 1899, Young accepted an offer from Dewey to supervise instruction at his school. Both Young and Dewey saw the rhetoric of the laboratory as useful, not just in their negotiations with Harper over the school's status, but also as a means of attracting students (that is, students' parents) and teachers to the project.[48]

"Dewey's School" (as it was known) was a laboratory in three senses. The first is the most familiar today: students at the school learned by doing—specifically by doing experiments. This was implicit in Dewey's early focus on construction, which paired mental and manual training. Students learned chemistry through cooking, for example, with "eggs serving as the material of experiment." For Dewey, experiments elicited something natural from within the children themselves: "For the child simply to desire to cook an egg, and accordingly drop it in water for three minutes, and take it out when he is told, is not educative. But for the child to realize his own impulse by recognizing the facts, materials and conditions involved, and then to regulate his impulse through that recognition, is educative." Self-direction was key to Dewey's definition of experiment: if a task involved no sense of purpose or instinctive attraction, it would not suffice. Not a single fact should be taught "except as the child sees that fact entering into and modifying his own acts and relationships." Experimentation, from kids cooking eggs to the discoveries of eminent physicists, was rooted in a reflective impulse, one a proper education in scientific thinking was designed to elicit from every student.[49]

The second sense of the laboratory was predicated on this connection: teachers, too, were supposed to be spontaneous, creative, and adaptive. After all, teachers were human; this way of thinking was supposed to come naturally. "The school," Dewey wrote early on, "has two sides, which of course are the obverse and reverse of the other—the one for the children, the other for the students in the University taking up pedagogical work." The line between student and teacher was a blurry one. When leading groups of eight-year-olds, the adults in the classroom were the teachers; when reflecting on what they had done and imagining ways to do it better, they were students. This blurriness started in classroom instruction, the methods of which Dewey saw as "ultimately reducible to the question of the order of development of the child's powers and interests." As a specific version of "child-centered teaching," this approach looked not to the whims of this or that child, but to child psychology, to guide pedagogy. "The law for presenting and treating material," Dewey wrote in *My Pedagogic Creed,* "is the law implicit within the child's own nature." Psychology was how that nature was elucidated; so, once again, prescription emerged from description. Because children experimented, and because they did so naturally, the teachers had to be experimental, too. Soon, the thinking went, it would be second nature.[50]

Like his redefinition of the reflex arc, Dewey's reorientation of the methods of teaching was predicated on mental unity. As he put it elsewhere, in an account of the relationship between children's imaginations and their ability to express them: "there are not two sides to the child, an image and its expression; the image is only in its expression, the expression is only the image moving, vitalizing itself." As with students' minds, so with the teachers': the task was reflective adaptation, learning to teach by engaging experimentally with students themselves. Training teachers without this kind of experience, without what he called "practical exhibition and testing" was, as Dewey told Harper in a pitch for the project, like "professing to give thorough training in a science and then neglecting to provide a laboratory for faculty and students to work in." Just as chemistry students learned in the laboratory, so pedagogy students—future teachers—had to experiment firsthand. Like the lessons they taught, classroom practices were "worked out by the teachers themselves cooperatively, with considerable use of the trial-and-error method." Everyone's mind followed the same patterns.[51]

Dewey's third sense of laboratory was the original one he proposed to Harper: the school was *his* laboratory, a place to bring ideas "to the test of practice." Dewey made this clear in early promotional efforts. "The conception underlying the school is that of a laboratory," Dewey wrote in the *University Record:* "It bears the same relation to the work in pedagogy that a laboratory bears to biology, physics, or chemistry." When he called it "a pedagogical experiment"—he meant it. Even the school's physical aspects were experimental in this sense. In his first budget report, for example, under "Permanent Equipment," Dewey crossed out "Library"—and wrote "Laboratory." As the century ended, Dewey's research reflected his engagement with students and teachers at the school more and more. *My Pedagogic Creed* and *The School and Society* were based on his work there, and both pointed toward integrating the psychological research in the "Reflex Arc" paper with advocacy of what would come to be called "progressive education" at the beginning of the next decade. The classroom was a laboratory, then, in the same sense—and for the same reasons—that Bruno Latour invoked when he wrote "Give me a Laboratory and I Will Raise the World." It had power—new, but real—beyond its walls.[52]

Dewey shared this sense of the laboratory's power with his friend Jane Addams, then doing her own experiments at Hull House in Chicago. While she rejected the name "sociology laboratory" for Hull House "because a Settlement should be something much more human and spontaneous than such a phrase connotes," she called her work "an experimental effort to aid in the solution of the social and industrial problems which are engendered by the modern conditions of life in a great city." Addams's projects, including with the Working People's Social Science Club, were all experimental: "The one thing to be dreaded in the Settlement is that it lose its flexibility, its power of quick adaptation, its readiness to change its methods as its environment may demand. It must be open to conviction and must have a deep and abiding sense of tolerance. It must be hospitable and ready for experiment. It should demand from its residents a scientific patience in the accumulation of facts and the steady holding of their sympathies as one of the best instruments for that accumulation." Like Dewey's school, Hull House was an organic whole. According to Addams, the idea emerged from "the desire to make the entire social organism democratic, to extend democracy beyond its political expression"—which is what Dewey was after as well. Science's experimental ethos, understood as spontaneous and natural, was the key not just to imagining a just and democratic society, but to making one possible.[53]

Addams and Dewey were joined in their "spontaneous" reform by George Herbert Mead, a former student who came to Chicago with Dewey in the 1890s. Mead, a founder of sociology and social psychology, did more than anyone to push Dewey's evolutionary theory from the individual to the social sphere. Active in local politics and the labor movement, Mead sought to fuse political and academic projects, "to establish the theory of social reform among the inductive sciences." To do this, he thought both science and society needed to be redefined as *hypothetical* pursuits. "In the social world we must recognize the working hypothesis as the form into which all theories must be cast as completely as in the natural sciences," he wrote, adding that "this is the attitude of the scientist in the laboratory, whether his work remains purely scientific or is applied immediately to conduct. His foresight does not go beyond the testing of his hypothesis." Method marked the way forward for a union of science and social reform, so long as the working hypothesis was central. In a way, Mead's methodological commitment closed a loop: a century before, hypotheses had been all but forbidden in the tradition to which he was an heir. By the eve of the twentieth century, he was making hypotheses not only central to science, but the key to social productivity and political activism.[54]

"The settlement is practical in its attitude," Mead wrote, "but inquiring and scientific in its method." Recognizing the scope of hypotheses, their power as well as their limitations, was already a key component of the experimental work being done by Addams, Dewey, and others. But using the scientific method in these social contexts entailed an extra complication: scientific studies of society *change* society, which in turn changes science. What Mead called "reflective consciousness" in the scientific method started a feedback loop through which science intervenes directly in its objects: "Reflective consciousness does not then carry us on to the world that is to be, but puts our own thought and endeavor into the very process of evolution, and evolution within consciousness that has become reflective has the advantage over other evolution in that the form does not tend to perpetuate himself as he is, but identifies himself with the process of development." The result was "an identification of our effort with the problem that presents itself," culminating in "the recognition and use of scientific method and control." Everything, from everyday life to social movements, led back to science—redefined as an evolved tendency inside all of us, a natural feature of adaptive, embodied cognition.[55]

What Addams discovered at Hull House and Mead found at labor meetings, Dewey learned in the classroom. For all three, science was something spontaneous, and it certainly was not limited to its familiar manifestations in the botanical garden or the chemistry laboratory. It was a quotidian practice, and not just that: it was a requirement for living in the modern world. What Addams would call a "subjective necessity" and Mead might dub the ultimate "working hypothesis" was the culmination of decades of thinking scientifically *about science,* about where everyday behaviors ended and scientific investigation began. Significantly, it was not enough for any of the three to say that these broader phenomena reduced to what went on in the mind. Addams, for example, did not oppose the "subjective necessity" of her famous piece on settlement houses to their "objective necessity" from a political perspective. The two needs were the same, individual and social sides of a single problem confronted in modern society. For Addams, as for her colleagues in Chicago, science was an attitude of mind geared toward fusing the individual and the social, toward solving political problems in a democratic way.[56]

At its root, scientific method *was* cognition. Unfolding at the intersection of evolution and experiment, of natural selection and hypothesis testing, it was indistinguishable from the way problems of survival were "solved" well beyond the laboratory—or beyond the human. "The experimental method is the method of making, of following the history of production," wrote Dewey. "And, as already intimated, the conception of evolution is no more the discovery of a general law of life than it is the generalization of all scientific method." This was both a natural history of science and a manual for living. "Democracy is possible only because of a change in intellectual conditions. It implies tools for getting at truth in detail, and day by day, as we go along." Evolutionary theory was the change in conditions Dewey had in mind, and psychology was the best tool for getting at it. Where, decades earlier, Darwin had forged an evolutionary theory out of an experimental ideal, Dewey now justified a broad experimental philosophy as the fruit of mental evolution. Together, mind and method were supposed to start a new era in politics and pedagogy. Or at least, that was the idea.[57]

HOW WE THINK

"Scientific method," Dewey told an audience of scientists in 1910, "is not just a method which it has been found profitable to pursue in this or that abstruse subject for purely technical reasons." Their subject was much bigger than that. "It is not a peculiar development of thinking for highly specialized ends; it *is* thinking so far as thought has become conscious of its proper ends and of the equipment indispensable for success in their pursuit." Scientific method was not just the right tool for the job—it was the only tool for any job. A few years earlier, Dewey had made the same point by illustrating the scientific method of "the plain man": "He assumes uninterrupted, free, and fluid passage from ordinary experience to abstract thinking, from thought to fact, from things to theories and back again. Observation passes into development of hypothesis; deductive methods pass to use in description of the particular; inference passes into action with no sense of difficulty save those found in the particular task in question." As Dewey had gleaned from James, our minds are streams, not trains; they overflow their banks and are as real as the rocks you walk on. The scientific method was not just the best way to solve those problems we labeled scientific; according to Dewey, it was *how we think*, full stop.[58]

The same year Dewey told scientists that their method was everyday thinking, he gave a wider audience an accessible introduction to the same topic. Though he was now at Columbia (he had left Chicago partly over an argument about his wife's employment status), Dewey was still drawing on the experiences he had accumulated there. The book in which he did so, *How We Think,* was an effort to distil the lessons of the Laboratory School into a general account. Dewey laid out his cards in the preface: "This book represents the conviction that the needed centralizing and steadying factor is found in adopting as the end of endeavor that attitude of mind, that habit of thought, which we call scientific . . . [and] that the native and unspoiled attitude of childhood, marked by ardent curiosity, fertile imagination, and love of experimental inquiry, is near, very near, to the attitude of the scientific mind." As the ensuing pages made clear, Dewey owed this insight to the teachers and students with whom he had worked at the Laboratory School. Having drawn on his experience in many articles and books, Dewey was taking a moment to step back from technical matters to reflect on thinking in general. As always, he felt children exhibited it in its purest form.[59]

At the center of *How We Think* was what Dewey called "the analysis of a complete act of thought." Illustrated with quotidian examples—doing the dishes, calculating a journey uptown—the idea was to draw out of individual acts of reflective thought a general pattern, something like what he had done when describing students' efforts in earlier books. In *How We Think,* this took the form of a numbered list, meant to summarize the book's claims in a convenient format:

> Upon examination, each instance reveals, more or less clearly, five logically distinct steps: (*i*) a felt difficulty; (*ii*) its location and definition; (*iii*) suggestion of possible solution; (*iv*) development by reasoning of the bearings of the suggestion; (*v*) further observation and experiment leading to its acceptance or rejection; that is, the conclusion of belief or disbelief.

On the surface, there was nothing special about Dewey's list. It looked like the account of "the plain man" he had published in *Studies in Logical Theory,* with its emphasis on hypotheses and the confrontation of everyday problems. But the inclusion of numbers set the "steps" apart from one another, paving the way for chopping the list into bits. Far from a stream, thinking now seemed like a ladder, planted in difficulty and reaching toward resolution.[60]

If the numbers in Dewey's list made acts of thought seem almost algorithmic, his focus on its individual character obscured the social side of learning and science. Studying "the free play of the children's communicative instinct" had led Dewey to recognize that we learn not only by doing, but also by talking—that is, by interacting with one another in a social setting. While that setting was absent from the logical analysis of Dewey's list, the rest of *How We Think* was full of "social stimuli" and "social conditions"—the indispensable foundations of thinking. On Dewey's view, the evolutionary history of our minds was resolutely social, an adaptive response not to problems we encountered individually, but to those we grapple with as a group. Mead took this to the next level, arguing that our very sense of self, our status as individuals, had emerged from our sense of *selves*, of others. In the same year as *How We Think*, Mead put this point in stark terms: "Other selves in a social environment logically antedate the consciousness of self which introspection analyzes." Studying others was a way to study ourselves; knowing how we might behave is only useful as a clue for how others will. Evolutionary theory, which so often seems to emphasize the lonely, harried, surviving individual, in fact revealed the broader context in which any adaptation might have originally helped individuals thrive socially.[61]

Science, seen as an adaptation, was a tool for controlling others—and, only because of that, a tool of self-control. As Mead put it in another essay from the same year: "Successful social conduct brings one into a field within which a consciousness of one's own attitudes helps toward the control of the conduct of others." Though it was far from Mead's (or Dewey's) intention, the language of success and failure, the product of an evolutionary logic, entailed its own surprising results for psychologists—whether they entered the classroom or not. A sense of *why* this or that trait proved advantageous, a focus on utility in the past or the present, seemed to lead to a new push for prescriptive, rather than simply descriptive, accounts of learning and problem solving. An interest in educational issues, especially those related to the role of science in social progress, only buttressed this shift. Methodological debates had been intertwined with ideas about teaching and learning since Herschel's formulation of a reform method, but the rise of child study toward the end of the century placed a new premium on models of mind, turning descriptions of science and its method into prescriptions for how to think scientifically. This transit from is to ought, from a tacit feature of earlier debates to the avowed goal of new ones, depended upon Dewey and Mead's social vision but also helped eclipse it. In the hands of their enthusiastic followers, the embodied cognition of early evolutionary psychologists became a disembodied manual for thinking scientifically.[62]

The broader shift from social to individual and from descriptive to prescriptive helps explain the reception of Dewey's numbered list. Both the start of the process ("a felt difficulty") and its goal ("the conclusion of belief") were *meant* to fit into a broader world. But it is easy to see them instead as introspective rather than interactive. Dewey saw no barrier between the two, as the rest of the book and his life's work made clear. Yet his five steps implied otherwise. The list's staccato format made it extractable, exportable, a kind of one-size-fits-all tool of the very sort Dewey was trying to fight in the classroom. Absent the evolutionary and experimental concerns in which the list had been forged and that it was meant to signal, the thinking process it represented seemed at once narrower and more general. What had started out as "a complete act of thought," an accessible account of *How We Think,* could now be read as a set of rules for right thinking. A description of what makes us human, of how we connect to the world around us and how we might do so more successfully, could become a prescriptive ritual, a method only. And that was precisely what happened.

8

A METHOD ONLY

John Dewey's *How We Think* was short and sweet. The book was meant to introduce readers to the lessons he had learned at the Laboratory School in simple, accessible language. Its title was supposed to say it all: *How We Think* was a guide to all thinking, from small tasks to scientific theories. If Dewey was right, these were two versions of the same thing: a basic process that only made sense in light of evolution and development, linking humans to other animals and science to all the ways we and other animals behave. Equipping readers with a unified view of cognition was about explaining not just *how* but *why* we think the way we do—and how we might think better. That was what the list of steps was for: to summarize, in pithy form, a vast natural history he had seen on display in classrooms, a miniature version of evolutionary development playing out in a wide range of courses. Thinking, for Dewey and many of his followers, was never *reducible* to those steps—the steps were an outline, to be filled in with details of actual behavior. Those details were not a distraction from deeper understanding; they *were* that understanding. Dewey's list pointed toward a view of thinking as a living, breathing, embodied activity.

At least, that was the idea. But *How We Think* quickly acquired an enthusiastic audience that Dewey had not quite intended: authors in the burgeoning science textbook industry. Looking for a handy account of what made science special, these authors seized on Dewey's list as a way to unite different scientific subfields regardless of the subject matter involved. For them, finding a common thread among the sciences was a publishing problem. As historian of education John Rudolph has shown, Dewey's list helped them solve it. Rudolph is careful to point out that *How We Think* did not give rise to the "new pragmatic spirit that infused and shaped the wider cultural configurations of science and engineering in the Progressive Era"; it was, instead, "a powerfully epitomized statement of this emerging definition of scientific method." In other words, Dewey's list was primed for mass education, even if he never intended science to be seen, or taught, as a set of steps. He even renamed them "phases" to try to avoid this simplified way of reading his list—but it was too late. Dewey's steps were meant to epitomize an organic account of thinking in general, but they were quickly turned into something else: a symbol of the separation of science from everyday thinking, a talisman of scientific exceptionalism.[1]

How did that happen? Here, Dewey himself provides a clue. A year before his list appeared, Dewey participated in a celebration of the man who started it all: Charles Darwin. 1909 marked the centenary of Darwin's birth (and the fiftieth anniversary of the *On the Origin of Species*), and followers used the occasion to rethink the legacies of his theory of natural selection. In Dewey's case, this meant, once more, bringing evolution and experiment together: "The influence of Darwin upon philosophy resides in his having freed the new logic for application to mind and morals by conquering the phenomena of life. When he said of species what Galileo had said of the earth, *e pur se muove*, he emancipated once for all genetic and experimental ideas as an organon of asking questions and looking for explanations in philosophy." Natural selection was not just a genetic theory. It was an exemplar of genetic reasoning, a tool for thinking that also justified its own application across a range of contexts. And in framing the theory this way, Dewey highlighted one of its key components: adaptability, or the ease with which it could be extracted, reformulated, and applied in new areas.[2]

Commemorations and creative appropriations crowded in, so much so that all the essays and arguments threatened to occlude the man being celebrated. The years leading up to 1909 have even been called "the eclipse of Darwinism," in which enthusiasm and alternatives threatened to "out-Darwin Darwin," according to one commentator. Every field got in on the action. "For the future historian of science," wrote one participant, the year's events "must for all time serve as a land-mark indicating the present stage of development of scientific doctrine in every department of human thought where science holds sway." Psychology was no exception. Like Dewey, leading figures in the field repurposed Darwin for their own pet projects, insisting on this or that interpretation of mental evolution and the best way to study it. Mind and method, in other words, were together again, with psychologists recasting their own disputes in evolutionary language. Lloyd Morgan, for example, insisted that the scientist only "refashions and rearranges the data afforded by previous inheritance and acquisition." The scientific method was a natural extension of evolution. "The history of human progress has been mainly the history of man's higher educability," Morgan concluded, "the products of which he has projected on to his environment." Science was evolution's ultimate achievement—and, paradoxically, a way of pushing past it.[3]

These reflections on Darwin's work reveal the blueprint by which science came to seem less natural and more artificial, its method less an organic description and more an instrumental guide. If science had evolved, it had some essence that was both identifiable and, importantly, useful. But what was that essence? Baldwin, another of Darwin's celebrants, made the point bluntly: "The method of knowledge is the now familiar Darwinian procedure of 'trial and error.'" It was a tool. The rise of evolutionary psychology had turned minds into instruments "for solving problems of adjustment and truth." Minds mirrored natural selection. "The thinker," Baldwin continued, "whether working in the laboratory with things or among the products of his own imaginative thought, *tries out* hypotheses; and only by trying out hypotheses does he establish truth." Here, a loop was closed: the mind was evolutionary because evolution was mental. Stanley Hall took it even further: "The present psychological situation cries out for a new Darwin of the mind," he predicted, "who shall break the persistent spell of theoretical problems incapable of scientific solution." Such a figure would bring psychology "back to the study of human life as it is lived out," turning it into "a true natural history of the soul." If Baldwin and Hall, bitter enemies on so many fronts, could agree on this prospect, maybe "a new Darwin of the mind" really was around the corner.[4]

But it was not to be. The idea of a single method binding disciplines together proved to be the undoing of work that had once sustained it. From the shared mind posited by evolutionary theory emerged a shared method that seemed less and less natural as the twentieth century wore on. Experiment, once taken to be an evolutionary triumph, *replaced* evolution as a guide to the study of the mind. To see how, one needs to briefly track method into the areas in which it held the most sway. One was pragmatism, a philosophy once presented as "a method only," a world of means rather than ends. The other was behaviorism, an approach to psychology that, in its extreme form, erased the mind. Overlapping and sometimes mutually reinforcing, both pragmatism and behaviorism put method first—and thereby helped undo the naturalism about science that had driven the story. Both culturally and politically, the language of pragmatism and behaviorism ruled for much of the twentieth century. What seemed like a success was also an undoing. This vision of science was rule-bound rather than wild, algorithmic rather than creative. At the end of the age of methods, the image of science as embodied cognition became something both constrained and constraining, a set of steps rather than a description of lived reality. The rise of "the scientific method" was, in this sense, less a success than a tragedy.[5]

A LIVE HYPOTHESIS

Nature's method peaked in 1898. That year, pragmatism was introduced as a new philosophy of method, the end point of a century of thinking about thinking as an embodied, human activity. In a lecture at the University of California, William James introduced the term and credited it to his friend Charles Peirce—"with whose very existence as a philosopher," he admitted, "I dare say many of you are unacquainted." Calling it "the principle of practicalism" and "the principle of Peirce, the principle of pragmatism," James offered pragmatism as a means of making our beliefs matter in the world. For James, believing was not a matter of private thought; it was a public concern. Paraphrasing Peirce, he argued: "Beliefs, in short, are really rules for action; and the whole function of thinking is but one step in the production of habits of action." Earlier, he had made a similar point in a controversial lecture on "The Will to Believe"—but articulated it in the language of scientific method: "Let us give the name of *hypothesis* to anything that may be proposed to our belief; and just as the electricians speak of live and dead wires, let us speak of any hypothesis as either *live* or *dead*. A live hypothesis is one that appeals as a real possibility to him to whom it is proposed." Significantly, "real possibility" is not just imagined: a person's belief is "measured by his willingness to act." In an important sense, belief *is* a person's action. Hypotheses are what we live by; they are the stuff of life and death. Belief culminated in method, James argued. In science and beyond, method was how our minds changed the world.[6]

When, a decade later, James expanded "the principle of pragmatism" into a series of lectures at the Lowell Institute in Boston, he underscored these connections between mind and method. Having sparred with Peirce since taking their idea public, he chose to subtitle the book that resulted *A New Name for Some Old Ways of Thinking* as a nod to his own interpretation of its lineage. Rather than trace it to Kant and post-Kantian philosophy, as Peirce did by that time, James wanted to ground pragmatism in British empiricism, starting with the book's dedication "to the memory of John Stuart Mill, from whom I first learned the pragmatic openness of mind and whom my fancy likes to picture as our leader were he alive to-day." Given James's clear reservations about Mill's theory of mind (specifically, the associationism he adapted from his father), he obviously meant something else by claiming Mill as the movement's leader. What he meant was method, specifically a means of balancing competing demands against one another. James called the need to strike such a balance "the present dilemma in philosophy," a "clash of human temperaments" between rationalists and empiricists, or those he deemed "the tender-minded" and "the tough-minded." Mill could help him cross the chasm between them.[7]

Most people wanted both sides, including James. "You want a system that will combine both things," he told his audience, "a scientific loyalty to facts and willingness to take account of them, the spirit of adaptation and accommodation, in short, but also the old confidence in human values and the resultant spontaneity, whether of the religious or of the romantic type." It was a dilemma—but James had an answer. "I offer the oddly-named thing pragmatism as a philosophy that can satisfy both kinds of demand. It can remain religious like the rationalisms, but at the same time, like the empiricisms, it can preserve the richest intimacy with facts." And it was here that he had Mill in mind. For Mill's own efforts, from his essays on Bentham and Coleridge to his mature theory of democratic governance, attempted to unite the Enlightenment ideals of his early education with the Romantic yearning in which he found solace when those ideals failed. Crediting Mill's "pragmatic openness of mind" was not a nod to his *System of Logic,* with its classification of how we think and systematization of costs and benefits. Instead, James meant the Mill of *On Liberty,* whose balance of competing political desires in a private equilibrium of mental states was essential to a functioning collective government.[8]

For the pragmatist, this was a mental give-and-take; each of us had to balance belief in a hypothesis against the willingness to test it. Individuals reflected society: in both, the goal was reaching an equilibrium. For James, nothing less than the truth was at stake: "Any idea that will carry us prosperously from any one part of our experience to any other part, linking things satisfactorily, working securely, simplifying, saving labor; is true for just so much, true in so far forth, true *instrumentally.*" Pragmatism, later conceived as a "mediator and reconciler," was to be a neutral party in the comparison of results. "She has in fact no prejudices whatever, no obstructive dogmas, no rigid canons of what shall count as proof. She is completely genial. She will entertain any hypothesis, she will consider any evidence." Given James's ongoing search for a scientific defense of religious belief, "her" capacity for reconciliation makes a lot of sense. But this adaptability was also a constraint: an apolitical pragmatism cannot defend itself against uses with which its proponents might disagree. Solving problems is not the same as choosing which problems to solve—and the latter, especially, is where politics come in. An apolitical method may have been desirable in 1906, but the inability to defend pragmatism or define its aims would become a liability for James and his followers.[9]

The trouble began almost immediately. On the tenth anniversary of James's Berkeley lecture and a year after *Pragmatism* appeared, the philosopher Arthur Lovejoy subjected the existing literature to an exhaustive analysis, listing (at least) thirteen definitions of pragmatism—which he saw as a fatal strike against its coherence. Turning one of James's metaphors against him, Lovejoy boiled down the pragmatic theory of truth (one of the thirteen pragmatisms) to "a system of deferred payments; it gets no cash down; and it is also a rule of this kind of finance that when the payments are finally made, they are always made in outlawed currency." This was a temporal problem: If truth is determined after the fact, then what use—in a pragmatic sense—is the idea itself, if we can only be sure whether it is true when it is too late? James had tried to address this issue in "The Will to Believe" by defining beliefs as hypotheses we test by living. And Dewey, in the very next issue, returned to the same point: pragmatism owed its method to "the dominant influence of experimental science: the method of treating conceptions, theories, etc., as working hypotheses, as directors for certain experiments and experimental observations." He even tried to connect James to Peirce, the estranged godfather of the pragmatic method, on these grounds. "Pragmatism as attitude," he wrote, "represents what Mr. Peirce has happily termed 'the laboratory habit of mind' extended into every area where inquiry may fruitfully be carried on." For Dewey, the question was not whether pragmatism had the right method—it was why it had taken philosophers so long to adopt it.[10]

Like Lovejoy, Dewey forecast pragmatism's fate in financial terms. "After a period of pools and mergers, the tendency is to return to the advantages of individual effort and responsibility," he wrote, adding: "Possibly 'pragmatism' as a holding company for allied, yet separate interests and problems, might be dissolved and revert to its original constituents." And dissolution is exactly what happened. While Lovejoy's critique was not fatal, James did not have long to fend it off: he died in 1910, from a long-standing heart ailment. In his absence, Dewey and others struggled to keep the "holding company" together in the face of an increasingly united front of enemies. Allies like psychologists and philosophers turned to their own disciplines, soon abandoning the territory carved out between their fields for pragmatic deliberation. As has been well documented, the Great War—and Dewey's controversial support of American entry into it—further tarnished pragmatism's allure; it became easy to frame the philosophy's apolitical core as permissive of, or responsible for, tyranny. Pragmatism's identity with its method, insisted upon by James and supported by Dewey, helped to undo it: in wartime, apolitical method was recast as weakness, not flexibility. The balancing act that James had recognized in Mill and had pursued himself for so long was not suited to every age; the method of pragmatism, much like Mill's liberalism, proved helpless against the forces pulling it apart.[11]

PREDICTION AND CONTROL

Pragmatism may have failed because it did not define its own ends, but behaviorism succeeded for the same reason. The two seem like two sides of the same coin, each a defender of a shared scientific vision in its respective discipline. And indeed, the division between philosophy and psychology, pragmatism and behaviorism, had not hardened when the core texts of either was written, meaning that their central ideas were even closer than we now recognize. To see how they emerged and then split apart, it is useful to adopt a pragmatic, even behavioristic, approach to the *effects* of these texts—the ways they were implemented by followers and condemned by opponents. Doing so means taking ideas seriously through reflexive application. The first thing such an endogenous analysis reveals is that the ideal of experimentalism and obsession with what you can *observe* was more of a caricature than an actual practice. But caricatures matter, and in the case of behaviorism it was the extreme version that won out. In the hands of those wielding it increasingly far from the laboratory, behaviorism's reductionist potential ruled.[12]

Behaviorism was a radical departure, but its basic premises were not. "Psychology as the behaviorist views it," began an essay of that name, "is a purely objective experimental branch of natural science." This statement alone, by behaviorism's controversial founder John Watson, was not itself radical. Indeed, by the time it was published in 1913, it was close to dogma in the field. Even figures opposed to Watson's scientism would have agreed. James, for example, had written something similar two decades earlier. In "A Plea for Psychology as a 'Natural Science,'" James anticipated Watson: "All natural sciences," he asserted, "aim at practical prediction and control, and in none of them is this more the case than in psychology to-day." What made Watson's essay polemical, if laboratories were ascendant and field observations comparatively ignored? Almost everyone agreed that psychology was a science, sciences were practical, and a focus on practice was the point of science in the twentieth century. For pragmatists like James and behaviorists like Watson, science was so powerful that it redefined, even reconstituted the world it studied; their focus on effects produced a *world* of effects, of organisms aimed at concrete goals.[13]

What made Watson stand out was just how far he took this point and how much weight he put on his key term: behavior. Yes, animals of all sorts were analyzed in behavioral terms, and had been for many years. But such behaviors were usually treated as windows into the mind. At least, that was how James had meant it in his earlier essay. "Whatever conclusions an ultimate criticism may come to about mental states," James pled, "they form a practically admitted sort of object whose habits of coexistence and succession and relations with organic conditions form an entirely definite subject of research." Watson abandoned "mental states" entirely. Rather than try to describe the minds of others, his scientific psychology was limited to what you could observe. Animal intelligence was reduced to animal behavior, the invisible or occult qualities of inner life replaced by what could be seen and, significantly, measured. Mind was not just set aside "for the sake of practical effectiveness," as James had put it. Instead, for Watson and his followers, the mind had ceased to exist. With mental life out of the picture, experimental method reigned.[14]

This was a strange victory for method. Watsonian behaviorism was experiment taken to the extreme, a kind of pragmatism on steroids. It took the anxieties that had shaped comparative psychology and indeed the age of methods as a whole—anxieties about evidence and inference, about what was going on in the mind of the scientist as much as in the minds being studied—and turned them into an argument. The "uniformity in experimental procedure and in the method of stating results" that Watson advocated was not just a methodological claim—it slipped across the transom to become a claim about the limits of the mind. This made Watson and his approach into a dark mirror of Darwin, almost a century earlier. Where Darwin had written his anxieties about hypothesis onto the natural world, turning experiment into evolution, Watson did the same—but to wildly different effect. Far from completing the circle Darwin started, Watson's insistence on methodological absolutes shut down the naturalistic study of science Darwin had helped inaugurate. In Watson's hands, method was a matter of limits, of warding off irrelevant inferences and—in a cruel twist—demonizing hypotheses. Behavior was *everything* in a specific and powerful sense: it was all that psychologists could study, and as a result—behavior was all that there was.[15]

It was soon customary to distinguish between two types of behaviorism. The first, associated with Watson, was methodological behaviorism. For Watson and his followers, we are restricted in what we say about other minds to what we can directly observe. Only behavior was fair game: "Any other hypothesis than that which admits the independent value of behavior material, regardless of any bearing such material may have upon consciousness, will inevitably force us to the absurd position of attempting to *construct* the conscious content of the animal whose behavior we have been studying." While methodological behaviorism went much further than Morgan's Canon, it is a tame alternative to the "radical behaviorism" championed a decade or two later by the Harvard psychologist B. F. Skinner. Watson had banished the mind in favor of behavior, only to have Skinner recolonize the mind in the *name* of behavior. After that, even the invisible phenomena of mind were measurable responses to external stimuli. To Skinner, this translation "restores some kind of balance" between introspection and behaviorism; the private world of the mind was recast in the language of the laboratory. Many disagreed, but Skinner's behaviorism marked the triumph of method over mind, a grim apotheosis of the long-standing desire to see science as an adaptive method for all to use.[16]

When Watson named the volume that resulted from his graduate work *Animal Education,* he was referencing Thorndike and others who came before him while also suggesting, subtly, an advance on their approach. As intelligence gave way to learning, and as the practical applications of animal studies became more obvious and more enticing, it was a short leap to claim that rats running in a maze were "educated" just like children in classrooms or scientists in laboratories. The analogy was not perfect—and Watson did not want it to be. After all, he supplemented his observational studies with invasive surgeries in order to experiment on the process, something he would not have done on children. Blinding, maiming, and burning rats in order to test the role of the senses in the learning process, Watson's work refracted the debates over animal cruelty that swirled around him. But his approach also revealed something deeper: a capacity for the kind of detachment he had urged in his behaviorist manifesto. The seeds of methodological absolutism were sowed early, in his ability to conduct experiments on cognition without any feeling for the organism and its experience. No wonder he eventually pushed minds out of the field.[17]

Just a few years later, Watson was gone. Well, not gone, but forced out of his position at the university after starting a relationship with a student. But thanks to the traction his views had gained, Watson had a soft landing: he was immediately hired by an advertising agency to put the principles of behaviorism to work selling products. "It can be just as thrilling," he later wrote, "to watch the growth of a sales curve of a new product as to watch the learning curve of animals or men." To Watson, it was all behavior. Rats in the laboratory and consumers in the supermarket amounted to the same sort of thing: what mattered was predicting and controlling the behavior one observed. Comparison of this sort continues today, as findings from animal models suggest new avenues for research and even treatment in humans. Watson's either-or attitude about organisms may have been protesting too much, but his insistence on the lack of a difference paved the way for broader applications of psychological research. The dominance of (defensive) concerns about method over substantive theoretical issues launched what one historian calls "the methodological imperative" in twentieth-century psychology. Once launched, there was no going back. What began as a warning about how to do science subsumed everything in its path.[18]

SCIENTIFIC MANAGEMENT

By the second decade of the twentieth century, well into what we now call the Progressive era, "science" had become even more of a watchword. Pragmatism and behaviorism were far from alone in flying the banner of science, and Dewey's claims for its links with democratic society were no longer the crying in the wilderness they had seemed to Peirce in the 1880s. These years were the heyday of what Andrew Jewett has called "scientific democracy," a celebration of the ethical life of scientific thinking. But of course, the veneration of science, especially scientific control, was not always what we would recognize today as progressive. Eugenics, too, flourished in these years, as scientists and politicians alike applied theories of inheritance to communities and nations. In hindsight such ambitions invite anxiety, if not rejection, but we have to recall that the aim of remaking populations on scientific principles was entertained by people with all sorts of political views. The impulse to use science in the service of governance defined a Progressive ethos, backing up ideals of change and retrenchment at once. These years saw scientific method turned into an instrument of political and social uplift as well as a defense of the status quo in a mirror image of its multiple uses in the age of the whig interpretation of science.[19]

Only in the postwar period, and especially under the influence of Richard Hofstadter's framing of the impact of evolutionary social theory, did our image of "social Darwinism" narrow into the view of Spencerian conservatism that it invokes today. Around 1900 it was still possible to proclaim science's emancipatory promise and the mind's organic evolution in the same breath, as W. E. B. Du Bois and others frequently did. Science remained a way of life for some, a way of imagining democratic society as a meeting of minds or as an experimental program. The practice of human experimentation did not yet carry the gruesome connotations it acquired following the Holocaust and the Tuskegee Syphilis Study. For better or worse, thinking of culture and the mind in scientific terms remained the stuff of grand visions well into the twentieth century, with all the utopian ideals and dystopian realities those visions entailed. The narrowing of a scientific social vision into its worst manifestations, horrible as they have been, has had the effect of occluding the political and intellectual landscape that preceded that narrowing—including the capacious sense of science that informed so much work in the age of methods.[20]

But even as science and its applications retained an all-encompassing allure, their method was increasingly understood in static and simplified terms. Dewey's approach found new life in a range of contexts, even as they were reinvented and compressed. Medicine, for example, was being remade in these years, at least rhetorically. A landmark survey called *Medical Education in the United States and Canada* (known as the "Flexner Report," after its author), published in the same year as *How We Think,* cited Dewey as an expert on scientific method. "Science," Flexner quoted Dewey as saying, "has been taught too much as an accumulation of ready-made material, with which students are to be made familiar, not enough as a method of thinking, an attitude of mind, after the pattern of which mental habits are to be transformed." Despite this wider view of what science was capable of, Flexner is remembered for the opposite: a narrow, technical view of scientific medicine that had been coalescing in the field for some time. Flexner's report was weaponized in an ongoing war over medical education and wielded by those who saw scientific medicine as a laboratory field. Dewey's steps were treated similarly. Flexner would question this equation: "Science resides in the intellect," he wrote in 1925, "not in the instrument." But the damage was done. Science was fast being reduced to a method only.[21]

The same was true in the world of business. Frederick Winslow Taylor's *Principles of Scientific Management,* which appeared in 1911, expressed the same goal in terms of industrial efficiency. Scientific management, like scientific medicine, was predicated on the existence of a single, superior method for attaining certain ends. Both movements seemed to prioritize practice over theory, phenomena you could observe over the data of personal experience. For Taylor, as for Flexner, science was the key to progress in an industrial age: "Now, among the various methods and implements used in each element of each trade there is always one method and one implement which is quicker and better than any of the rest. And this one best method and best implement can only be discovered or developed through a scientific study and analysis of all of the methods and implements in use, together with accurate, minute, motion and time study." As Taylor contributed to the redefinition of science as a form of instrumental reason, his work made room for a new kind of expertise, based not so much on doing the work involved (a specific trade, say) as *managing* that work using the tools of experimental science. The scientific method may have been accessible to anyone in principle, but in practice it had become an elite tool of bureaucratic control. Science, once conceived as a balancing act, was shifting to one side of the balance, a counterweight to what was now conceived of as a nonscientific society.[22]

The unlikely prophet of this transformation was Walter Lippmann, a student of James's who in the 1920s clashed publicly and prominently with Dewey over the prospect of democratic scientific culture. Lippmann had begun, a decade earlier, in basic agreement with his mentor. His early writings decried mechanical routine and celebrated organic flexibility: "We have hoped for machine regularity," Lippmann wrote in 1913's *Preface to Politics,* "when we needed human initiative and leadership, when life was crying that its inventive abilities should be freed." Like Dewey, he saw Darwin's gift as his method—"the new twist he gave to science"—and like so many progressives, he saw politics as a choice about "the method by which change is to come about." *Drift and Mastery,* which Lippmann published the following year, made his choice clear: "Science is the irreconcilable foe of bogeys, and therefore, a method of laying the conflicts of the soul. It is the unfrightened, masterful and humble approach to reality—the needs of our natures and the possibilities of the world. The scientific spirit is the discipline of democracy, the escape from drift, the outlook of a free man." Science was mastery, for Lippmann. It meant "the careful application of administrative technique, the organization and education of the consumer for control, the discipline of labor for an increasing share of the management." In Lippmann, James met Taylor: science was mental, the mind was a matter of method, and method was increasingly instrumental, a tool for controlling aimless drift.[23]

A decade later, Lippmann's concerns about drift—and his faith in mastery—had become a skepticism about the public's capacity to reason. This was a new way to think about method: as an instrument for stemming the tide of popular sentiment. In *Public Opinion* and *The Phantom Public,* Lippmann dismissed an earlier faith in deliberation in favor of top-down control. Society, he argued, had grown too complex for grassroots politics. "This whole development," Lippmann wrote in 1922, "has been the work, not so much of a spontaneous creative evolution, as of blind natural selection." Scientific expertise, on this view, was managerial expertise. Dewey disagreed. In *The Public and Its Problems,* he rose to democracy's defense while acknowledging that "the public" was a contradictory, even incoherent basis for a political system. At issue was the way Lippmann divided the world in two, setting drift against mastery, nature against nurture. Dewey saw this division as artificial, not least because science was not so much a weight on one side of a balance as the balance itself. Method meant both, just as it had for Darwin. If natural selection was blind, it was also creative. For Dewey, scientific method was not how we mastered drift—it was how we lived with it. Science required a balance of drift and mastery at once.[24]

Or at least, science had once required both. By the third decade of the twentieth century, the nature of scientific thinking had traveled far from its meaning in the age of methods. At a superficial level, this transformation was not that obvious: science was more powerful than ever, in terms of both its capacity to alter the natural and social worlds and the authority it commanded as something to which people could appeal. To see how it had changed, we must look closer: at the aspects of the scientific process being invoked, how those aspects were presented in schools and before the public. At that level, and especially in the human sciences that had helped to make "the scientific method" possible, the differences were stark. What was recently a natural history of cognition now felt mechanistic; the evolutionary theory of mind that underwrote method for decades was being replaced, bit by bit, with something algorithmic. An adaptive process once seen as central to human nature was now cordoned off from everyday thinking—even potentially implementable without the human mind entirely.[25]

Today, champions and critics alike think science is in trouble. Its authority is undermined by a "replication crisis" crippling confidence from within, while deniers of scientific consensus on issues ranging from climate change to vaccine schedules have compromised its image in the wider world. Debates over the nature and causes of these problems often center on the scientific method: on whether it is biased or a cure for bias, free from politics or fundamentally political, for better or worse. But such debates tend to obscure an important historical lesson: Science has not always been "a method only," nor is method necessarily as flat as it seems. It is possible, as it was in the age of methods, to think of science as the flawed, fallible activity of some imperfect, evolving creatures *and* as a worthy, even noble pursuit. Just over a century ago, science was a politics of equilibrium, something that encompassed rather than denied political difference. The crises we now face may stem from the collapse of that way of thinking, from a new myth we tell about science and ourselves. But if history teaches us anything, it is that those myths can change. They do not do so easily, but it is possible to imagine them changing again. If they do, and if we loop back to the age of methods, science could once again be nothing more—and nothing less—than how we think.

NOTES

1. Age of Methods

1. Not all scientists agree that the scientific method does not exist—but, without weighing in on the matter, it is clear that their disagreement reinforces "the scientific method" as an object of debate. The touchstone of methodological skepticism in the history and philosophy of science remains Paul Feyerabend, *Against Method: Outline of an Anarchistic Theory of Knowledge* (Atlantic Highlands, NJ: Humanities Press, 1975). For a recent reconsideration, see Matthew J. Brown and Ian James Kidd, "Introduction: Reappraising Paul Feyerabend," in *Reappraising Feyerabend:* Special Issue of *Studies in History and Philosophy of Science Part A* 57 (June 2016): 1–8.
2. Steven Shapin, *The Scientific Revolution* (Chicago: University of Chicago Press, 1996), 95, 3–4. This book's opening line is justifiably famous: "There was no such thing as the Scientific Revolution, and this is a book about it." Like Shapin, I take seriously a myth in which most historians of science no longer believe—and, by doing so, reveal the means by which explicit attention to myths of method had a real impact on the scientific practices about which those myths were told. On the idea of methodology as "mythic speech," see John A. Schuster, "Methodologies as Mythic Structures: A Preface to the Future Historiography of Method," *Metascience* 1 / 2 (1984): 15–36.

3. John Dewey, *How We Think* (Boston: D. C. Heath and Co., 1910), 72. For Rudolph's account of the steps and their impact, see John L. Rudolph, "Epistemology for the Masses: The Origins of 'The Scientific Method' in American Schools," *History of Education Quarterly* 45, no. 3 (Fall 2005): 341–376. On the broader context of American science education, see John L. Rudolph, *How We Teach Science: What's Changed, and Why It Matters* (Cambridge, MA: Harvard University Press, 2019). My argument is intended to complement Rudolph's—showing how the history of education intersected with the rise of the human sciences, which helped make possible the methodological narrowing he describes.
4. On "methods discourse," see Jutta Schickore, *About Method: Experimenters, Snake Venom, and the History of Writing Scientifically* (Chicago: University of Chicago Press, 2017), esp. 23–27. Schickore identifies three "layers" of methods discourse that interact in scientific writing, an account I buttress in this book by insisting on the interpenetration of those layers in the case of psychology and the significance of methodological concepts in the construction of the sciences of mind.
5. These nineteenth-century theorists were not unique in their anxiety about experimental and other methodological matters. For an account of anxiety's important role in a later period, for example, see Jill Morawski, "Epistemological Dizziness in the Psychology Laboratory: Lively Subjects, Anxious Experimenters, and Experimental Relations, 1950–1970," *Isis* 106, no. 3 (September 2015): 567–597.
6. John Dewey, "Darwin's Influence upon Philosophy," *Popular Science Monthly*, July 1909, 90–98, here 93–94.
7. James alluded to this genealogy in his public introduction of pragmatism. See William James, "Philosophical Conceptions and Practical Results," *University of California Chronicle* 4 (September 1898): 287–310, esp. 307–308. A decade later he underscored pragmatism's roots again in a series of lectures—to which he gave a telling subtitle. See James, *Pragmatism: A New Name for Some Old Ways of Thinking* (New York: Longmans, Green, and Co., 1907).
8. On the relationship between the history of science and the history of knowledge in general, see Jürgen Renn, "From the History of Science to the History of Knowledge—and Back," *Centaurus* 57, no. 1 (February 2015): 37–53.
9. Thomas S. Kuhn, "The Essential Tension: Tradition and Innovation in Scientific Research," in *The Essential Tension: Selected Studies in Scientific Tradition and Change* (Chicago: University of Chicago Press, 1977), 225–239, here 227. Originally delivered in 1959, Kuhn's "Essential Tension" lecture alludes to the

larger project that was occupying him at the time: the account of "normal" and "revolutionary" science that would become his book *The Structure of Scientific Revolutions* (Chicago: University of Chicago Press, 1962). Kuhn's work in these years reflected the concerns of the age of methods, insofar as his account of science was indebted to contemporary psychological research. On Kuhn's psychological debts, see David Kaiser, "Thomas Kuhn and the Psychology of Scientific Revolution," in *Kuhn's Structure of Scientific Revolutions at Fifty: Reflections on a Science Classic,* ed. Robert J. Richards and Lorraine Daston (Chicago: University of Chicago Press, 2016), 71–95.

10. The role of affective anxiety in helping to produce slippages between theory and method in these years is discussed case-by-case in the chapters that follow. My discussion is modeled on the role of *anticipatory* affect in scientific work examined in Vincanne Adams, Michelle Murphy, and Adele E. Clarke, "Anticipation: Technoscience, Life, Affect, Temporality," *Subjectivity* 28, no. 1 (September 2009): 246–265. Anxiety, like anticipation, encodes a temporal dimension in the subjective side of scientific practice, which is only heightened by the explicit attention given to methodology by the figures examined below.

11. Postwar reforms in science education and policy were as much about consolidating authority and projecting values as they were about classroom learning and public understanding. On educational reforms as professional consolidation, see John L. Rudolph, *Scientists in the Classroom: The Cold War Reconstruction of American Science Education* (New York: Palgrave Macmillan, 2002). On science as a way to project American values, see Audra J. Wolfe, *Freedom's Laboratory: The Cold War Struggle for the Soul of Science* (Baltimore: Johns Hopkins University Press, 2018). See also Audra J. Wolfe, *Competing with the Soviets: Science, Technology, and the State in Cold War America* (Baltimore: Johns Hopkins University Press, 2013). On scientific "open-mindedness" as a political virtue, see Jamie Cohen-Cole, *The Open Mind: Cold War Politics and the Sciences of Human Nature* (Chicago: University of Chicago Press, 2014).

12. On the intersections of psychology and utility in this period, see Cathy Gere, *Pain, Pleasure, and the Greater Good: From the Panopticon to the Skinner Box and Beyond* (Chicago: University of Chicago Press, 2017). Linking utility, psychology, and evolution was not confined to the nineteenth century. The decline of cybernetics and the (second) rise of evolutionary psychology at the end of the twentieth century saw a similar efflorescence. See, for example, Donald T. Campbell, "Rationality and Utility from the Standpoint of Evolutionary Biology," *Journal of Business* 59, no. 4 (October 1986): S355–S364. On the continued role of utility and instrumentalism in science education, see John L.

Rudolph, "Inquiry, Instrumentalism, and the Public Understanding of Science," *Science Education* 89, no. 5 (September 2005): 803–821.

13. This emphasis on evolution and fallibility was echoed in Karl Popper's work on "conjectures and refutations" a few decades later. For a comprehensive account, see Popper, *Objective Knowledge: An Evolutionary Approach* (Oxford: Oxford University Press, 1972).

14. Charles S. Peirce, "Introductory Lecture on the Study of Logic," *Johns Hopkins University Circular* 2, no. 19 (November 1882): 11–12, here 11, emphasis in the original. Within a decade, Peirce described semiotics—to which his logic and pragmatism led—in the same terms. See Roberta Kevelson, *Charles S. Peirce's Method of Methods* (Amsterdam: John Benjamins, 1987). I should note that the view of logic as a theory of inquiry was not standard at the time, and Peirce's formulation influenced several versions that followed. See, for example, John Dewey, *Logic: The Theory of Inquiry* (New York: Henry Holt and Co., 1938), 9. For Peirce, fallibility was not anathema to a logical view of science: it was central to it. Here, Peirce was articulating a view of science—as a fallible human endeavor—that others in his cohort shared in broad strokes.

15. Daniel J. Wilson has narrated this rise-and-fall in terms of a half-century of attempts by American philosophers to keep up with the changing meaning of science—and indeed, to remake themselves in its image. See Daniel J. Wilson, *Science, Community, and the Transformation of American Philosophy, 1860–1930* (Chicago: University of Chicago Press, 1990). On the "crisis of confidence" in particular, see Daniel J. Wilson, "Science and the Crisis of Confidence in American Philosophy, 1870–1930," *Transactions of the Charles S. Peirce Society* 23, no. 2 (April 1987): 235–262. I aim to complement Wilson by emphasizing concrete practices in psychology and the overarching importance of methodological issues beyond philosophy, especially insofar as the "crisis" was felt in scientific fields as well.

16. On scientific languages, see Michael D. Gordin, *Scientific Babel: How Science Was Done before and after Global English* (Chicago: University of Chicago Press, 2015). On the special relationship (especially regarding empiricism and pragmatism), see Robert J. Roth, *British Empiricism and American Pragmatism: New Directions and Neglected Arguments* (New York: Fordham University Press, 1993). For an example of a mentorship network, see Laura Otis, *Müller's Lab: The Story of Jakob Henle, Theodor Schwann, Emil Du Bois-Reymond, Hermann von Helmholtz, Rudolf Virchow, Robert Remak, Ernst Haeckel, and Their Brilliant, Tormented Advisor* (Oxford: Oxford University Press, 2007). For an example of networks around publication venue, see Christopher D. Green, Ingo Feinerer, and Jeremy T. Burman, "Searching for the Structure of Early American

Psychology: Networking *Psychological Review*, 1894–1908," *History of Psychology* 18, no. 1 (February 2015): 15–31. On the analysis of networks structured around beliefs or ideas, see, for example, Andrei Boutyline and Stephen Vaisey, "Belief Network Analysis: A Relational Approach to Understanding the Structure of Attitudes," *American Journal of Sociology* 122, no. 5 (March 2017): 1371–1447. These can all be read as versions of what was once practiced as "prosopography." See Steven Shapin and Arnold Thackray, "Prosopography as a Research Tool in History of Science: The British Scientific Community 1700–1900," *History of Science* 12, no. 1 (March 1974): 1–28.

17. The literature on these transformations is vast and, at least in part, reviewed in later chapters. One recent book provides an overview of key aspects. See Ruth Barton, *The X Club: Power and Authority in Victorian Science* (Chicago: University of Chicago Press, 2018). See also Barton, "'Men of Science': Language, Identity, and Professionalization in the Mid-Victorian Scientific Community," *History of Science* 41, no. 1 (March 2003): 73–119; and Barton, "'Huxley, Lubbock, and Half a Dozen Others': Professionals and Gentlemen in the Formation of the X Club, 1851–1864," *Isis* 89, no. 3 (September 1998): 410–444. On professionalization, see especially Paul Lucier, "The Professional and the Scientist in Nineteenth-Century America," *Isis* 100, no. 4 (December 2009): 699–732. On the transformation in science's meaning, see Richard R. Yeo, *Defining Science: William Whewell, Natural Knowledge, and Public Debate in Early Victorian Britain* (Cambridge: Cambridge University Press, 1993). My argument builds on Yeo's by attending to psychology and extending method's genealogy into the twentieth century. On "scientist," see Sydney Ross, "Scientist: The Story of a Word," *Annals of Science* 18, no. 2 (June 1962): 65–85. On the rise of the scientific journal, see Alex Csiszar, *The Scientific Journal: Authorship and the Politics of Knowledge in the Nineteenth Century* (Chicago: University of Chicago Press, 2018).
18. Lorraine Daston and Peter Galison, *Objectivity* (New York: Zone Books, 2007). See also Daston and Galison, "The Image of Objectivity," *Representations*, no. 40 (Autumn 1992): 81–128. For an alternative history of the rise of objectivity in this period, see Theodore M. Porter, *Trust in Numbers: The Pursuit of Objectivity in Science and Public Life* (Princeton: Princeton University Press, 1995). Porter's disagreement with Daston and Galison is documented in Porter, "The Objective Self," *Victorian Studies* 50, no. 4 (July 2008): 641–647. I am deeply indebted to both views and have attempted to integrate them by exploring a shared politics of method that reveals the virtues and vices of objectivity in its intellectual and political dimensions.

19. There are many histories of methodology as a millennium-long philosophical problem. They begin with Ralph M. Blake, Curt J. Ducasse, and Edward H. Madden, *Theories of Scientific Method: The Renaissance through the Nineteenth Century,* ed. Edward H. Madden (Seattle: University of Washington Press, 1960). A few years later, an essay by Larry Laudan upended some of that book's assumptions, including the idea that such a long history was possible. See Laurens Laudan, "Theories of Scientific Method from Plato to Mach: A Bibliographical Review," *History of Science* 7, no. 1 (January 1968): 1–63. See also the essays collected in Larry Laudan, *Science and Hypothesis: Historical Essays on Scientific Methodology* (Dordrecht: D. Reidel, 1981). On method from a philosophical perspective, see Ronald N. Giere and Richard S. Westfall, eds., *Foundations of Scientific Method: The Nineteenth Century* (Bloomington: Indiana University Press, 1973); John A. Schuster and Richard R. Yeo, eds., *The Politics and Rhetoric of Scientific Method: Historical Studies* (Dordrecht: Reidel, 1986); Barry Gower, *Scientific Method: A Historical and Philosophical Introduction* (London: Routledge, 1996); Peter Achinstein, *Science Rules: A Historical Introduction to Scientific Methods* (Baltimore: Johns Hopkins University Press, 2004); and Steven Gimbel, *Exploring the Scientific Method: Cases and Questions* (Chicago: University of Chicago Press, 2011). On Aristotle, see G. E. R. Lloyd, *Greek Science after Aristotle* (New York: W. W. Norton, 1975). On the reinvention of ancient methodology in the Renaissance, see Walter J. Ong, *Ramus, Method, and the Decay of Dialogue: From the Art of Discourse to the Art of Reason* (Cambridge, MA: Harvard University Press, 1958). On Bacon, see Lisa Jardine, *Francis Bacon: Discovery and the Art of Discourse* (Cambridge: Cambridge University Press, 1974). See also Peter Dear, "From Truth to Disinterestedness in the Seventeenth Century," *Social Studies of Science* 22, no. 4 (November 1992): 619–631. On Descartes, see Daniel Garber, *Descartes Embodied: Reading Cartesian Philosophy through Cartesian Science* (Cambridge: Cambridge University Press, 2000). Isaac Newton's legacy is discussed in greater detail in Chapter 2.
20. My accounts of positivism's impact has benefited from a collection on its general contours in the human sciences since the nineteenth century: see George Steinmetz, ed., *The Politics of Method in the Human Sciences: Positivism and Its Epistemological Others* (Durham, NC: Duke University Press, 2005). On Bergson's impact in the United States, see Larry McGrath, "Bergson Comes to America," *Journal of the History of Ideas* 74, no. 4 (2013): 599–620.
21. On how such debates were refracted through local and regional concerns in the early nineteenth century, see Denise Phillips, *Acolytes of Nature: Defining Natural Science in Germany, 1770–1850* (Chicago: University of Chicago Press, 2012).

On the untranslatability of *Wissenschaft* in particular, see Denise Phillips, "Francis Bacon and the Germans: Stories from When 'Science' Meant 'Wissenschaft,'" *History of Science* 53, no. 4 (December 2015): 378–394. On debates over popular science later in the century, see Lynn K. Nyhart, *Modern Nature: The Rise of the Biological Perspective in Germany* (Chicago: University of Chicago Press, 2009). In a separate article, Nyhart argues for attention to methodological disputes in this period—including, in passing, the debates with which this book begins. See Nyhart, "Wissenschaft and Kunde: The General and the Special in Modern Science," *Osiris* 27, no. 1 (January 2012): 250–275.

22. In an essay focused mostly on German *Wissenschaftsideologien* but with comparative attention to the cases of France and Britain, Lorraine Daston has sketched the divergence of European academies of science during the eighteenth and (especially) nineteenth centuries, with resulting differences in how members of those societies came to understand the meaning of science and disciplines. See Daston, "The Academies and the Unity of Knowledge: The Disciplining of the Disciplines," *Differences: A Journal of Feminist Cultural Studies* 10, no. 2 (Summer 1998): 67–86.
23. Changes in the human sciences were not alone in altering science's meaning, of course. The shift from object to process mirrored broader changes in economic and social orders in the period as well. See, for example, Jonathan Levy, "Capital as Process and the History of Capitalism," *Business History Review* 91, no. 3 (Autumn 2017): 483–510.
24. Ian Hacking, "The Looping Effects of Human Kinds," in *Causal Cognition: A Multidisciplinary Debate,* ed. Dan Sperber, David Premack, and Ann James Premack (Oxford: Oxford University Press, 1995), 351–394, here 352, 367, 369. See also Ian Hacking, "Making Up People," *London Review of Books,* August 17, 2006. On Hacking's "looping" analytic and its application to later work in the human sciences, see Joel Isaac, "Tangled Loops: Theory, History, and the Human Sciences in Modern America," *Modern Intellectual History* 6, no. 2 (June 2009): 397–424. For another example of the same kind of application, see Sarah E. Igo, *The Averaged American: Surveys, Citizens, and the Making of a Mass Public* (Cambridge, MA: Harvard University Press, 2007). I am indebted to both Isaac and Igo for discussions of my treatment of looping effects.
25. On scientists as their own case studies, see Jamie Cohen-Cole, "The Reflexivity of Cognitive Science: The Scientist as Model of Human Nature," *History of the Human Sciences* 18, no. 4 (November 2005): 107–139. See also Michael Pettit, "The Con Man as Model Organism: The Methodological Roots of Erving Goffman's Dramaturgical Self," *History of the Human Sciences* 24, no. 2 (April 2011): 138–154. For more on such "kinds," see Laura Stark, "Out of Their Depths:

'Moral Kinds' and the Interpretation of Evidence in Foucault's Modern Episteme," *History and Theory* 55, no. 4 (December 2016): 131–147. See also Nancy D. Campbell and Laura Stark, "Making Up 'Vulnerable' People: Human Subjects and the Subjective Experience of Medical Experiment," *Social History of Medicine* 28, no. 4 (November 2015): 825–848.

26. Hacking, "Looping Effects," 392. On Whewell's invention of this way of talking about kinds, see Ian Hacking, "Natural Kinds: Rosy Dawn, Scholastic Twilight," *Royal Institute of Philosophy Supplements* 61 (October 2007): 203–239. See also Roger Smith, "Does Reflexivity Separate the Human Sciences from the Natural Sciences?," *History of the Human Sciences* 18, no. 4 (November 2005): 1–25.

27. Haraway reminds us that, as historians, we are implicated in looping effects—by describing them. See Donna Haraway, "Situated Knowledges: The Science Question in Feminism and the Privilege of Partial Perspective," *Feminist Studies* 14, no. 3 (Autumn 1988): 575–599, here 592. My argument also draws on the complementary view put forward in Helen E. Longino, "Cognitive and Non-Cognitive Values in Science: Rethinking the Dichotomy," in *Feminism, Science, and the Philosophy of Science*, ed. Lynn Hankinson Nelson and Jack Nelson (Dordrecht: Kluwer, 1996), 39–58. I am also inspired by Samia Khatun, *Australianama: the South Asian Odyssey in Australia* (Oxford: Oxford University Press, 2019).

28. For Richards's "natural selection model," see Robert J. Richards, *Darwin and the Emergence of Evolutionary Theories of Mind and Behavior* (Chicago: University of Chicago Press, 1987), 559–593. See also Richards, "The Natural Selection Model of Conceptual Evolution," *Philosophy of Science* 44, no. 3 (September 1977): 494–501. For a similar view, see David L. Hull, *Science as a Process: An Evolutionary Account of the Social and Conceptual Development of Science* (Chicago: University of Chicago Press, 1988). See also Thomas Nickles, "The Strange Story of Scientific Method," in *Models of Discovery and Creativity* (Dordrecht: Springer, 2009), 167–207. On "-emic" analysis, see Nick Jardine, "Etics and Emics (Not to Mention Anemics and Emetics) in the History of the Sciences," *History of Science* 42, no. 3 (September 2004): 261–278. On "exportable analysis," see D. Graham Burnett, *The Sounding of the Whale: Science and Cetaceans in the Twentieth Century* (Chicago: University of Chicago Press, 2012), 673–675. On "endogenous analysis," see Henry M. Cowles, "Review of *Life Atomic: A History of Radioisotopes in Science and Medicine*, by Angela N. H. Creager," *Medical History* 60, no. 4 (October 2016): 563–565.

29. Alfred North Whitehead, *Science and the Modern World* (New York: Macmillan, 1925), 141. On Whitehead (through a Deweyan lens), see William T. Myers,

"Dewey and Whitehead on the Starting Point and Method," *Transactions of the Charles S. Peirce Society* 37, no. 2 (April 2001): 243–255. For "instruments," see James, *Pragmatism*, 53, emphasis in the original. My attention to the concrete effects theorizing is also a nod to to James.

30. Of course, the idea that a concept or term has a politics does not mean that those who invoke it always agree, or that they will act in the same ways. Far from it. The point of linking ideas and action in language is that we too often hold them apart, privileging one over the other. Daniel Rodgers makes this case succinctly: "The making of words is indeed an act, not a business distinct from the hard, behavioral part of politics but a thing people do. . . . The old dichotomy between behavior and ideas, intellectual history and the history of politics, shopworn with use, never in truth made much sense." See Daniel T. Rodgers, *Contested Truths: Keywords in American Politics since Independence* (New York: Basic Books, 1987), 5. On thinking of political concepts as *verbs*, see Rodgers, "Thinking in Verbs," *Intellectual History Newsletter* 18 (1996): 21–23.
31. See James A. Secord, *Visions of Science: Books and Readers at the Dawn of the Victorian Age* (Oxford: Oxford University Press, 2014). I aim to complement Secord's analysis by exploring how science's meaning shifted in response to the sciences of mind in this and subsequent periods. On the role of such histories in cementing the idea of "discovery," see Simon Schaffer, "Scientific Discoveries and the End of Natural Philosophy," *Social Studies of Science* 16, no. 3 (August 1986): 387–420. See also Andrew Cunningham, "Getting the Game Right: Some Plain Words on the Identity and Invention of Science," *Studies in History and Philosophy of Science Part A* 19, no. 3 (September 1988): 365–389. Schaffer and Cunningham charge these histories with moving in the *wrong* direction, but I am interested in what they got right (or at least enabled others to get right) about science's human nature.
32. On the links between these movements in intellectual and political history, see James T. Kloppenberg, *Uncertain Victory: Social Democracy and Progressivism in European and American Thought, 1870–1920* (Oxford: Oxford University Press, 1986). See also Robert H. Wiebe, *The Search for Order, 1877–1920* (New York: Hill and Wang, 1967). Both Kloppenberg and Wiebe build on the argument of Morton Gabriel White, *Social Thought in America: The Revolt against Formalism* (New York: Viking Press, 1949). On Kloppenberg's relationship to White, see Kloppenberg, "Morton White's Social Thought in America," *Reviews in American History* 15, no. 3 (September 1987): 507–519. My own reading of this moment is indebted to Daniel T. Rodgers, "In Search of Progressivism," *Reviews in American History* 10, no. 4 (December 1982): 113–132. Specifically, I read Rodgers's insistence that we attend to the self-understanding of historical

figures as complementary to Kloppenberg's approach. I argue that drawing analytical terms endogenously from the actors being described is a way to take that self-understanding seriously without sacrificing theoretical scope.

2. Hypothesis Unbound

1. [Mary Wollstonecraft Shelley], *Frankenstein; or, The Modern Prometheus* (London: Lackington, Hughes, Harding, Mavor and Jones, 1818), 97. Shelley's husband, Percy, was engaged in a similar rewriting of Prometheus at the same time, also framed in terms of balance. See Percy Bysshe Shelley, *Prometheus Unbound: A Lyrical Drama in Four Acts with Other Poems* (London: C. and J. Ollier, 1820), esp. xiii. For more on these themes as they relate to the history of science, see Henry M. Cowles, "History Comes to Life," *Modern Intellectual History* 16, no. 1 (April 2019): 309–320.
2. See Samuel Taylor Coleridge, *Biographia Literaria: Or, Biographical Sketches of My Literary Life and Opinions,* vol. 2 (London: R. Fenner, 1817), 11–12.
3. For "spontaneous overflow," see William Wordsworth, *Lyrical Ballads, with Pastoral and Other Poems,* 3rd ed., vol. 1 (London: T. N. Longman and O. Rees, 1802), x–xi. For Mill's definition, see [John Stuart Mill], "What Is Poetry?," *Monthly Repository* 7, no. 73 (January 1833): 60–70, here 63. For "springs of thought," see Percy Bysshe Shelley, *Shelley Memorials,* ed. Jane Shelley (London: Smith, Elder and Co., 1859), 120–122, here 121. This line of Shelley's is quoted in M. H. Abrams, *Natural Supernaturalism: Tradition and Revolution in Romantic Literature* (New York: Norton, 1971), 11. On the imagination in science and the lengths gone to constrain it, see Lorraine Daston, "Fear and Loathing of the Imagination in Science," *Daedalus* 127, no. 1 (January 1998): 73–95.
4. See Percy Bysshe Shelley, "A Defence of Poetry," in Shelley, *Essays, Letters from Abroad, Translations and Fragments.,* ed. Mary Shelley, vol. 1 (London: Edward Moxon, 1840), 47. For Wordsworth's amendment and "vivid sensation," see Wordsworth, *Lyrical Ballads,* ii, emphasis in the original. For "metrical arrangement," see xxiv; for "carrying sensation," see xxxviii. On the meaning of experiment to these authors, see Robert Mitchell, *Experimental Life: Vitalism in Romantic Science and Literature* (Baltimore: Johns Hopkins University Press, 2013). See also Maureen N. McLane, *Romanticism and the Human Sciences: Poetry, Population, and the Discourse of the Species* (Cambridge: Cambridge University Press, 2000).
5. For "heterocosm," see M. H. Abrams, *The Mirror and the Lamp: Romantic Theory and the Critical Tradition* (Oxford: Oxford University Press, 1953), 272; see 283

on similar distinctions between German and English critics in the context of literature.

6. The dictum, which Newton added to the end of the second edition of his *Principia Mathematica* in 1713, appears in Isaac Newton, *The Principia: Mathematical Principles of Natural Philosophy,* trans. I. Bernard Cohen and Anne Miller Whitman (Berkeley: University of California Press, 1999), 943. On Newton's hypotheses, see George E. Smith, "The Methodology of the Principia," in *The Cambridge Companion to Newton,* ed. I. Bernard Cohen and George E. Smith (Cambridge: Cambridge University Press, 2002), 138–173. See also Alan E. Shapiro, "Newton's 'Experimental Philosophy,'" *Early Science and Medicine* 9, no. 3 (January 2004): 185–217. Richard Yeo and Laura Snyder vividly evoke the methodological inheritances of the early nineteenth century. See Richard R. Yeo, *Defining Science: William Whewell, Natural Knowledge, and Public Debate in Early Victorian Britain* (Cambridge: Cambridge University Press, 1993); and Laura J. Snyder, *Reforming Philosophy: A Victorian Debate on Science and Society* (Chicago: University of Chicago Press, 2006). See also Daniel Patrick Thurs, "Scientific Methods," in *Wrestling with Nature: From Omens to Science,* ed. Peter Harrison, Ronald L. Numbers, and Michael H. Shank (Chicago: University of Chicago Press, 2011), 307–335, esp. 311–315.
7. The meaning of "Whig" was—and remains—contested. See J. G. A. Pocock, "The Varieties of Whiggism from Exclusion to Reform: A History of Ideology and Discourse," in *Virtue, Commerce, and History: Essays on Political Thought and History, Chiefly in the Eighteenth Century* (Cambridge: Cambridge University Press, 1985), 215–310. I expand on Pocock's conclusion by emphasizing the wider resonances of "reform" in these decades. For the political background, I have drawn on both Boyd Hilton, *A Mad, Bad, and Dangerous People? England, 1783–1846* (Oxford: Oxford University Press, 2008), esp. 309–371, 439–492; and Linda Colley, *Britons: Forging the Nation, 1707–1837* (New Haven: Yale University Press, 1992), esp. 321–363. On whiggism in politics and science in this period, see Joe Bord, *Science and Whig Manners: Science and Political Style in Britain, c. 1790–1850* (New York: Palgrave Macmillan, 2009). On the heyday of Whig histories, see Joseph Hamburger, *Macaulay and the Whig Tradition* (Chicago: University of Chicago Press, 1976); and especially J. W. Burrow, *A Liberal Descent: Victorian Historians and the English Past* (Cambridge: Cambridge University Press, 1981).
8. On "fondness," see Thomas Brown, *Lectures on the Philosophy of the Human Mind,* vol. 1 (Edinburgh: W. and C. Tait, 1820), 159 and 175. For an overview of the (contested) status of hypotheses, see Laura J. Snyder, "Hypotheses in 19th Century British Philosophy of Science: Herschel, Whewell, Mill," in *The*

Significance of the Hypothetical in Natural Science, ed. Michael Heidelberger and Gregor Schiemann (Berlin: Walter de Gruyter, 2009), 59–76. My interpretation builds on Snyder's by emphasizing how mind and brain were reimagined as being, by nature, hypothetical.

9. For "Age of Machinery," see [Thomas Carlyle], "Signs of the Times," *Edinburgh Review* 49 (June 1829): 439–459, here 442–443. On Carlyle's visit to the Royal Society, see a letter to Robert Mitchell printed in Thomas Carlyle, *Early Letters of Thomas Carlyle,* ed. Charles Eliot Norton, vol. 1 (London: Macmillan and Co., 1886), 207–215, here 211–212. On Carlyle, see Fred Kaplan, *Thomas Carlyle: A Biography* (Ithaca, NY: Cornell University Press, 1983). On the fate of the Society, see David Philip Miller, "Between Hostile Camps: Sir Humphry Davy's Presidency of the Royal Society of London, 1820–1827," *British Journal for the History of Science* 16, no. 1 (March 1983): 1–47.
10. [Carlyle], "Signs of the Times," 443, 458, 448, 444.
11. See Matthew L. Jones, *Reckoning with Matter: Calculating Machines, Innovation, and Thinking about Thinking from Pascal to Babbage* (Chicago: University of Chicago Press, 2016). Biographical details about Babbage are drawn from Anthony Hyman, *Charles Babbage: Pioneer of the Computer* (Princeton: Princeton University Press, 1982). On the difficulties of the Difference Engine, see Doron Swade, *The Difference Engine: Charles Babbage and the Quest to Build the First Computer* (New York: Penguin Books, 2002). Carlyle's insults are from an oft-quoted letter to his brother, written in 1840. See Charles Richard Sanders, ed., *The Collected Letters of Thomas and Jane Welsh Carlyle,* vol. 12 (Durham, NC: Duke University Press, 1985), 335–336. On the enmity of Carlyle and Babbage, see Jon Agar, *The Government Machine: A Revolutionary History of the Computer* (Cambridge, MA: MIT Press, 2003), 37–44.
12. Charles Babbage, *On the Economy of Machinery and Manufactures* (London: Charles Knight, 1832), iii, 316. Much of the book had already appeared as an entry in a volume of Coleridge's *Encyclopaedia Metropolitana,* published in the same year as "Signs of the Times." Carlyle coined "Condition-of-England" in Thomas Carlyle, *Chartism* (London: James Fraser, 1840), esp. 1–5. On Carlyle's place in this debate, see Catherine Gallagher, *The Industrial Reformation of English Fiction: Social Discourse and Narrative Form, 1832–1867* (Chicago: University of Chicago Press, 1985), esp. 187–218. On Babbage's, see Maxine Berg, *The Machinery Question and the Making of Political Economy, 1815–1848* (Cambridge: Cambridge University Press, 1982), esp. 179–202.
13. Babbage, *On the Economy of Machinery and Manufactures,* 15, 307, emphasis in the original. For "universal language," see Charles Babbage, "On a Method of Expressing by Signs the Action of Machinery," *Philosophical Transactions of the*

Royal Society of London 116, no. 3 (1826): 250–265. On its impact, see Simon Schaffer, "Babbage's Intelligence: Calculating Engines and the Factory System," *Critical Inquiry* 21, no. 1 (October 1994): 203–227. On the links between this way of seeing things and Herschel, for example, see William J. Ashworth, "Memory, Efficiency, and Symbolic Analysis: Charles Babbage, John Herschel, and the Industrial Mind," *Isis* 87, no. 4 (December 1996): 629–653.

14. Charles Babbage, *Reflections on the Decline of Science in England: And on Some of Its Causes* (London: B. Fellowes, 1830), 1, 11. My treatment of Babbage's *Decline* as well the general discussion of scientific optimism in this chapter are indebted to James A. Secord, *Visions of Science: Books and Readers at the Dawn of the Victorian Age* (Oxford: Oxford University Press, 2014), esp. 52–79.
15. For "practical use," see Babbage, *Decline of Science,* 18. For mill-ponds, see [Carlyle], "Signs of the Times," 451.
16. [John Wade], *The Black Book: Or, Corruption Unmasked!* (London: J. Fairburn, 1820). On this genre and its place in the politics of the period, see Philip Harling, *The Waning of "Old Corruption": The Politics of Economical Reform in Britain, 1779–1846* (Oxford: Clarendon Press, 1996).
17. For "clerisy," see Samuel Taylor Coleridge, *On the Constitution of the Church and State, according to the Idea of Each* (London: Hurst, Chance, and Co., 1830), 47–48. On its relationship to the BAAS, see Jack Morrell and Arnold Thackray, *Gentlemen of Science: Early Years of the British Association for the Advancement of Science* (Oxford: Oxford University Press, 1981), 17–29. For Babbage's remarks on the matter, see Babbage, *Decline of Science,* 9–14. For "aristocracy of talent," see Thomas Carlyle, *Past and Present* (London: Chapman and Hall, 1843), 23–28, 205–216, here 215.
18. For the "Parliament of Science" nickname, see the address of Thomas Stewart Traill, published in British Association for the Advancement of Science, *Report of the Annual Meeting* (London: J. Murray, 1838), xlii. See also the essays in Roy M. MacLeod and Peter Collins, eds., *The Parliament of Science: The British Association for the Advancement of Science, 1831–1981* (London: Science Reviews, 1981). For "motley," see [Augustus De Morgan], "Reports of the British Association. Vols. I. and II. London: 1833 and 1834," *British and Foreign Review* 1, no. 1 (July 1835): 134–157, here 154.
19. On the Sedition Acts, see the exchange between Ian Inkster and Paul Weindling in *The British Journal for the History of Science:* Ian Inkster, "London Science and the Seditious Meetings Act of 1817," *British Journal for the History of Science* 12, no. 2 (July 1979): 192–196; Paul Weindling, "Science and Sedition: How Effective Were the Acts Licensing Lectures and Meetings, 1795–1819?," *British Journal for the History of Science* 13, no. 2 (July 1980): 139–153; and Ian Inkster,

"Seditious Science: A Reply to Paul Weindling," *British Journal for the History of Science* 14, no. 2 (July 1981): 181–187. For "gigantic development," see J. M. Baernreither, *English Associations of Working Men* (London: Swann Sonneschein, 1891), 11. For more on this passage, see Roy MacLeod, "Introduction: On the Advancement of Science," in MacLeod and Collins, *The Parliament of Science,* 17–42.

20. On Liberal Anglicans in this period, see Richard Brent, *Liberal Anglican Politics: Whiggery, Religion, and Reform, 1830–1841* (Oxford: Clarendon Press, 1987). On the persistence of aristocratic power, see Peter Mandler, *Aristocratic Government in the Age of Reform: Whigs and Liberals, 1830–1852* (Oxford: Clarendon Press, 1990). On Liberal Anglicanism in general, see Duncan Forbes, *The Liberal Anglican Idea of History* (Cambridge: Cambridge University Press, 1952). On the "inner core," see also Morrell and Thackray, *Gentlemen of Science,* 109–128, 298–308.
21. For "united exertions," see William Vernon Harcourt, "Report of the First Annual Meeting," *Proceedings of the British Association for the Advancement of Science* 1 (1831): 1–44, here 28. On long-standing worries about France, see (for example) Colley, *Britons,* esp. 364–375.
22. [William Whewell], review of Mary Somerville's *On the Connexion of the Physical Sciences, Quarterly Review* 51, no. 101 (March 1834): 54–68, here 59. For "peripatetic character," see Hugh Miller, *First Impressions of England and Its People* (London: J. Johnstone, 1847), 249–250. For Herschel's imagery, see John F. W Herschel, "Whewell on Inductive Sciences," *Quarterly Review* 68, no. 135 (June 1841): 177–238, here 186.
23. [Whewell], review of Somerville, 59–60, emphasis in the original. The classic account of Whewell's coinage (including references to its American afterlife) is Sydney Ross, "Scientist: The Story of a Word," *Annals of Science* 18, no. 2 (June 1962): 65–85. For a comprehensive list of his coinages, see P. J. Wexler, "The Great Nomenclator: Whewell's Contributions to Scientific Terminology," *Notes and Queries* 8, no. 1 (January 1961): 27–32. On the German debates, see Denise Phillips, *Acolytes of Nature: Defining Natural Science in Germany, 1770–1850* (Chicago: University of Chicago Press, 2012).
24. Mary Somerville, *On the Connexion of the Physical Sciences,* 2nd ed. (London: John Murray, 1835), 2, 1–2. On Somerville's status as a woman of science, see Kathryn A. Neeley, *Mary Somerville: Science, Illumination, and the Female Mind* (Cambridge: Cambridge University Press, 2001). On the persistence of the term "man of science," see Ruth Barton, "'Men of Science': Language, Identity, and Professionalization in the Mid-Victorian Scientific Community," *History of Science* 41, no. 1 (March 2003): 73–119. For more general context, see Ruth

Barton, *The X Club: Power and Authority in Victorian Science* (Chicago: University of Chicago Press, 2018).

25. [Whewell], review of Somerville, 65; [John Herschel], "Mechanism of the Heavens," *Quarterly Review* 47, no. 94 (July 1832): 537–559, here 548. On the gendered nature of scientific knowing, see Londa L. Schiebinger, *Nature's Body: Gender in the Making of Modern Science* (Boston: Beacon Press, 1993).
26. J. C. Clerk-Maxwell, "The Correlation of Physical Forces, Sixth Edition," *Nature* 10, no. 251 (August 20, 1874): 302–304, here 303. On this misprision, see Richard R. Yeo, "Scientific Method and the Image of Science, 1831–1891," in MacLeod and Collins, *The Parliament of Science*, 65–88, esp. 69–70. On these varieties of positivism, see Trevor Pearce, "'Science Organized': Positivism and the Metaphysical Club, 1865–1875," *Journal of the History of Ideas* 76, no. 3 (July 2015): 441–465.
27. [John Stuart Mill], "The Spirit of the Age," *The Examiner*, no. 1197 (January 9, 1831): 20–21; no. 1199 (January 23, 1831): 50–52; no. 1201 (February 6, 1831): 82–84; no. 1209 (April 3, 1831): 210–211; no. 1216 (May 15, 1831): 307; and no. 1218 (May 29, 1831): 339–341. On Carlyle's (mis-)reading of Mill, see Emery Neff, *Carlyle and Mill: An Introduction to Victorian Thought*, 2nd ed. (New York: Columbia University Press, 1926). On Mill's relationship to Comte, see Mary Pickering, *Auguste Comte: An Intellectual Biography* (Cambridge: Cambridge University Press, 2009), 1:505–538, 2:70–113. On their disagreement over method, see David Lewisohn, "Mill and Comte on the Methods of Social Science," *Journal of the History of Ideas* 33, no. 2 (April–June 1972): 315–324.
28. [Mill], "Spirit No. 1," 20; [Mill], "Spirit No. 5 (Part II)," 341, 340. On the political aesthetics of Mill's essays, see David Russell, "Aesthetic Liberalism: John Stuart Mill as Essayist," *Victorian Studies* 56, no. 1 (October 2013): 7–30.
29. [John Stuart Mill], "The Works of Jeremy Bentham," *London and Westminster Review* 29 (August 1838): 467–506, here 468, 473; [Mill], "The Literary Remains of Samuel Taylor Coleridge," *London and Westminster Review* 33 (March 1840): 139–163, here 139, 154, 140–141. For details on Mill's life, I have relied on (in order) John Stuart Mill, *Autobiography* (London: Longmans, Green, Reader, and Dyer, 1873); Alan Ryan, *The Philosophy of John Stuart Mill* (London: Macmillan, 1970); Nicholas Capaldi, *John Stuart Mill: A Biography* (Cambridge: Cambridge University Press, 2004); and Richard Reeves, *John Stuart Mill: Victorian Firebrand* (London: Atlantic Books, 2007).
30. [Mill], "Spirit No. 2," 51.
31. [John Stuart Mill], review of Herschel's *Preliminary Discourse on the Study of Natural Philosophy*, *The Examiner*, no. 1207 (March 20, 1831): 179–180; [Henry

Peter Brougham], *A Discourse of the Objects, Advantages, and Pleasures of Science* (London: Baldwin, Cradock, and Joy, 1827), 6 , emphasis in the original. On Brougham, see G. N. Cantor, "Henry Brougham and the Scottish Methodological Tradition," *Studies in History and Philosophy of Science Part A* 2, no. 1 (May 1971): 69–89. On the Society, see Alan Rauch, *Useful Knowledge: The Victorians, Morality, and the March of Intellect* (Durham, NC: Duke University Press, 2001), 22–59.

32. [Brougham], *Discourse,* 45. For Herschel's threat, see Morrell and Thackray, *Gentlemen of Science,* 48. Babbage was not persuaded by Herschel's reading, and in fact quoted Herschel in his preface as though to suggest that Herschel had endorsed Babbage's book. See Babbage, *Decline of Science,* vii–ix.
33. John F. W. Herschel, *A Preliminary Discourse on the Study of Natural Philosophy* (London: Longman, Rees, Orme, Brown, and Green, 1830), 219, 26, 9.
34. [Mill], review of Herschel, 179, emphasis in the original. On "science of science," see Mill to John Sterling, October 20–22, 1831, in John Stuart Mill, *Collected Works of John Stuart Mill,* vol. 12 (London: Routledge, 1996), 78–79. Cited in Snyder, *Reforming Philosophy,* 95.
35. [Mill], review of Herschel, 179. On boundary objects, see Susan Leigh Star and James R. Griesemer, "Institutional Ecology, 'Translations' and Boundary Objects: Amateurs and Professionals in Berkeley's Museum of Vertebrate Zoology, 1907–39," *Social Studies of Science* 19, no. 3 (August 1989): 387–420. Star later reconsidered the definition. See Susan Leigh Star, "This Is Not a Boundary Object: Reflections on the Origin of a Concept," *Science, Technology, & Human Values* 35, no. 5 (September 2010): 601–617. Although hypotheses are too abstract to meet her later definition, the concept proves useful in making sense of efforts to *materialize* the hypothesis as a feature of organic, embodied, even material thinking.
36. Herschel, *Preliminary Discourse,* 174–175, emphasis in the original. This neutrality was related to, but not the same as, the "mechanical objectivity" traced by Lorraine Daston and Peter Galison in their recent work. See Daston and Galison, *Objectivity* (New York: Zone Books, 2007).
37. Herschel, *Preliminary Discourse,* 104, 72. On the rhetorical power of Baconianism in this period, see Richard Yeo, "An Idol of the Market-Place: Baconianism in Nineteenth Century Britain," *History of Science* 23, no. 61 (September 1985): 251–298. On debates over Newton's genius, see Richard Yeo, "Genius, Method, and Morality: Images of Newton in Britain, 1760–1860," *Science in Context* 2, no. 2 (October 1988): 257–284. See also Richard Bellon, "There Is Grandeur in This View of Newton: Charles Darwin, Isaac Newton and Victorian Conceptions of Scientific Virtue," *Endeavour* 38, no. 3 (September 2014): 222–234. Whether or not Herschel was being strategic or genuine (or both) in his appeals

to these forebears, the effect was the same: to ease the way for his reintroduction of hypotheses.

38. Newton, *Principia,* 794. For "figments," see Herschel, *Preliminary Discourse,* 144. On Herschel's Baconianism, see [William Whewell], review of Herschel's *Preliminary Discourse on the Study of Natural Philosophy, Quarterly Review* 45, no. 90 (July 1831): 374–407. On Whewell's understanding of himself as a Baconian, see Laura J. Snyder, "Renovating the Novum Organum: Bacon, Whewell and Induction," *Studies in History and Philosophy of Science Part A* 30, no. 4 (December 1999): 531–557.
39. Herschel, *Preliminary Discourse,* 149, 196, 204, 186. Herschel was far from the first to see analogy, hypothesis, and imagination as significant and potentially problematic in this way. See, for example, Katharine Park, Lorraine J. Daston, and Peter L. Galison, "Bacon, Galileo, and Descartes on Imagination and Analogy," *Isis* 75, no. 2 (June 1984): 287–289 (and the essays that follow). Herschel differed from his predecessors in *how* he allowed imagination into scientific work amid heightened anxiety about its place in proper inquiry.
40. Herschel, *Preliminary Discourse,* 198–199, 208, 164, 190–191.
41. Ibid., 69. For Mill's reference to unity, see [Mill], review of Herschel, 179.
42. [Whewell], review of Herschel, 376–377. On reviews of Herschel's book, see Richard Yeo, "Reviewing Herschel's *Discourse,*" *Studies in History and Philosophy of Science Part A* 20, no. 4 (December 1989): 541–552. On Herschel's reputation, see Susan Faye Cannon, "John Herschel and the Idea of Science," *Journal of the History of Ideas* 22, no. 2 (April 1961): 215–239.
43. [Whewell], review of Herschel, 390, 399–400. Herschel, *Preliminary Discourse,* 164–165. On Whewell's psychologism in this period, see John F. Metcalfe, "Whewell's Developmental Psychologism: A Victorian Account of Scientific Progress," *Studies in History and Philosophy of Science Part A* 22, no. 1 (March 1991): 117–139.
44. For more on these looping effects and their role in scientific naturalism, see Henry M. Cowles, "History Naturalized," *Historical Studies in the Natural Sciences* 47, no. 1 (February 2017): 107–116.
45. [Mill], review of Herschel, 179. The quotation is from Herschel, *Preliminary Discourse,* 72–73.
46. The classic account of Whewell's life is Isaac Todhunter, *William Whewell, D.D., Master of Trinity College, Cambridge: An Account of His Writings with Selections from His Literary and Scientific Correspondence* (London: Macmillan and Co., 1876). The letter to Jones is quoted at 2:115, emphasis in the original. Whewell quoted Herschel in his review of Herschel, 406–407. The original context was Herschel, *Preliminary Discourse,* 7–8.

47. In addition to the work of Yeo and Snyder, another source for my account of Whewell is Menachem Fisch, *William Whewell, Philosopher of Science* (Oxford: Clarendon Press, 1991). See also Menachem Fisch and Simon Schaffer, eds., *William Whewell: A Composite Portrait* (Oxford: Clarendon Press, 1991). My argument complements Fisch's work by situating Whewell's project in the broader age of methods, both chronologically and geographically.
48. For "lookers on," see Todhunter, *Whewell,* 2:29. On Whewell's scientific work, see Michael S. Reidy, *Tides of History: Ocean Science and Her Majesty's Navy* (Chicago: University of Chicago Press, 2008); Steffen Ducheyne, "Whewell's Tidal Researches: Scientific Practice and Philosophical Methodology," *Studies in History and Philosophy of Science Part A* 41, no. 1 (March 2010): 26–40; and Ducheyne, "Fundamental Questions and Some New Answers on Philosophical, Contextual and Scientific Whewell: Some Reflections on Recent Whewell Scholarship and the Progress Made Therein," *Perspectives on Science* 18, no. 2 (Summer 2010): 242–272. See also Laura J. Snyder, "Whewell and the Scientists: Science and Philosophy of Science in 19th Century Great Britain," in *History of Philosophy of Science: New Trends and Perspectives,* ed. Michael Heidelberger and Friedrich Stadler (Dordrecht: Kluwer, 2002), 81–94.
49. Todhunter, *Whewell,* 2:286–287. For Brewster's review, David Brewster, "Whewell's Philosophy of the Inductive Sciences," *Edinburgh Review* 74, no. 150 (January 1842): 265–306, here 269.
50. William Whewell, "The Philosophy of the Pure Sciences," as cited in Henry M. Cowles, "The Age of Methods: William Whewell, Charles Peirce, and Scientific Kinds," *Isis* 107, no. 4 (December 2016): 722–737, here 727. For "giving rise," see William Whewell, *Of Induction: With Especial Reference to Mr. J. Stuart Mill's* System of Logic (London: J. W. Parker and Son, 1849), 30–31. For the sum, see William Whewell, *The Philosophy of the Inductive Sciences, Founded upon Their History,* vol. 2 (London: J. W. Parker and Son, 1840), 250. On Whewell's Kantianism, see Robert E. Butts, "Induction as Unification: Kant, Whewell, and Recent Developments," in *Kant and Contemporary Epistemology,* ed. Paolo Parrini (Dordrecht: Kluwer Academic, 1994), 273–289. See also Laura J. Snyder, "Discoverers' Induction," *Philosophy of Science* 64, no. 4 (December 1997): 580–604.
51. William Whewell, *History of the Inductive Sciences, from the Earliest to the Present Times,* vol. 1 (London: J. W. Parker and Son, 1837), 9–10. For "penetrated," see William Whewell, "Of the Transformation of Hypotheses in the History of Science," *Transactions of the Cambridge Philosophical Society* 9, no. 2 (1851): 139–146, here 139, 146.

52. See Todhunter, *Whewell,* 1:90, on "Prospects," and 1:186–187 on "guessing." Whewell was not alone in incorporating guessing into an account of truth in this period. A decade earlier, his friend Julius Hare had published a book that made the same point. See [Augustus William Hare and Julius Charles Hare], *Guesses at Truth* (London: J. Taylor, 1827).
53. For Whewell's early doubts, see Todhunter, *Whewell,* 2:124–126. For "barren speculation," see William Whewell, *The Philosophy of the Inductive Sciences, Founded upon Their History,* 2nd ed., vol. 2 (London: J. W. Parker and Son, 1847), 19. On the "process," see Whewell, *Of Induction,* 22. For the "method," see Todhunter, *Whewell,* 2:416. For the "reflex sciences," see Whewell, "The Philosophy of the Pure Sciences" and "The Inductive Philosophy," as quoted in Cowles, "Age of Methods," 728, emphasis in the original.
54. Whewell, *History,* 3:430–431. For "irresistibly," see Whewell, *Philosophy,* 71.
55. Whewell, *History,* 3:623–624.
56. The authoritative account is Snyder, *Reforming Philosophy.* Snyder is sympathetic to both adversaries, though she sides with Whewell against those (starting with Mill) who have misread his intentions. To this end she adopts Whewell's, not Mill's, definition of induction. As someone more sympathetic to Mill in general, I have complemented Snyder's thick description with a consideration of the legacies of both men later in the century.
57. On these disputes, see Laura J. Snyder, "The Mill-Whewell Debate: Much Ado about Induction," *Perspectives on Science* 5, no. 2 (Summer 1997): 159–198.
58. James Mill, *Analysis of the Phenomena of the Human Mind* (London: Baldwin and Cradock, 1829). For an example of the reception of these ideas, see the critical apparatus in the posthumous edition of James Mill's *Analysis of the Phenomena of the Human Mind,* ed. John Stuart Mill (London: Longmans, Green, Reader, and Dyer, 1869).
59. James Mill, *Analysis,* 56. The history of associationism remains to be written. For an early account that was close to an official history, see Howard C. Warren, *A History of the Association Psychology* (New York: Charles Scribner's Sons, 1921).
60. The first edition of Combe's work was George Combe, *The Constitution of Man Considered in Relation to External Objects* (Edinburgh: J. Anderson, 1828). On phrenology's popularity in the period, see Roger Cooter, *The Cultural Meaning of Popular Science: Phrenology and the Organization of Consent in Nineteenth-Century Britain* (Cambridge: Cambridge University Press, 1985). On its progressive implications, see R. J. Cooter, "Phrenology: The Provocation of Progress," *History of Science* 14, no. 4 (December 1976): 211–234. For a more recent account of phrenology in a wider context, see James Poskett, *Materials of the*

Mind: Phrenology, Race, and the Global History of Science, 1815–1920 (Chicago: University of Chicago Press, 2019).

61. On the rise of experimental psychology, see Kurt Danziger, *Constructing the Subject: Historical Origins of Psychological Research* (Cambridge: Cambridge University Press, 1990). On the role of method with respect to theoretical advances, see Danziger, "On Theory and Method in Psychology," in *Recent Trends in Theoretical Psychology,* ed. W. Baker et al. (New York: Springer, 1988), 87–94.
62. Herbert Butterfield, *The Whig Interpretation of History* (London: G. Bell and Sons, 1931), v. On whiggism in the history of science, see Nick Jardine, "Whigs and Stories: Herbert Butterfield and the Historiography of Science," *History of Science* 41, no. 2 (June 2003): 125–140. For a recent defense, see Hasok Chang, "We Have Never Been Whiggish (about Phlogiston)," *Centaurus* 51, no. 4 (November 2009): 239–264. For a response, see Michael D. Gordin, "The Tory Interpretation of History," *Historical Studies in the Natural Sciences* 44, no. 4 (September 2014): 413–423.
63. On "whole box of tools," see Laura J. Snyder, "'The Whole Box of Tools': William Whewell and the Logic of Induction," in *Handbook of the History of Logic,* vol. 4: *British Logic in the Nineteenth Century,* ed. Dov M. Gabbay and John Woods (Amsterdam: North Holland, 2008), 163–228.
64. William Whewell, *On the Principles of English University Education* (London: J. W. Parker, 1837), 46–53. On curricular reforms, see John R. Gibbins, "'Old Studies and New': The Organisation of Knowledge in University Curriculum," in *The Organisation of Knowledge in Victorian Britain,* ed. Martin Daunton (Oxford: British Academy, 2005), 235–262. The phrase "whig interpretation of science" has been used to indict the sins Butterfield noted—but not to link science and reform in the 1830s. See Theodore M. Porter and Dorothy Ross, "Introduction: Writing the History of Social Science," in *The Cambridge History of Science,* vol. 7: *The Modern Social Sciences,* ed. Theodore M. Porter and Dorothy Ross (Cambridge: Cambridge University Press, 2003), 1–10, here 5.
65. Whewell, *English University Education,* 138–139. The phrase "scientific whigs" was also used by Duncan Forbes to describe a related evolution in (Scottish) Enlightenment thought, linking David Hume, Adam Smith, and others to later figures such as the Mills. See Duncan Forbes, "Scientific Whiggism: Adam Smith and John Millar," *Cambridge Journal* 7 (1953–1954): 643–670.

3. Nature's Method

1. For "walks with Henslow," see *Autobiography of Charles Darwin, 1809–1882, with Original Omissions Restored,* ed. Nora Barlow (New York: Harcourt, Brace,

1958), 64. Details about Darwin's life were gathered from his autobiography and from Adrian Desmond and James Moore, *Darwin: The Life of a Tormented Evolutionist* (New York: W. W. Norton, 1991); and Janet Browne, *Charles Darwin: A Biography,* 2 vols. (Princeton: Princeton University Press, 1996, 2003).

2. For "*finished* Naturalist," see John Stevens Henslow to Darwin, August 24, 1831, in *The Correspondence of Charles Darwin,* ed. Frederick Burkhardt et al., vol. 1: *1821–1836* (Cambridge: Cambridge University Press, 1985), 128–129, here 129. On his early training, see James A. Secord, "The Discovery of a Vocation: Darwin's Early Geology," *British Journal for the History of Science* 24, no. 2 (June 1991): 133–157. On Darwin's geology in general, see Sandra Herbert, *Charles Darwin, Geologist* (Ithaca, NY: Cornell University Press, 2005). See also Alistair Sponsel, "An Amphibious Being: How Maritime Surveying Reshaped Darwin's Approach to Natural History," *Isis* 107, no. 2 (June 2016): 254–281.
3. On the myth, see Frank J. Sulloway, "Darwin and His Finches: The Evolution of a Legend," *Journal of the History of Biology* 15, no. 1 (Spring 1982): 1–53. See also Sulloway, "Darwin's Conversion: The *Beagle* Voyage and Its Aftermath," *Journal of the History of Biology* 15, no. 3 (Autumn 1982): 325–396. On the value of Darwin's notebooks, see Jonathan Hodge, "The Notebook Programmes and Projects of Darwin's London Years," in *The Cambridge Companion to Darwin,* 2nd ed., ed. Jonathan Hodge and Gregory Radick (Cambridge: Cambridge University Press, 2009), 44–72.
4. Much has been written on Darwin's early philosophical reading. See especially Michael T. Ghiselin, *The Triumph of the Darwinian Method* (Chicago: University of Chicago Press, 1969); David L. Hull, "Charles Darwin and Nineteenth Century Philosophies of Science," in *Foundations of Scientific Method: The Nineteenth Century,* ed. Ronald N. Giere and Richard S. Westfall (Bloomington: Indiana University Press, 1973), 115–132; Michael Ruse, "Darwin's Debt to Philosophy: An Examination of the Influence of the Philosophical Ideas of John F. W. Herschel and William Whewell on the Development of Charles Darwin's Theory of Evolution," *Studies in History and Philosophy of Science Part A* 6, no. 2 (June 1975): 159–181; and David L. Hull, "Darwin's Science and Victorian Philosophy of Science," in Hodge and Radick, *Cambridge Companion to Darwin,* 173–196. On Darwin's anxieties about authorship, see Alistair Sponsel, *Darwin's Evolving Identity: Adventure, Ambition, and the Sin of Speculation* (Chicago: University of Chicago Press, 2018).
5. On "Darwin's metaphor," see Robert M. Young, "Darwin's Metaphor: Does Nature Select?," *Monist* 55, no. 3 (July 1971): 442–503. This essay was famously republished in Robert M. Young, *Darwin's Metaphor: Nature's Place in Victorian*

Culture (Cambridge: Cambridge University Press, 1985). See also Michael Ruse, "Charles Darwin and Artificial Selection," *Journal of the History of Ideas* 36, no. 2 (April 1975): 339–350; James A. Secord, "Nature's Fancy: Charles Darwin and the Breeding of Pigeons," *Isis* 72, no. 2 (June 1981): 163–186; and L. T. Evans, "Darwin's Use of the Analogy between Artificial and Natural Selection," *Journal of the History of Biology* 17, no. 1 (Spring 1984): 113–140. On Darwin's "logical vocabulary" as a strategy, see Edward Manier, "'External Factors' and 'Ideology' in the Earliest Drafts of Darwin's Theory," *Social Studies of Science* 17, no. 4 (November 1987): 581–609.

6. On "vocabularies of method," see Henry M. Cowles, "On the Origin of Theories: Charles Darwin's Vocabulary of Method," *American Historical Review* 122, no. 4 (October 2017): 1079–1104. On the importance of terms, see John Beatty, "What's in a Word? Coming to Terms in the Darwinian Revolution," *Journal of the History of Biology* 15, no. 2 (Summer 1982): 215–239. See also Beatty, "Speaking of Species: Darwin's Strategy," in *The Darwinian Heritage*, ed. David Kohn (Princeton: Princeton University Press, 1985), 265–281. My argument here owes a great deal to Gillian Beer, *Darwin's Plots: Evolutionary Narrative in Darwin, George Eliot, and Nineteenth-Century Fiction* (London: Routledge, 1983). See also George Levine, *Darwin and the Novelists: Patterns of Science in Victorian Fiction* (Cambridge, MA: Harvard University Press, 1988); Stephen G. Alter, *Darwinism and the Linguistic Image: Language, Race, and Natural Theology in the Nineteenth Century* (Baltimore: Johns Hopkins University Press, 1999); and David Amigoni, *Colonies, Cults and Evolution: Literature, Science and Culture in Nineteenth-Century Writing* (Cambridge: Cambridge University Press, 2007).
7. Charles Kingsley, "How to Study Natural History," in *The Works of Charles Kingsley*, vol. 19 (London: Macmillan, 1880), 289–310, here 297, 296, 294, 308, 299. On the lecture's original context, see Bernard V. Lightman, *Victorian Popularizers of Science: Designing Nature for New Audiences* (Chicago: University of Chicago Press, 2007), 71–81, esp. 73–74. On Kingsley's life, see J. M. I. Klaver, *The Apostle of the Flesh: A Critical Life of Charles Kingsley* (Boston: Brill, 2006). On his science, see Will Abberley, "Animal Cunning: Deceptive Nature and Truthful Science in Charles Kingsley's Natural Theology," *Victorian Studies* 58, no. 1 (Autumn 2015): 34–56.
8. Charles Kingsley, *The Water-Babies: A Fairy-Tale for a Land-Baby* (London: Macmillan and Co., 1863), 249–250. On *Water-Babies* and the history of science, see John Beatty and Piers J. Hale, "Water Babies: An Evolutionary Parable," *Endeavour* 32, no. 4 (December 2008): 141–146.
9. Kingsley, *Water Babies*, 316. On Kingsley's Baconianism, see Piers J. Hale, "Monkeys into Men and Men into Monkeys: Chance and Contingency in the

Evolution of Man, Mind and Morals in Charles Kingsley's 'Water Babies,'" *Journal of the History of Biology* 46, no. 4 (Winter 2013): 551–597, e.g. 562, 567–568, and 574. Spencer's assertion is from Herbert Spencer, *Education: Intellectual, Moral, and Physical* (London: Williams and Norgate, 1861), 76. On Spencer's recapitulation, see Jessica Straley, "Of Beasts and Boys: Kingsley, Spencer, and the Theory of Recapitulation," *Victorian Studies* 49, no. 4 (Summer 2007): 583–609.

10. For "primal forms," see Charles Kingsley to Charles Darwin, November 18, 1859, in *Correspondence of Charles Darwin*, vol. 7: *1858–1859* (Cambridge: Cambridge University Press, 1992), 379–380, here 380. Mother Carey's line appears in *Water-Babies,* 284. For the later formulation, see Charles Kingsley, "The Natural Theology of the Future," in *Works of Charles Kingsley,* 19:313–336, here 332. For "Darwin's other bulldog," see Piers J. Hale, "Darwin's Other Bulldog: Charles Kingsley and the Popularisation of Evolution in Victorian England," *Science & Education* 21, no. 7 (July 2012): 977–1013.
11. Charles Darwin, *On the Origin of Species by Means of Natural Selection, or, The Preservation of Favoured Races in the Struggle for Life,* 5th ed. (London: John Murray, 1869), ii. The Whewell quotation is from William Whewell, *Astronomy and General Physics Considered with Reference to Natural Theology* (London: William Pickering, 1833), 356. On the *Bridgewater Treatises,* see Jonathan Topham, "Science and Popular Education in the 1830s: The Role of the 'Bridgewater Treatises,'" *British Journal for the History of Science* 25, no. 4 (December 1992): 397–430; and Topham, "Beyond the 'Common Context': The Production and Reading of the Bridgewater Treatises," *Isis* 89, no. 2 (June 1998): 233–262.
12. For evidence of the request, see Charles Darwin to Charles Kingsley, November 30 [1859], in *Correspondence of Charles Darwin,* 7:407. For the insertion, see Charles Darwin, *On the Origin of Species by Means of Natural Selection, Or, The Preservation of Favoured Races in the Struggle for Life,* 2nd ed. (London: John Murray, 1860), 481. "Creator" appears on 490. On balancing science and religion, see John Hedley Brooke, "Natural Theology and the Plurality of Worlds: Observations on the Brewster-Whewell Debate," *Annals of Science* 34, no. 3 (May 1977): 221–286. For a fuller expression, see John Hedley Brooke, *Science and Religion: Some Historical Perspectives* (Cambridge: Cambridge University Press, 1991). For his reading of the *Origin,* see John Hedley Brooke, "'Laws Impressed on Matter by the Creator'? The *Origin* and the Question of Religion," in *The Cambridge Companion to the "Origin of Species,"* ed. Michael Ruse and Robert J. Richards (Cambridge: Cambridge University Press, 2008), 256–274. The final line appears, in both the 1859 and 1860 editions of the *Origin,* on 490.

13. Darwin, *Autobiography,* 59. On Paley's curricular place, see Aileen Fyfe, "The Reception of William Paley's 'Natural Theology' in the University of Cambridge," *British Journal for the History of Science* 30, no. 3 (September 1997): 321–335. On Darwin's engagement, see Jonathan Topham, "Biology in the Service of Natural Theology: Paley, Darwin, and the Bridgewater Treatises," in *Biology and Ideology from Descartes to Dawkins,* ed. Denis R. Alexander and Ronald L. Numbers (Chicago: University of Chicago Press, 2010), 88–113.
14. William Paley, *Natural Theology, Or, Evidences of the Existence and Attributes of the Deity* (London: R. Faulder, 1802), 1–2. On his argument, see Adam R. Shapiro, "William Paley's Lost 'Intelligent Design,'" *History and Philosophy of the Life Sciences* 31, no. 1 (January 2009): 55–77. On its reception, see Aileen Fyfe, "Publishing and the Classics: Paley's *Natural Theology* and the Nineteenth-Century Scientific Canon," *Studies in History and Philosophy of Science Part A* 33, no. 4 (December 2002): 729–751.
15. Charles Darwin, *Charles Darwin's Beagle Diary,* ed. R. D. Keynes (Cambridge: Cambridge University Press, 2001), 443, 42. For his compliment, see Charles Darwin to Joseph Hooker, [February 10, 1845], in *Correspondence of Charles Darwin,* vol. 3: *1844–1846* (Cambridge: Cambridge University Press, 1988), 139–140, here 140. On Darwin's debts to Humboldt, see Robert J. Richards, *The Romantic Conception of Life: Science and Philosophy in the Age of Goethe* (Chicago: University of Chicago Press, 2002), 514–554. For the controversy, see Michael Ruse, "The Romantic Conception of Robert J. Richards," *Journal of the History of Biology* 37, no. 1 (March 2004): 3–23; and Richards, "Michael Ruse's Design for Living," *Journal of the History of Biology* 37, no. 1 (March 2004): 25–38. See also Phillip R. Sloan, "'The Sense of Sublimity': Darwin on Nature and Divinity," *Osiris* 16 (2001): 251–269, esp. 252–256; and Paul White, "Darwin, Concepción, and the Geological Sublime," *Science in Context* 25, no. 1 (March 2012): 49–71.
16. For "intense pleasure," see Darwin to W. D. Fox, 19 [September 1831], in *Correspondence of Charles Darwin,* 1:162–164, here 164. For "read and reread," see Darwin to J. S. Henslow, [July 11, 1831], in ibid., 125–126, here 125. For "hardly sit" (and "tropical glow"), see Darwin to Caroline Darwin, [April 28, 1831], in ibid., 121–122, here 122. See Darwin, *Beagle Diary,* for "Nothing could be better" (18), "enthusiasm" (20, 48), "disappointment" (23). For "french expressions," see Caroline Darwin to Charles Darwin, October 28 [1833], in *Correspondence of Charles Darwin,* 1:345–346, here 345. For his note to Henslow, see Darwin to J. S. Henslow, May 18–June 16, 1832, in ibid., 1:236–239.
17. Humboldt's later interpretation appears in Alexander von Humboldt, *Cosmos: A Sketch of the Physical Description of the Universe,* vol. 1, trans. E. C. Otté (London: Henry G. Bohn, 1849), 1–2. Richards interprets this as evidence of Humboldt's

aesthetic appreciation of unity. See Richards, *Romantic Conception of Life,* 520–521. On "Humboldtian science," see Susan Faye Cannon, *Science in Culture: The Early Victorian Period* (New York: Science History Publications, 1978), 73–110. See also Michael Dettelbach, "Humboldtian Science," in *Cultures of Natural History,* ed. Nicholas Jardine, James A. Secord, and E. C. Spary (Cambridge: Cambridge University Press, 1996), 287–304.

18. For "expression," see Louis Agassiz, *Lake Superior: Its Physical Character, Vegetation, and Animals, Compared with Those of Other and Similar Regions* (Boston: Gould, Kendall and Lincoln, 1850), 145, emphasis in the original. For "succession," see Louis Agassiz and A. A. Gould, *Principles of Zoology: Touching the Structure, Development, Distribution, and Natural Arrangement of the Races of Animals, Living and Extinct* (Boston: Gould, Kendall and Lincoln, 1848), 206. For "Divine Intelligence," see Louis Agassiz, *An Essay on Classification* (London: Longman, Brown, Green, Longmans and Roberts, 1859), 8. On Agassiz's life, see Edward Lurie, *Louis Agassiz: A Life in Science* (Chicago: University of Chicago Press, 1960). On Agassiz and Gray, see C. George Fry and Jon Paul Fry, *Congregationalists and Evolution: Asa Gray and Louis Agassiz* (Lanham, MD: University Press of America, 1989).
19. Richard Owen, *On the Nature of Limbs: A Discourse Delivered on Friday, February 9, at an Evening Meeting of the Royal Institution of Great Britain* (London: J. Van Voorst, 1849), 2, 86. Whether Owen meant his claims about Platonic ideas is contested. For an argument that he did, see Adrian Desmond, *Archetypes and Ancestors: Palaeontology in Victorian London, 1850–1875* (Chicago: University of Chicago Press, 1986). For an argument that the claims were rhetorical, see Nicolaas A. Rupke, *Richard Owen: Biology without Darwin* (Chicago: University of Chicago Press, 2009). My argument holds either way.
20. Darwin, *Autobiography,* 87. On Owen's delicate balance, see Nicolaas A. Rupke, *Richard Owen: Victorian Naturalist* (New Haven: Yale University Press, 1994).
21. Charles Darwin, *The Descent of Man, and Selection in Relation to Sex,* vol. 1 (London: John Murray, 1871), 153. For Kingsley's statement, see Kingsley, "Natural Theology of the Future," 335.
22. For "language," see Adam Sedgwick, "Anniversary Address to the Geological Society, 1831," in *Proceedings of the Geological Society of London,* vol. 1: *November 1826 to June 1833* (London: Richard Taylor, 1834), 281–316, here 315. For "slow," see Sedgwick, "Presidential Address to the Geological Society, 1830," in *Proceedings,* 187–212, here 207. See also Sedgwick, *A Discourse on the Studies of the University* (London: J. W. Parker, 1833). On Sedgwick's rhetoric, see Richard Bellon, "The Moral Dignity of Inductive Method and the Reconciliation of Science and Faith in Adam Sedgwick's *Discourse,*" *Science & Education* 21, no. 7 (July 2012): 937–958.

23. William Whewell, *On the Philosophy of Discovery: Chapters Historical and Critical* (London: J. W. Parker and Son, 1860), 379, 381, 387, 395. On Whewell's divine mind, see Richard Yeo, "William Whewell, Natural Theology and the Philosophy of Science in Mid-Nineteenth Century Britain," *Annals of Science* 36, no. 5 (1979): 493–516.
24. Whewell, *Astronomy and General Physics,* 335.
25. Charles Babbage, *The Ninth Bridgewater Treatise: A Fragment* (London: J. Murray, 1837), 23–24. The relationship between Babbage's treatise and the series was complex. See J. R. Topham, "'An Infinite Variety of Arguments': The Bridgewater Treatises and British Natural Theology in the 1830s" (PhD diss., University of Lancaster, 1993), 74–75. On the book's effect on his relationship with Whewell, see Laura J. Snyder, *The Philosophical Breakfast Club: Four Remarkable Friends Who Transformed Science and Changed the World* (New York: Broadway Books, 2011), 204–210.
26. Babbage, *Ninth Bridgewater Treatise,* 30–31, 34, 225–227. For the letter (and a discussion), see Susan Faye Cannon, "The Impact of Uniformitarianism: Two Letters from John Herschel to Charles Lyell, 1836–1837," *Proceedings of the American Philosophical Society* 105, no. 3 (June 1961): 301–314. On Paley and Babbage, see Tamara Ketabgian, "Prosthetic Divinity: Babbage's Engine, Spiritual Intelligence, and the Senses," *Victorian Review* 35, no. 2 (Fall 2009): 33–36.
27. For "amusement," see Darwin, *Autobiography,* 120; for "zeal," see 67–68. His book list is Darwin, *"Books to Be Read" and "Books Read" Notebook* [1838–1851], 3v–4r and 4v–5r, in Peter J. Vorzimmer, "The Darwin Reading Notebooks (1838–1860)," *Journal of the History of Biology* 10, no. 1 (Spring 1977): 107–153, here 120. For a separate list, see Charles Darwin, *Charles Darwin's Notebooks, 1836–1844: Geology, Transmutation of Species, Metaphysical Enquiries,* ed. Paul H. Barrett et al. (Ithaca, NY: Cornell University Press, 1987), 237–328, here 318–328.
28. For "Hurrah," see Charles Darwin, *Notebook E* [Transmutation of Species (10.1838–7.1839)], 59, in Darwin, *Notebooks,* 413. There is some disagreement about whether Darwin referred to the letter's transcription in the second edition of Babbage's *Ninth Bridgewater Treatise* or to a version circulating in London at the time. See Silvan S. Schweber, "The Origin of the 'Origin' Revisited," *Journal of the History of Biology* 10, no. 2 (October 1977): 229–316, here 286n124. See also Charles Darwin, *The Red Notebook of Charles Darwin,* ed. Sandra Herbert (London: British Museum [Natural History], 1980), 86n40. For "Arguing," see Darwin, *Notebook N* [Metaphysics and Expression (1838–1839)], 49, in Darwin, *Notebooks,* 576–577. On Darwin's relationship to Herschel, see Michael Ruse, "Darwin and Herschel," *Studies in History and Philosophy of Science Part A* 9, no. 4 (December 1978): 323–331; and S. S. Schweber, "John Herschel and

Charles Darwin: A Study in Parallel Lives," *Journal of the History of Biology* 22, no. 1 (Spring 1989): 1–71.

29. For "love of the deity," see Darwin, *Notebook C,* 166, in Darwin, *Notebooks,* 291. For "cold water," see Darwin, *Notebook M* [Metaphysics on Morals and Speculations on Expression (1838)], 19, in ibid., 524. For "avoid," see Darwin, *Notebook M,* 57, in ibid., 532–533. On Darwin's ambivalence, see David Kohn, "Darwin's Ambiguity: The Secularization of Biological Meaning," *British Journal for the History of Science* 22, no. 2 (July 1989): 215–239. On the importance of these topics to his larger project, see Sandra Herbert, "The Place of Man in the Development of Darwin's Theory of Transmutation: Part I," *Journal of the History of Biology* 7, no. 2 (October 1974): 217–258; and Herbert, "The Place of Man in the Development of Darwin's Theory of Transmutation: Part II," *Journal of the History of Biology* 10, no. 2 (October 1977): 155–227.
30. For "young geologists," see Darwin to Charles Henry Lardner Wood, March 4, 1850, in *Correspondence of Charles Darwin,* vol. 4: *1847–1850* (Cambridge: Cambridge University Press, 1989), 316–317. The earlier letter is Darwin to Adolf von Morlot, October 10 [1844], in *Correspondence of Charles Darwin,* 3:64–66, here 64. On Glen Roy, see Martin Rudwick, "Darwin and Glen Roy: A 'Great Failure' in Scientific Method?," *Studies in History and Philosophy of Science Part A* 5, no. 2 (August 1974): 97–185. On corals, see Sponsel, *Darwin's Evolving Identity.* The very existence of "Darwin's delay" is in dispute. See John van Wyhe, "Mind the Gap: Did Darwin Avoid Publishing His Theory for Many Years?," *Notes and Records of the Royal Society of London* 61, no. 2 (May 2007): 177–205; and Roderick D. Buchanan and James Bradley, "'Darwin's Delay': A Reassessment of the Evidence," *Isis* 108, no. 3 (September 2017): 529–552.
31. Charles Darwin, *Journal of Researches into the Geology and Natural History of the Various Countries Visited by H.M.S. Beagle: Under the Command of Captain FitzRoy, R.N., from 1832 to 1836* (London: Henry Colburn, 1839), 607–608. For "speculative," see John F. W Herschel, "Whewell on Inductive Sciences," *Quarterly Review* 68, no. 135 (June 1841): 177–238. Darwin's notes on Herschel appear in Darwin, *"Books to Be Read" and "Books Read" Notebook,* 15v, emphasis in the original.
32. Charles Darwin, *Notebook D* [Transmutation of Species (7–10 1838)], 25–26, 49, in Darwin, *Notebooks,* 339–340, 347. On "affinity" and "analogy," see Stephen Jay Gould, "Evolution and the Triumph of Homology, or Why History Matters," *American Scientist* 74, no. 1 (January 1986): 60–69. On "unwillingness," see Darwin, *Notebook M,* 154e, in Darwin, *Notebooks,* 559.
33. Charles Darwin, *Notebook B* [Transmutation of Species (1837–1838)], 3–4 and 18, in Darwin, *Notebooks,* 167–236, here 171 and 175.

34. For "late work," see Darwin to Charles Lyell, July 30, 1837, in *Correspondence of Charles Darwin,* vol. 2: *1837–1842,* 32. For "confessing a murder," see Darwin to Joseph Dalton Hooker, January 11, 1844, in *Correspondence of Charles Darwin,* 3:1–3, here 2.
35. Charles Darwin, "Macculloch. Attrib of Deity" [Essay on Theology and Natural Selection], 58v–59, in Darwin, *Notebooks,* 638. For "grains of sand," see Darwin, *Notebook M,* 31, in ibid., 527.
36. For Darwin's letter to Gray, see Charles Darwin and Alfred Russel Wallace, "On the Tendency of Species to Form Varieties; and on the Perpetuation of Varieties and Species by Natural Means of Selection," *Journal of the Proceedings of the Linnaean Society of London* 3 (July 1858): 45–62, here 50–53. Darwin's first published discussion of "unconscious selection" appears in Darwin, *Origin,* 34–40, here 36. For more on "unconscious selection," see Stephen G. Alter, "Separated at Birth: The Interlinked Origins of Darwin's Unconscious Selection Concept and the Application of Sexual Selection to Race," *Journal of the History of Biology* 40, no. 2 (October 2006): 231–258. On the closing of the gap, see D. Graham Burnett, "Savage Selection: Analogy and Elision in *On the Origin of Species,*" *Endeavour* 33, no. 4 (December 2009): 121–126.
37. The reference to naturalists' "unconsciously seeking" is on 420 of *Origin* and is echoed on 425.
38. For "pure hypothesis," see Darwin, *Notebook D,* 58, in Darwin, *Notebooks,* 352. For "Cuidado," see Darwin, *Notebook B,* 44, in Darwin, *Notebooks,* 171. For his warnings, see Darwin to Morlot, October 10 [1844], in *Correspondence of Charles Darwin,* 3:64–66, here 65. For "line of argument," see Darwin, *Notebook D,* 117–118, in Darwin, *Notebooks,* 370–371. For "fools' experiments," see E. R. Lankester, "Charles Robert Darwin," in *Library of the World's Best Literature Ancient and Modern,* vol. 2, ed. C. D. Warner (New York: R. S. Peale and J. A. Hill, 1896), 4385–4393, here 4391. On Darwin's use of the phrase, see Ralph Colp, "Notes on Charles Darwin's 'Autobiography,'" *Journal of the History of Biology* 18, no. 3 (Autumn 1985): 357–401, esp. 399–400. On his debts to Herschel, see Peter Gildenhuys, "Darwin, Herschel, and the Role of Analogy in Darwin's Origin," *Studies in History and Philosophy of Science Part C: Studies in History and Philosophy of Biological and Biomedical Sciences* 35, no. 4 (December 2004): 593–611. See also Cowles, "On the Origin of Theories."
39. Darwin, *Notebook D,* 118, 26, in Darwin, *Notebooks,* 371, 340. On instinct, see Robert J. Richards, "Instinct and Intelligence in British Natural Theology: Some Contributions to Darwin's Theory of the Evolution of Behavior,"

Journal of the History of Biology 14, no. 2 (Autumn 1981): 193–230. See also Robert J. Richards, "Lloyd Morgan's Theory of Instinct, 12–32.

40. While the "Sketch" of 1842 was only rediscovered (in a cupboard beneath the stairs) after Emma's death, the "Essay" appeared earlier. Both were republished in 1909. See Charles Darwin, *The Foundations of the Origin of Species,* ed. Francis Darwin (Cambridge: Cambridge University Press, 1909). For his letter to Hooker, see *Correspondence of Charles Darwin,* 3:1–3. On these publication arrangements, see Darwin to Emma Darwin, July 5, 1844, in ibid., 3:43–45, here 43–44.
41. For Huxley's recollection, see Charles Darwin, *The Life and Letters of Charles Darwin, Including an Autobiographical Chapter,* ed. Francis Darwin, vol. 2 (London: John Murray, 1887), 14. The pieces from the "Sketch" appear in Darwin, *Foundations:* 1 ("habits of life"), 10 ("contrast with Lamarck"), and 17 ("partly habit"); "consensual" appears on 57. On Darwin and Lamarck, see Jessica Riskin, *The Restless Clock: A History of the Centuries-Long Argument over What Makes Living Things Tick* (Chicago: University of Chicago Press, 2016), 214–249. On distinctions between earlier and later versions, see Dov Ospovat, *The Development of Darwin's Theory: Natural History, Natural Theology, and Natural Selection, 1838–1859* (Cambridge: Cambridge University Press, 1981), esp. 83–86.
42. Charles Darwin, *The Variation of Animals and Plants under Domestication,* vol. 2 (London: John Murray, 1868), 374, 389, 395. On pangenesis, see Gerald L. Geison, "Darwin and Heredity: The Evolution of His Hypothesis of Pangenesis," *Journal of the History of Medicine and Allied Sciences* 24, no. 4 (October 1969): 375–411. See also Rasmus G. Winther, "Darwin on Variation and Heredity," *Journal of the History of Biology* 33, no. 3 (Winter 2000): 425–455.
43. Darwin, *Variation,* 1:8, 9, 3. On the reception of pangenesis, see Kate Holterhoff, "The History and Reception of Charles Darwin's Hypothesis of Pangenesis," *Journal of the History of Biology* 47, no. 4 (Winter 2014): 661–695. For a related view, see Ghiselin, *Triumph,* 181–186.
44. Darwin, *Variation,* 2:357. On his "well-abused hypothesis," see Darwin, *Autobiography,* 130.
45. Darwin to Asa Gray, July 20 [1857], in *Correspondence of Charles Darwin,* vol. 6: *1856–1857* (Cambridge: Cambridge University Press, 1990), 431–433, here 432. For his recollection about hypothesizing, see Darwin, *Autobiography,* 141.
46. Darwin, *Notebook D,* 135e (excised sheets), in Darwin, *Notebooks,* 339–340. On Darwin's reading of Malthus, see Dov Ospovat, "Darwin after Malthus," *Journal of the History of Biology* 12, no. 2 (Autumn 1979): 211–230. See also Ospovat, *Development of Darwin's Theory,* 60–86. On orchids and chance, see John

Beatty, "Chance Variation: Darwin on Orchids," *Philosophy of Science* 73, no. 5 (December 2006): 629–641. The role of chance remained a vexed issue for Darwin—and indeed, still vexes his readers today.

47. Darwin, *Autobiography,* 120. On Malthus's impact, see Peter Vorzimmer, "Darwin, Malthus, and the Theory of Natural Selection," *Journal of the History of Ideas* 30, no. 4 (October 1969): 527–542. See also David Kohn, "Theories to Work By: Rejected Theories, Reproduction, and Darwin's Path to Natural Selection," *Studies in History of Biology* 4 (1980): 67–170.
48. For "uphill," see Darwin to Joseph Dalton Hooker, 15 [May 1860], in *Correspondence of Charles Darwin,* vol. 8: *1860* (Cambridge: Cambridge University Press, 1993), 210–212, here 211. For "fight my best," see Darwin to Asa Gray, May 18 [1860], in ibid., 216–217, here 216. For "fight publicly," see Darwin to Hooker, [July 2, 1860], in ibid., 272–273, here 272. For "flank movement," see Gray to Darwin, July 2–3, 1862, in ibid., vol. 10: *1862,* 291–294, here 292; and Darwin to Gray, July 23[–24] [1862], in ibid., 330–334, here 331. On the "flank movement" for attention, see Kathryn Tabb, "Darwin at Orchis Bank: Selection after the Origin," *Studies in History and Philosophy of Science Part C: Studies in History and Philosophy of Biological and Biomedical Sciences* 55 (February 2016): 11–20, here 19.
49. For "long battle," see Darwin to G. H. K. Thwaites, October 20 [1860], in *Correspondence of Charles Darwin,* 8:440–441, here 441. For Babbage's emphasis on error, see Babbage, *Ninth Bridgewater Treatise,* 27–28.
50. The line "red in tooth and claw" famously appeared in Alfred, Lord Tennyson's poem "In Memoriam," published in 1850. For an example of its later use to emphasize competition, see Richard Dawkins, *The Selfish Gene* (Oxford: Oxford University Press, 1976), 2. Michael Ruse's "science red in tooth and claw" describes the theory's *reception,* not its production. See Michael Ruse, *The Darwinian Revolution: Science Red in Tooth and Claw,* 2nd ed. (Chicago: University of Chicago Press, 1999).
51. Asa Gray, "Notice [of Robert Brown and Charles Darwin]," *American Naturalist* 8, no. 8 (1874): 475–479, here 479. On teleology, see James G. Lennox, "Darwin Was a Teleologist," *Biology and Philosophy* 8, no. 4 (October 1993): 409–421; Michael T. Ghiselin, "Darwin's Language May Seem Teleological, but His Thinking Is Another Matter," *Biology and Philosophy* 9, no. 4 (October 1994): 489–492; and Lennox, "Teleology by Another Name: A Reply to Ghiselin," *Biology and Philosophy* 9, no. 4 (October 1994): 493–495. See also Richard England, "Natural Selection, Teleology, and the Logos: From Darwin to the Oxford Neo-Darwinists, 1859–1909," *Osiris* 16 (2001): 270–287.
52. Gray, "Notice [of Robert Brown and Charles Darwin]," 477. For "imaginary illustrations," see Charles Darwin, *Origin,* 90.

53. For "virtual witnessing," see Steven Shapin, "Pump and Circumstance: Robert Boyle's Literary Technology," *Social Studies of Science* 14, no. 4 (November 1984): 481–520. See also Steven Shapin and Simon Schaffer, *Leviathan and the Air-Pump: Hobbes, Boyle, and the Experimental Life* (Princeton: Princeton University Press, 1985). On "Darwinian thought experiments," see James G. Lennox, "Darwinian Thought Experiments: A Function for Just-So Stories," in *Thought Experiments in Science and Philosophy,* ed. Tamara Horowitz and Gerald J. Massey (Savage, MD: Rowman and Littlefield, 1991), 223–246. For "cognitive strategy," see Lennox, "Darwin's Methodological Evolution," *Journal of the History of Biology* 38, no. 1 (Spring 2005): 85–99.
54. Charles Darwin, *Notebook B,* 36, 49–50, in Darwin, *Notebooks,* 180, 182. On Darwin's use of trees, see Heather Brink-Roby, "Natural Representation: Diagram and Text in Darwin's 'On the Origin of Species,'" *Victorian Studies* 51, no. 2 (January 2009): 247–273.
55. Charles Darwin, *On the Various Contrivances by Which British and Foreign Orchids Are Fertilised by Insects, and on the Good Effects of Intercrossing* (London: J. Murray, 1862), 351.
56. For "brutes," see Kingsley, "How to Study," 299–300. For "fanatics," see Kingsley, *Water-Babies,* 290. On his "backward" orientation, see Stanwood S. Walker, "'Backwards and Backwards Ever': Charles Kingsley's Racial-Historical Allegory and the Liberal Anglican Revisioning of Britain," *Nineteenth-Century Literature* 62, no. 3 (December 2007): 339–379.
57. For "division of labour," see Charles Kingsley to Charles Darwin, June 14, 1865, in *Correspondence of Charles Darwin,* vol. 13: *1865* (Cambridge: Cambridge University Press, 2002), 183–184, here 183. For "final causes," see Kingsley, "Natural Theology of the Future," 329. For "mistress," see Charles Kingsley, *Madam How and Lady Why; Or, First Lessons in Earth Lore for Children* (London: Strahan and Co., 1869), 3. On theology as balance, see David M. Levy and Sandra J. Peart, "Charles Kingsley and the Theological Interpretation of Natural Selection," *Journal of Bioeconomics* 8, no. 3 (December 2006): 197–218.
58. Darwin, *Autobiography,* 144–145 ("complex" and "unbounded patience"), 119 ("note-book"). On his faulty account, see Janet Browne, "Presidential Address: Commemorating Darwin," *British Journal for the History of Science* 38, no. 3 (September 2005): 251–274.
59. Darwin to Asa Gray, September 5 [1857], in *Correspondence of Charles Darwin,* 6:445–450, here 449. For "progress of opinion," see Darwin to *Athenaeum,* May 5 [1863], in ibid., vol. 11: *1863,* 380–381, here 380.

4. Mental Evolution

1. For "absolutely splendid," see Friedrich Engels to Karl Marx, [Manchester, December 11 or 12, 1859], in *The Collected Works of Karl Marx and Frederick Engels*, vol. 40: *Letters, 1856–1859*, 550–551. For "basis for our views," see Marx to Engels, December 19, 1860, in ibid., vol. 41: *Letters, 1860–1864*, 231–233, here 232. For "class struggle," see Marx to Ferdinand Lassalle, (London: Lawrence & Wishart, 1983) January 16, 1861, in ibid., 41:245–247, here 246–247. For "beasts and plants," see Marx to Engels, June 18, 1862, in ibid., 40:380–382, here 381. For the disproof of an offer by Marx to dedicate *Capital* to Darwin, see Lewis S. Feuer, P. Thomas Carroll, and Ralph Colp Jr., "On the Darwin-Marx Correspondence," *Annals of Science* 33, no. 4 (July 1976): 383–394. For a summary of the relationship between the two, see Jonathan Sperber, *Karl Marx: A Nineteenth-Century Life* (New York: Liveright, 2013), 393–397.
2. On Darwin's Malthusian debts and their political-economic context, see Piers J. Hale, *Political Descent: Malthus, Mutualism, and the Politics of Evolution in Victorian England* (Chicago: University of Chicago Press, 2014). See also Silvan S. Schweber, "Darwin and the Political Economists: Divergence of Character," *Journal of the History of Biology* 13, no. 2 (Summer 1980): 195–289.
3. For "Darwin's dangerous idea," see Daniel Dennett, *Darwin's Dangerous Idea: Evolution and the Meanings of Life* (New York: Simon and Schuster, 1995). On how the Huxley-Wilberforce debate has flattened the history of science and religion, see Frank A. J. L James, "An 'Open Clash between Science and the Church'? Wilberforce, Huxley and Hooker on Darwin at the British Association, Oxford, 1860," in *Science and Beliefs: From Natural Philosophy to Natural Science, 1700–1900*, ed. David M. Knight and Matthew Eddy (Burlington, VT: Ashgate, 2005), 171–193.
4. For "inversion," see [Robert MacKenzie Beverley], *The Darwinian Theory of the Transmutation of Species* (London: J. Nisbet, 1867), 295, 306–307. See also Daniel Dennett, "Darwin's 'Strange Inversion of Reasoning,'" *Proceedings of the National Academy of Sciences of the United States of America* 106 (2009): 10061–10065. For "adaptation," see [Thomas Henry Huxley], "Criticisms on 'The Origin of Species,'" *Natural History Review* 4, no. 16 (October 1864): 566–580, here 568. On Huxley's self-presentation, see Paul White, *Thomas Huxley: Making the "Man of Science"* (Cambridge: Cambridge University Press, 2003). On his status as an agnostic, see Bernard Lightman, "Huxley and Scientific Agnosticism: The Strange History of a Failed Rhetorical Strategy," *British Journal for the History of Science* 35, no. 3 (September 2002): 271–289.

See also Lightman, *The Origins of Agnosticism: Victorian Unbelief and the Limits of Knowledge* (Baltimore: Johns Hopkins University Press, 1987).

5. On the rise of "naturalism," see Gowan Dawson and Bernard V. Lightman, eds., *Victorian Scientific Naturalism: Community, Identity, Continuity* (Chicago: University of Chicago Press, 2014); Bernard V. Lightman and Michael S. Reidy, eds., *The Age of Scientific Naturalism: Tyndall and His Contemporaries* (London: Pickering and Chatto, 2014); and Matthew Stanley, *Huxley's Church and Maxwell's Demon: From Theistic Science to Naturalistic Science* (Chicago: University of Chicago Press, 2014). For an overview of this literature, see Henry M. Cowles, "History Naturalized," *Historical Studies in the Natural Sciences* 47, no. 1 (February 2017): 107–116.
6. Charles Darwin, *On the Origin of Species by Means of Natural Selection, or, The Preservation of Favoured Races in the Struggle for Life* (London: John Murray, 1859), 488. Darwin's inclusion of these lines has been the object of historical disagreement. See Carl J. Bajema, "Charles Darwin on Man in the First Edition of the 'Origin of Species,'" *Journal of the History of Biology* 21, no. 3 (Autumn 1988): 403–410; Peter J. Bowler, "Darwin on Man in the 'Origin of Species': A Reply to Carl Bajema," *Journal of the History of Biology* 22, no. 3 (Autumn 1989): 497–500; and Kathy J. Cooke, "Darwin on Man in the 'Origin of Species': An Addendum to the Bajema-Bowler Debate," *Journal of the History of Biology* 23, no. 3 (Autumn 1990): 517–521. My account of psychology in the period is indebted to Robert M. Young, *Mind, Brain, and Adaptation in the Nineteenth Century: Cerebral Localization and Its Biological Context from Gall to Ferrier* (Oxford: Oxford University Press, 1990).
7. [John Stuart Mill], "Bain's Psychology," *Edinburgh Review* 110, no. 224 (October 1, 1859): 287–320, here 287. For "compounded," see James Mill, *Analysis of the Phenomena of the Human Mind,* vol. 1 (London: Baldwin and Cradock, 1829), 1. On British psychology in these years, see Rick Rylance, *Victorian Psychology and British Culture, 1850–1880* (Oxford: Oxford University Press, 2000). See also Peter Garratt, *Victorian Empiricism: Self, Knowledge, and Reality in Ruskin, Bain, Lewes, Spencer, and George Eliot* (Madison, NJ: Fairleigh Dickinson University Press, 2010). On the relationship between the Mills, see Bruce Mazlish, *James and John Stuart Mill: Father and Son in the Nineteenth Century* (New York: Basic Books, 1975). See also Fred Wilson, *Psychological Analysis and the Philosophy of John Stuart Mill* (Toronto: University of Toronto Press, 1990).
8. [Mill], "Bain's Psychology," 288–289. For evidence of Bain and Mill's complicated relationship to each other and to James Mill, see James Mill, *Analysis of the Phenomena of the Human Mind,* ed. John Stuart Mill (London: Longmans, Green, Reader, and Dyer, 1869).

9. On Bain's life, see Alexander Bain, *Autobiography,* ed. William Leslie Davidson (London: Longmans, Green, and Co., 1904). On the relationship between Bain and Mill, see Cathy Gere, *Pain, Pleasure, and the Greater Good: From the Panopticon to the Skinner Box and Beyond* (Chicago: University of Chicago Press, 2017), 133–138. For Bain's take on Mill's *System,* see Alexander Bain, review of Mill's *A System of Logic, Ratiocinative and Inductive; Being a Connected View of the Principles of Evidence, and the Methods of Scientific Investigation, Westminster Review* 39, no. 2 (May 1843): 412–456, here 414. On Bain's impact, see William L. Davidson, "Professor Bain," *Mind* 13, no. 49 (January 1904): 151–155. On *Mind,* see Christopher D. Green, "The Curious Rise and Fall of Experimental Psychology in *Mind,*" *History of the Human Sciences* 22, no. 1 (February 2009): 37–57.
10. Alexander Bain, *The Senses and the Intellect* (London: J. W. Parker and Son, 1855), vi ("laws of association" and "subdivision"), 120 ("reduced"), 315 ("various faculties" and "thinking portion"). The main points are repeated in [Bain], "On Toys," *Westminster Review* 37, no. 1 (January 1842): 97–121. On Mill's championing of Bain, see Rylance, *Victorian Psychology,* 159–162.
11. Bain, *Senses and Intellect,* 520, 540–541. On his associationism, see Benjamin Morgan, *The Outward Mind: Materialist Aesthetics in Victorian Science and Literature* (Chicago: University of Chicago Press, 2017), 91–99.
12. Bain, *Senses and Intellect,* 331, 436. His discussion of scientific labor followed an anonymous essay critiquing science's "abuses of language." See [Alexander Bain], "On the Abuse of Language, in Science and in Common Life," *Fraser's Magazine* 35, no. 206 (February 1847): 139–140.
13. The chapter on "emotions of the intellect" appears in Alexander Bain, *The Emotions and the Will* (London: J. W. Parker and Son, 1859), 199–227. For "suffering" and "revulsion," see 205. On Bain's argument about emotions in general, see Thomas Dixon, *From Passions to Emotions: The Creation of a Secular Psychological Category* (Cambridge: Cambridge University Press, 2003), esp. 150–159. See also Thomas Dixon, "'Emotion': The History of a Keyword in Crisis," *Emotion Review* 4, no. 4 (October 2012): 338–344.
14. For "rush of delight," practical bearing," and "attainment of truth," see Bain, *Emotions and Will,* 201; for "lacerates" and "operations of daily life," see 206.
15. For Bain's footnote, see Mill, *Analysis,* 102n31. For his critique of the son, see Alexander Bain, *John Stuart Mill: A Criticism, with Personal Recollections* (London: Longmans, Green, and Co., 1882), esp. 146–147, 121.
16. Bain, *Mill,* 121. For "one among many trains," see Bain, *Senses and Intellect,* 562.
17. For "circumference," see Bain, *Emotions and Will,* 595.
18. Bain, *Senses and Intellect,* 409, 575.

19. Bain, *Emotions and Will,* 353, 636.
20. [Huxley], "Criticisms," 568. See also Thomas Henry Huxley, *Lay Sermons, Addresses, and Reviews* (London: Macmillan Co., 1870). On "trial and error," see Henry M. Cowles, "Hypothesis Bound: Trial and Error in the Nineteenth Century," *Isis* 106, no. 3 (September 2015): 635–645.
21. The canonical statement of Bain's "interstitial" status is Edwin Garrigues Boring, *A History of Experimental Psychology,* 2nd ed. (New York: Appleton-Century-Crofts, 1950), 223–231. Lorraine Daston has called Bain's position a "curious compromise." See Lorraine J. Daston, "The Theory of Will versus the Science of Mind," in *The Problematic Science: Psychology in Nineteenth-Century Thought,* ed. Mitchell Ash and William Woodward (New York: Praeger, 1982), 88–115, esp. 96–98.
22. [Herbert Spencer], "Bain on the Emotions and the Will," *British and Foreign Medico-Chirurgical Review* 25, no. 49 (January 1860): 58–72, here (including previous paragraph) 59, 64, 65. Bain and Spencer are often treated as a pair. See, for example, Rylance, *Victorian Psychology,* 177–186. My view is that Bain's approach is distinguished from Spencer's by its *internal* emphasis.
23. On Spencer's intellectual biography, see Mark Francis, *Herbert Spencer and the Invention of Modern Life* (Stocksfield: Acumen, 2007). On organism-environment interaction as the key to his writings, see Trevor Pearce, "From 'Circumstances' to 'Environment': Herbert Spencer and the Origins of the Idea of Organism–Environment Interaction," *Studies in History and Philosophy of Biological and Biomedical Sciences* 41, no. 3 (September 2010): 241–252.
24. [Spencer], "Bain on the Emotions," 58. On Spencer's impact, see Mark Francis and Michael Taylor, eds., *Herbert Spencer: Legacies* (London: Routledge, 2015), esp. 1–15. See also Bernard V. Lightman, ed., *Global Spencerism: The Communication and Appropriation of a British Evolutionist* (Leiden: Brill, 2016). On Spencer's place in nineteenth-century thought, see Robert J. Richards, *Darwin and the Emergence of Evolutionary Theories of Mind and Behavior* (Chicago: University of Chicago Press, 1987), 243–294.
25. Bain, *Senses and Intellect,* vi.
26. Francis Herbert Bradley, *Ethical Studies* (London: Henry S. King, 1876), 35–36. On (older) theories of the mind as a collection, see Sean Silver, *The Mind Is a Collection: Case Studies in Eighteenth-Century Thought* (Philadelphia: University of Pennsylvania Press, 2015). On idealism, see W. J. Mander, *British Idealism: A History* (Oxford: Oxford University Press, 2011). On Bradley, see Phillip Ferreira, *Bradley and the Structure of Knowledge* (Albany: SUNY Press, 1999).
27. Bain, *Emotions and Will,* 23–25n.
28. [Spencer], "Bain on the Emotions," 63.

29. Herbert Spencer, *An Autobiography,* vol. 2 (London: Williams and Norgate, 1904), 7. As usual, one must treat this autobiographical account with caution. See, for example, Paul Elliott, "Erasmus Darwin, Herbert Spencer, and the Origins of the Evolutionary Worldview in British Provincial Scientific Culture, 1770–1850," *Isis* 94, no. 1 (March 2003): 1–29. For the "conceivability" critique, see [Spencer], "The Development Hypothesis," *The Leader* 3, no. 104 (March 20, 1852): 280–281. On the development of Spencer's theory, see Young, *Mind, Brain, and Adaptation,* 167–172. On the centrality of psychology to Spencer's formulation of "evolution," see Peter J. Bowler, "The Changing Meaning of 'Evolution,'" *Journal of the History of Ideas* 36, no. 1 (January 1975): 95–114; and Peter J. Bowler, "Herbert Spencer and 'Evolution'—An Additional Note," *Journal of the History of Ideas* 36, no. 2 (April 1975): 367.
30. [Spencer], "Development Hypothesis," 280, 281. For authority not being habitual, see Herbert Spencer, *The Life and Letters of Herbert Spencer,* vol. 2, ed. David Duncan (London: Methuen, 1908), 306. For speculative thinking, see, for example, Spencer, *Autobiography,* 1:166, 198–199. For his habitual roots, see Spencer, *Life and Letters,* 2:319.
31. [Herbert Spencer], "The Universal Postulate," *Westminster Review* 60, no. 118 (October 1853): 513–550, here 514, 520, 531.
32. Spencer's quotations of Mill appear on 522. For another view on Spencer's "universal postulate," see Francis, *Herbert Spencer,* 171–186.
33. [Spencer], "Universal Postulate," 527, 529, 538, 549, 550.
34. Herbert Spencer, "The Genesis of Science," *British Quarterly Review* 20, no. 39 (July 1854): 108–162, here 108, 110, emphasis in the original.
35. For "division of labour," see Spencer, "Genesis of Science," 126; see 129 for "psychological process." On Spencer's relationship to Comte, see, for example, Sydney Eisen, "Herbert Spencer and the Spectre of Comte," *Journal of British Studies* 7, no. 1 (November 1967): 48–67; and Robert Alun Jones, "Comte and Spencer: A Priority Dispute in Social Science," *Journal of the History of the Behavioral Sciences* 6, no. 3 (July 1970): 241–254.
36. Herbert Spencer, *The Principles of Psychology* (London: Longmans, Green, and Co., 1855), 339, 343.
37. Ibid., 344, 340.
38. On Spencer's "lionization," see Francis, *Herbert Spencer,* 144–156. For Darwin's first inclusion of the phrase "survival of the fittest," see Darwin, *Origin,* 5th ed., 72; see also the (new) title for chapter 4: "Natural Selection; or the Survival of the Fittest."
39. Darwin's amendment to the lines about psychology appears in Charles Darwin, *On the Origin of Species,* 6th ed. (London: John Murray, 1872), 428. For Bain's

chapter on evolution, see Alexander Bain, *The Emotions and the Will,* 3rd ed. (London: Longmans, Green, 1875), 47–68.

40. On Carpenter's place in Victorian science, see Alison Winter, "The Construction of Orthodoxies and Heterodoxies in the Early Victorian Life Sciences," in *Victorian Science in Context,* ed. Bernard V. Lightman (Chicago: University of Chicago Press, 1997), 24–50. On Carpenter's relationship with Laycock, see L. S. Jacyna, "The Physiology of Mind, the Unity of Nature, and the Moral Order in Victorian Thought," *British Journal for the History of Science* 14, no. 2 (July 1981): 109–132. See also Jacyna, "Moral Fibre: The Negotiation of Microscopic Facts in Victorian Britain," *Journal of the History of Biology* 36, no. 1 (Spring 2003): 39–85.
41. William Benjamin Carpenter, *Principles of Human Physiology: With Their Chief Applications to Psychology, Pathology, Therapeutics, Hygiène, and Forensic Medicine,* 5th ed. (Philadelphia: Blanchard and Lea, 1853), 791. For more on Carpenter, see Morgan, *The Outward Mind,* 69–72.
42. For "ignorance," see Owen Whooley, *On the Heels of Ignorance: Psychiatry and the Politics of Not Knowing* (Chicago: University of Chicago Press, 2019), 34. Carpenter's "unconscious cerebration" shows how the themes Whooley documents were projected onto the objects under study.
43. Alison Winter, *Mesmerized: Powers of Mind in Victorian Britain* (Chicago: University of Chicago Press, 1998), 287–305, here 287. For Carpenter's footnote, see Carpenter, *Principles of Human Physiology,* 789–790n1.
44. John Hughlings Jackson, "On Epilepsies and on the After Effects of Epileptic Discharges (Todd and Robertson's Hypothesis)," *West Riding Lunatic Asylum Medical Reports* 6, no. 2530 (1876): 266–309, here 287. On his "Spencerian medicine," see Kiersten Feil, "Evolution and the Developmental Perspective in Medicine: The Historical Precedent and Modern Rationale for Explaining Disorder and Normality with Evolutionary Processes" (PhD diss., University of Chicago, 2006), 43–91. On Hughlings Jackson's life, see Macdonald Critchley and Eileen A. Critchley, *John Hughlings Jackson: Father of English Neurology* (Oxford: Oxford University Press, 1998), esp. 53–60.
45. For "tigers," see J. Hughlings Jackson, "Remarks on the Diagnosis and Treatment of Diseases of the Brain," *British Medical Journal* 2 (July 14, 1888): 59–63, here 59. On "evolutionary neurology," see Elizabeth A. Franz and Grant Gillett, "John Hughlings Jackson's Evolutionary Neurology: A Unifying Framework for Cognitive Neuroscience," *Brain* 134, no. 10 (October 2011): 3114–3120.
46. J. Hughlings Jackson, "On the Anatomical & Physiological Localisation of Movements in the Brain," *Lancet* 101, no. 2577 (January 18, 1873): 84–85, here 84. For "bed-side," see Michael Foster, *A Text Book of Physiology* (London: Macmillan Co., 1877), 439.

47. John Hughlings Jackson, *Selected Writings of John Hughlings Jackson,* vol. 2 (London: Hodder and Stoughton, 1931), 354. On his relationship with philosophy, see L. Stephen Jacyna, "Process and Progress: John Hughlings Jackson's Philosophy of Science," *Brain* 134, no. 10 (October 2011): 3121–3126. See also Stephen Casper, *The Neurologists: A History of a Medical Specialty in Modern Britain, c.1789–2000* (Manchester: Manchester University Press, 2014).
48. Edward Burnett Tylor, *Primitive Culture: Researches into the Development of Mythology, Philosophy, Religion, Art, and Custom,* vol. 1 (London: J. Murray, 1871), 1. My account draws on the lifetime of scholarship by George Stocking, especially George W. Stocking, *Race, Culture, and Evolution: Essays in the History of Anthropology* (New York: Free Press, 1968); and Stocking, *Victorian Anthropology* (New York: Free Press, 1987). See also J. W. Burrow, *Evolution and Society: A Study in Victorian Social Theory* (Cambridge: Cambridge University Press, 1966).
49. For "self-examination," see Henrika Kuklick, *The Savage Within: The Social History of British Anthropology, 1885–1945* (Cambridge: Cambridge University Press, 1991), 1. See also Efram Sera-Shriar, *The Making of British Anthropology, 1813–1871* (London: Pickering and Chatto, 2013).
50. On "armchair anthropology," see Efram Sera-Shriar, "What Is Armchair Anthropology? Observational Practices in 19th-Century British Human Sciences," *History of the Human Sciences* 27, no. 2 (April 2014): 26–40. My thoughts here are indebted to conversations with Sarah Pickman. On fieldwork as a rite of passage, see Henrika Kuklick, "Personal Equations: Reflections on the History of Fieldwork, with Special Reference to Sociocultural Anthropology," *Isis* 102, no. 1 (March 2011): 1–33.
51. Stocking, *Victorian Anthropology,* 109. On "mechanical objectivity," see Lorraine Daston and Peter Galison, *Objectivity* (New York: Zone Books, 2007), 115–190.
52. Edward Burnett Tylor, *Researches into the Early History of Mankind and the Development of Civilization* (London: John Murray, 1865), 365. On this transformation, see George W. Stocking, "Matthew Arnold, E. B. Tylor, and the Uses of Invention," *American Anthropologist* 65, no. 4 (August 1963): 783–799. For "pebbles," see Tylor, *Primitive Culture,* 1:59. On Tylor's interpretation of instinct, see George W. Stocking, "'Cultural Darwinism' and 'Philosophical Idealism' in E. B. Tylor: A Special Plea for Historicism in the History of Anthropology," *Southwestern Journal of Anthropology* 21, no. 2 (Summer 1965): 130–147.
53. For counting, see Tylor, *Primitive Culture,* 1:218–246. For "unconscious hypothesis," see Herbert Spencer, *The Principles of Sociology,* vol. 1 (London: Williams and Norgate, 1876), 137. On "savage numbers," see Michael J. Barany,

"Savage Numbers and the Evolution of Civilization in Victorian Prehistory," *British Journal for the History of Science* 47, no. 2 (June 2014): 239–255.

54. Sir Edward Burnett Tylor, *Anthropology: An Introduction to the Study of Man and Civilization* (London: Macmillan and Co., 1881), 439.
55. Ibid., 54–55. On the rise of this form of expertise in the human sciences, see, for example, John Carson, *The Measure of Merit: Talents, Intelligence, and Inequality in the French and American Republics, 1750–1940* (Princeton: Princeton University Press, 2007).
56. For "savage and barbaric science," see Tylor, *Primitive Culture,* 2:64. For "vain," see Franz Boas, "The Mind of Primitive Man," *Journal of American Folklore* 14, no. 52 (1901): 1–11, here 7. On the shift from "savage" to "primitive," see Peter J. Bowler, "From 'Savage' to 'Primitive': Victorian Evolutionism and the Interpretation of Marginalized Peoples," *Antiquity* 66, no. 252 (September 1992): 721–729.
57. Tylor, *Primitive Culture,* 2:410.
58. Kuklick, *The Savage Within.* On related tensions of interpretation a few decades later, see Henrika Kuklick, "'Humanity in the Chrysalis Stage': Indigenous Australians in the Anthropological Imagination, 1899–1926," *British Journal for the History of Science* 39, no. 143 (December 2006): 535–568. The most famous interrogation of "the savage" in anthropology remains Claude Lévi-Strauss, *The Savage Mind* (Chicago: University of Chicago Press, 1966). On the ironic consequences of imperial psychology, see Erik Linstrum, *Ruling Minds: Psychology in the British Empire* (Cambridge, MA: Harvard University Press, 2016). See also Erik Linstrum, "The Politics of Psychology in the British Empire, 1898–1960," *Past & Present,* no. 215 (May 2012): 195–233. For more on unintended consequences, see Rebecca Lemov, *Database of Dreams: The Lost Quest to Catalog Humanity* (New Haven: Yale University Press, 2015).
59. See Stocking, *Victorian Anthropology,* xi. For a similar parallelism, see Curtis M. Hinsley, *Savages and Scientists: The Smithsonian Institution and the Development of American Anthropology, 1846–1910* (Washington, DC: Smithsonian Institution Press, 1981). For example, Hinsley argues that early work on American Indians was "an exercise in self-study by Americans who sensed but were unable to confront directly the tragic dimensions of their culture and of their own lives" (8).
60. On "nature's method," see Stephen Tomlinson, "The Method of Nature: Herbert Spencer and the Education of the Adaptive Mind," in *Herbert Spencer: Legacies,* ed. Mark Francis and Michael Taylor (London: Routledge, 2015), 16–39. For the longer history of this relationship, see Peter Dear, "Method and the Study of Nature," in *The Cambridge History of Seventeenth-Century Philosophy,* ed. Daniel

Garber and Michael Ayers, vol. 1 (Cambridge: Cambridge University Press, 2003), 147–177.

61. For "common sense," see Huxley, *Lay Sermons,* 72–93, here 77. On a very different use of a similar idea in the same period, see James Elwick, "Herbert Spencer and the Disunity of the Social Organism," *History of Science* 41, no. 1 (March 2003): 35–72. For Lewes, see George Henry Lewes, *Problems of Life and Mind,* vol. 1 (London: Trübner and Co., 1874), 89. See also Garratt, *Victorian Empiricism,* 102–126.
62. William Stanley Jevons, *The Principles of Science: A Treatise on Logic and Scientific Method,* vol. 1 (London: Macmillan Co., 1874), 167. Beyond the *Principles,* Jevons was best known for his work in economics. See Margaret Schabas, *A World Ruled by Number: William Stanley Jevons and the Rise of Mathematical Economics* (Princeton: Princeton University Press, 1990); and Harro Maas, *William Stanley Jevons and the Making of Modern Economics* (Cambridge: Cambridge University Press, 2005).
63. Thomas Henry Huxley, "The Progress of Science [1887]," in *Method and Results: Essays* (London: Macmillan and Co., 1893), 42–129, here 61. For one interpretation of the "paradox" of Huxley's approach, see George Levine, "Paradox: The Art of Scientific Naturalism," in Dawson and Lightman, *Victorian Scientific Naturalism,* 79–100. The passage from Huxley is quoted on 85.
64. Stanley, *Huxley's Church and Maxwell's Demon,* 248. On "common sense" as a rhetorical and political tool, see Sophia Rosenfeld, *Common Sense: A Political History* (Cambridge, MA: Harvard University Press, 2011), 250.

5. A Living Science

1. For "My Psychology," see William James to Théodule Ribot, May 13, 1888, in *The Correspondence of William James,* ed. Ignas K. Skrupskelis and Elizabeth M. Berkeley, vol. 6 (Charlottesville: University Press of Virginia, 1992–), 409. For "staggered," see Henry Holt to William James, June 8, 1878, in ibid., 5:14. Competing books included James McCosh, *Psychology: The Cognitive Powers* (New York: Charles Scribner's Sons, 1886); George Trumbull Ladd, *Elements of Physiological Psychology: A Treatise of the Activities and Nature of the Mind from the Physical and Experimental Point of View* (New York: Charles Scribner's Sons, 1887); and John Dewey, *Psychology* (New York: Harper and Bros., 1887). On the anxiety provoked by this wave, see William James to George Trumbull Ladd, April 4, 1887, in *Correspondence,* 6:216.
2. On James's life, see Ralph Barton Perry, *The Thought and Character of William James, as Revealed in Unpublished Correspondence and Notes, Together with His*

Published Writings (Boston: Little, Brown, and Co., 1935); Linda Simon, *Genuine Reality: A Life of William James* (New York: Harcourt Brace, 1998); and Robert D. Richardson, *William James: In the Maelstrom of American Modernism; A Biography* (Boston: Houghton Mifflin, 2006). The standard account of the rise of scientific psychology remains Kurt Danziger, *Constructing the Subject: Historical Origins of Psychological Research* (Cambridge: Cambridge University Press, 1990). See also Roger Smith, *Between Mind and Nature: A History of Psychology* (London: Reaktion Books, 2013). On psychology's growth in the United States (relative to philosophy), see Bruce Kuklick, *The Rise of American Philosophy: Cambridge, Massachusetts, 1860–1930* (New Haven: Yale University Press, 1977), 127–214. See also Deborah J. Coon, "Standardizing the Subject: Experimental Psychologists, Introspection, and the Quest for a Technoscientific Ideal," *Technology and Culture* 34, no. 4 (October 1993): 757–783.

3. For "dropsical mass," see William James to Henry Holt, May 9, 1890, in *Correspondence,* 7:24, emphasis in the original. Points of pride included the book's size ("the biggest book on psychology in any language except Wundt's, Rosmini's, and Daniel Greenleaf Thompson's!") and its "positivistic point of view." See William James to Alice Howe Gibbens James, May 24, 1890, in *Correspondence,* 7:38. On James's reception, see the essays in John J. Stuhr, ed., *100 Years of Pragmatism: William James's Revolutionary Philosophy* (Bloomington: Indiana University Press, 2010).
4. On the construction of the *Principles,* see James, *The Principles of Psychology,* vol. 3: *Notes, Appendixes, Apparatus, General Index* (Cambridge, MA: Harvard University Press, 1981), 1532–1579. For examples of critical reviews, see [G. Stanley Hall], review of James's *Principles of Psychology, American Journal of Psychology* 3, no. 4 (February 1891): 578–591; [Charles Sanders Peirce], "James's Psychology–1," *The Nation,* July 2, 1891, 15; [Peirce], "James's Psychology–2," *The Nation,* July 9, 1891, 32–33. See also Dominic W. Massaro, "A Century Later: Reflections on *The Principles of Psychology* by William James and on the Review by G. Stanley Hall," *American Journal of Psychology* 103, no. 4 (Winter 1990): 539–545.
5. The canonical account of "the new psychology" as a German import was Edwin Garrigues Boring, *A History of Experimental Psychology* (New York: Century Co., 1929). Though subsequent work has complicated Boring's history, the basic vectors remain. See, for example, the essays in Robert W. Rieber and David Keith Robinson, eds., *Wilhelm Wundt in History: The Making of a Scientific Psychology* (New York: Kluwer Academic, 2001). For "new psychologies," see Lorraine J. Daston, "The Theory of Will versus the Science of Mind," in *The Problematic Science: Psychology in Nineteenth-Century Thought,* ed. Mitchell Ash and William

Woodward (New York: Praeger, 1982), 88–115, here 90. On the rhetoric of "new psychology," see David E. Leary, "Telling Likely Stories: The Rhetoric of the New Psychology, 1880–1920," *Journal of the History of the Behavioral Sciences* 23, no. 4 (October 1987): 315–331. See also Elissa N. Rodkey, "Last of the Mohicans? James McCosh and Psychology 'Old' and 'New,'" *History of Psychology* 14, no. 4 (November 2011): 335–355. On "the laboratory revolution," see Andrew Cunningham and Perry Williams, eds., *The Laboratory Revolution in Medicine* (Cambridge: Cambridge University Press, 1992). On the "brass age," see Robert C. Davis, "The Brass Age of Psychology," *Technology and Culture* 11, no. 4 (October 1970): 604–612.

6. On James's complicated relationship with "experiment," see Rand B. Evans, "William James, 'The Principles of Psychology,' and Experimental Psychology," *American Journal of Psychology* 103, no. 4 (December 1990): 433–447. See also Paul Jerome Croce, *Science and Religion in the Era of William James* (Chapel Hill: University of North Carolina Press, 1995); and Paul Jerome Croce, *Young William James Thinking* (Baltimore: Johns Hopkins University Press, 2018).
7. Julie A. Reuben, *The Making of the Modern University: Intellectual Transformation and the Marginalization of Morality* (Chicago: University of Chicago Press, 1996). See also Jon H. Roberts and James Turner, *The Sacred and the Secular University* (Princeton: Princeton University Press, 2000). For "Octopus," see William James, "The Ph.D. Octopus," *Harvard Monthly,* March 1903, 1–9. On James's boundary-work, see Francesca Bordogna, *William James at the Boundaries: Philosophy, Science, and the Geography of Knowledge* (Chicago: University of Chicago Press, 2008).
8. For "launching," see Robert V. Bruce, *The Launching of Modern American Science, 1846–1876* (New York: Knopf, 1987). For critical engagement with Bruce's thesis, see Stuart W. Leslie, "Communities of Nineteenth-Century Science and Technology," *Reviews in American History* 17, no. 2 (June 1989): 232–237; and Paul Lucier, "The Professional and the Scientist in Nineteenth-Century America," *Isis* 100, no. 4 (December 2009): 699–732.
9. On reforms in this period, see Laurence R. Veysey, *The Emergence of the American University* (Chicago: University of Chicago Press, 1965). See also Louis Menand, Paul Reitter, and Chad Wellmon, eds., *The Rise of the Research University: A Sourcebook* (Chicago: University of Chicago Press, 2017).
10. On the role of science, see Andrew Jewett, *Science, Democracy, and the American University: From the Civil War to the Cold War* (Cambridge: Cambridge University Press, 2012). On France and Germany as destinations, see John Harley Warner, *Against the Spirit of System: The French Impulse in Nineteenth-Century American Medicine* (Princeton: Princeton University Press, 1998), esp. 291–329.

11. Charles Eliot, "The New Education, I," *Atlantic Monthly,* February 1869, 203–221; Charles Eliot, "The New Education, II," *Atlantic Monthly,* March 1869, 358–367. The quotation appears in part I, 206; Eliot emphasizes "experiment" throughout, but see 219 for an especially telling example. On Eliot's impact, see Hugh Hawkins, *Between Harvard and America: The Educational Leadership of Charles W. Eliot.* (Oxford: Oxford University Press, 1972).
12. For "globe" and "drill," see Eliot, "New Education I," 218. For "multiply," see Charles Eliot, "Inaugural Address by President Eliot," in *Addresses at the Inauguration of Charles William Eliot as President of Harvard College, Tuesday, October 19, 1869* (Cambridge, MA: Sever and Francis, 1869), 29–65, here 40.
13. Today this view of the mind is most often associated with "toolkit theory" in sociology. See Ann Swidler, "Culture in Action: Symbols and Strategies," *American Sociological Review* 51, no. 2 (April 1986): 273–286. For a critical discussion, see Omar Lizardo and Michael Strand, "Skills, Toolkits, Contexts and Institutions: Clarifying the Relationship between Different Approaches to Cognition in Cultural Sociology," *Poetics* 38, no. 2 (April 2010): 205–228.
14. "Pre-disciplinary matrix" is a double reference to Thomas Kuhn, pointing to his acknowledgment of "pre-paradigmatic" sciences and to the "disciplinary matrix" within which paradigms take hold. See Thomas S. Kuhn, *The Structure of Scientific Revolutions,* 2nd ed. (Chicago: University of Chicago Press, 1970). Joel Isaac describes Harvard (later on) as an "interstitial academy." See Joel Isaac, *Working Knowledge: Making the Human Sciences from Parsons to Kuhn* (Cambridge, MA: Harvard University Press, 2012), esp. 31–62.
15. On the Club, see Louis Menand, *The Metaphysical Club: A Story of Ideas in America* (New York: Farrar, Straus and Giroux, 2001). Even its existence has been doubted. See, for example, Philip P. Wiener, "Peirce's Metaphysical Club and the Genesis of Pragmatism," *Journal of the History of Ideas* 7, no. 2 (April 1946): 218–233; and Max H. Fisch, "Was There a Metaphysical Club in Cambridge?," in *Studies in the Philosophy of Charles Sanders Peirce,* ed. Edward C. Moore and Richard S. Robin (Amherst: University of Massachusetts Press, 1964), 3–32. See also Mark Mendell, "The Problem of the Origin of Pragmatism," *History of Philosophy Quarterly* 12, no. 1 (January 1995): 111–131.
16. Menand, *The Metaphysical Club,* 369. See also his introduction to Louis Menand, ed., *Pragmatism: A Reader* (New York: Vintage Books, 1997).
17. On pragmatism and evolution, see Philip P. Wiener, *Evolution and the Founders of Pragmatism* (Gloucester, MA: Smith, 1969). See also Lucas McGranahan, *Darwinism and Pragmatism: William James on Evolution and Self-Transformation* (London: Routledge, 2017). I am indebted to Wiener's pioneering work and to Trevor Pearce for many conversations on this subject.

18. On Chauncey Wright, see Edward H. Madden, *Chauncey Wright and the Foundations of Pragmatism* (Seattle: University of Washington Press, 1963). See also Cheryl Misak, *The American Pragmatists* (Oxford: Oxford University Press, 2013), 14–25. Peirce's account appears in Charles Sanders Peirce, *Collected Papers of Charles Sanders Peirce,* vol. 5 (Cambridge, MA: Harvard University Press, 1974), 6–9, here 8. For "Big Tent Positivism," see Trevor Pearce, "'Science Organized': Positivism and the Metaphysical Club, 1865–1875," *Journal of the History of Ideas* 76, no. 3 (July 2015): 441–465.
19. The correspondence between Wright and Darwin began in 1871. For "will of man," see Charles Darwin to Chauncey Wright, June 3 [1872], in *The Correspondence of Charles Darwin,* ed. Frederick Burkhardt et al., vol. 20: *1872* (Cambridge: Cambridge University Press, 2013), 241–242, here 241. For Wright's response, see Wright to Darwin, August 29, 1872, in ibid., 379–382, here 381. For "great advantage," see Wright to Darwin, September 9, 1872, in ibid., 398–400, here 398. On the interactions between Wright and Darwin, see Philip P. Wiener, "Chauncey Wright's Defense of Darwin and the Neutrality of Science," *Journal of the History of Ideas* 6, no. 1 (January 1945): 19–45. See also Wiener, *Evolution and the Founders of Pragmatism,* 152–171.
20. For "more like an essay," see Wright to Darwin, August 29, 1872, in *Correspondence of Charles Darwin,* 20:379–382, here 382. For "potential existence," see Chauncey Wright, "Evolution of Self-Consciousness," *North American Review,* April 1873, 245–310, here 248. On Wright's evolutionism, see Serge Grigoriev, "Chauncey Wright: Theoretical Reason in a Naturalist Account of Human Consciousness," *Journal of the History of Ideas* 73, no. 4 (October 2012): 559–582.
21. On Bain's influence, see Max H. Fisch, "Alexander Bain and the Genealogy of Pragmatism," *Journal of the History of Ideas* 15, no. 3 (June 1954): 413–444; and Cheryl J. Misak, *Cambridge Pragmatism: From Peirce and James to Ramsey and Wittgenstein* (Oxford: Oxford University Press, 2016), 18–20.
22. Alexander Bain, *The Emotions and the Will* (London: J. W. Parker and Son, 1859), 568. The summary is from Peirce, *Collected Papers,* 5:7. See also Menand, *The Metaphysical Club,* 225–226.
23. For the lines on Green, see [Charles Sanders Peirce], "Nicholas St. John Green," *Proceedings of the American Academy of Arts and Sciences* 12 (1876): 289–291, here 290.
24. Menand, *Metaphysical Club,* 232–234.
25. Newcomb was more interested in method as a set of *rules* than as an empirical description of the mind. See Albert E. Moyer, *A Scientist's Voice in American Culture: Simon Newcomb and the Rhetoric of Scientific Method* (Berkeley: University of California Press, 1992), esp. 47–51.

26. Simon Newcomb, "Exact Science in America," *North American Review,* October 1874, 286–308, here 307, 287, 308. On the essay's context, see Moyer, *A Scientist's Voice,* 82–86.
27. Simon Newcomb, "Abstract Science in America, 1776–1876," *North American Review,* January 1876, 88–123, here 122. For Adams's request, see Moyer, *A Scientist's Voice,* 86–89. For his own diagnosis, see [Henry Adams], *Democracy: An American Novel* (New York: Henry Holt and Co., 1880). On Adams's famous ambivalence, see William Wasserstrom, *The Ironies of Progress: Henry Adams and the American Dream* (Carbondale: Southern Illinois University Press, 1984).
28. Newcomb, "Abstract Science," 123. Of course, visions for implementation took many forms. On Newcomb's, see Jewett, *Science, Democracy,* 70–71n36. On his debts to Mill, see Moyer, *A Scientist's Voice,* 41–45.
29. On Benjamin Peirce, see Edward R. Hogan, *Of the Human Heart: A Biography of Benjamin Peirce* (Bethlehem, PA: Lehigh University Press, 2008). On his impact, see Sven R. Peterson, "Benjamin Peirce: Mathematician and Philosopher," *Journal of the History of Ideas* 16, no. 1 (January 1955): 89–112. For his pedagogical vision, see Benjamin Peirce, "The Intellectual Organization of Harvard University," *Harvard Register* 1, no. 5 (April 1880), 77. On his son's life, see Joseph Brent, *Charles Sanders Peirce: A Life,* rev. ed. (Bloomington: Indiana University Press, 1998). See also Cheryl Misak, "Charles Sanders Peirce (1839–1914)," in *The Cambridge Companion to Peirce,* ed. Cheryl Misak (Cambridge: Cambridge University Press, 2004), 1–26.
30. On the Institute, see Margaret W. Rossiter, "Benjamin Silliman and the Lowell Institute: The Popularization of Science in Nineteenth-Century America," *New England Quarterly* 44, no. 4 (December 1971): 602–626. See also Donald M. Scott, "The Profession That Vanished: Public Lecturing in Mid-Nineteenth-Century America," in *Professions and Professional Ideologies in America,* ed. Gerald L. Geison (Chapel Hill: University of North Carolina Press, 1983), 12–28. On science in this context, see David A. Hollinger, "Inquiry and Uplift: Late Nineteenth-Century American Academics and the Moral Efficacy of Scientific Practice," in *The Authority of Experts: Studies in History and Theory,* ed. Thomas L. Haskell (Bloomington: Indiana University Press, 1984), 142–156.
31. For the lectures, see Charles S. Peirce, "The Logic of Science; or, Induction and Hypothesis" (Lowell Lectures of 1866), in *Writings of Charles S. Peirce: A Chronological Edition,* vol. 1, ed. Max Harold Fisch et al. (Bloomington: Indiana University Press, 1982), 357–504. For Mill's definition, see John Stuart Mill, *An Examination of Sir William Hamilton's Philosophy and of the Principal Philosophical Questions Discussed in His Writings* (London: Longman, Green, Longman, Roberts and Green, 1865), 388. On Mill's psychologism, see David Godden,

"Psychologism in the Logic of John Stuart Mill: Mill on the Subject Matter and Foundations of Ratiocinative Logic," *History and Philosophy of Logic* 26, no. 2 (May 2005): 115–143.

32. Peirce, *Writings,* 1:162–175, here 164. The unfinished draft appears on 1:305–321. For more distinctions between forms of logic, see 1:322, 361–362.
33. [Charles Sanders Peirce], "The English Doctrine of Ideas," *The Nation,* November 25, 1869, 460–461, here 461. Peirce's mixed metaphors—logic impregnates, infects, immerses, and is useful all at once—illustrate the protean nature of "method" in the period.
34. Peirce, "Whewell," in "Lectures on British Logicians," in *Writings,* 2:337–345, here 339. On Peirce's debts to Whewell, see Henry M. Cowles, "The Age of Methods: William Whewell, Charles Peirce, and Scientific Kinds," *Isis* 107, no. 4 (December 2016): 722–737.
35. Peirce, "Whewell," 339, 342. On the psychologism of both, see John Losee, "Whewell and Mill on the Relation between Philosophy of Science and History of Science," *Studies in History and Philosophy of Science Part A* 14, no. 2 (June 1983): 113–126.
36. For "doubt," see Charles Sanders Peirce, "Chapter 1: Of the Difference between Doubt & Belief," in "Toward a Logic Book, 1872–73," in *Writings,* 3:22–23, here 22. For "feeling," see Charles Sanders Peirce, "The Fixation of Belief," *Popular Science Monthly,* November 1877, 1–15, here 5. For an anti-psychologistic reading, see Jeff Kasser, "Peirce's Supposed Psychologism," *Transactions of the Charles S. Peirce Society* 35, no. 3 (July 1999): 501–526. For an overview of the term's meanings, see Martin Kusch, *Psychologism: A Case Study in the Sociology of Philosophical Knowledge* (London: Routledge, 1994).
37. On the "Illustrations" and their reception, see Brent, *Peirce,* 115–119. Brent sees these as "the first formal expression of the scientific method as the logic of science" (117). On their place in Peirce's work, see Peter Ochs, "A Pragmatic Method of Reading Confused Philosophic Texts: The Case of Peirce's 'Illustrations,'" *Transactions of the Charles S. Peirce Society* 25, no. 3 (July 1989): 251–291.
38. Peirce, "Fixation," 2–3. Peirce later expanded his views on evolution. See Carl R. Hausman, *Charles S. Peirce's Evolutionary Philosophy* (Cambridge: Cambridge University Press, 1993).
39. Peirce, "Fixation," 5, 8, 9, 11, 14, 1, 15. On "fixing" belief, see David Wiggins, "Reflections on Inquiry and Truth Arising from Peirce's Method for the Fixation of Belief," in Misak, *Cambridge Companion to Peirce,* 87–126.
40. Charles Sanders Peirce, "How to Make Our Ideas Clear," *Popular Science Monthly,* January 1878, 286–302, here 291, 294, 293. James was the first to credit Peirce. See William James, "Philosophical Conceptions and Practical Results," *University*

of California Chronicle 4 (September 1898): 287–310, here 290. On Peirce's role, see Brent, *Peirce,* 83–89.

41. Charles Sanders Peirce, "Deduction, Induction, and Hypothesis," *Popular Science Monthly,* August 1878, 470–482, here 481–482. On hypotheses in particular, see Robert Sharpe, "Induction, Abduction, and the Evolution of Science," *Transactions of the Charles S. Peirce Society* 6, no. 1 (January 1970): 17–33, esp. 19–21.
42. For Peirce's example, see Peirce, "Deduction, Induction, and Hypothesis," 472. For "explanatory hypotheses," see Charles Sanders Peirce, "Lecture 6: Three Types of Reasoning," in *The Collected Papers of Charles Sanders Peirce,* vol. 5, ed. Charles Hartshorne and Paul Weiss (Cambridge, MA: Harvard University Press, 1931), 94–111, here 106. For an overview of the literature on abduction, see Daniel J. McKaughan, "From Ugly Duckling to Swan: C. S. Peirce, Abduction, and the Pursuit of Scientific Theories," *Transactions of the Charles S. Peirce Society* 44, no. 3 (July 2008): 446–468. For "inference to the best explanation," see Gilbert H. Harman, "The Inference to the Best Explanation," *Philosophical Review* 74, no. 1 (January 1965): 88–95. See also Peter Lipton, *Inference to the Best Explanation* (London: Routledge, 1991). For the view that abduction is *not* the seed of such inferences, see Daniel G. Campos, "On the Distinction between Peirce's Abduction and Lipton's Inference to the Best Explanation," *Synthese* 180, no. 3 (June 2011): 419–442.
43. Peirce's line comes from a lecture he delivered in 1903. For a discussion, see Peter Skagestad, "C. S. Peirce on Biological Evolution and Scientific Progress," *Synthese* 41, no. 1 (May 1979): 85–114.
44. For "a real science of man," see William James to Charles William Eliot, December 2, 1875, in *Correspondence,* 4:528. On James's teaching, see Ignas K. Skrupskelis, "Introduction," in William James, *Manuscript Lectures,* ed. Ignas K. Skrupskelis et al. (Cambridge, MA: Harvard University Press, 1988), xvii–lxiii. On psychology's increased popularity, see Jeffrey Sklansky, *The Soul's Economy: Market Society and Selfhood in American Thought, 1820–1920* (Chapel Hill: University of North Carolina Press, 2002), esp. 137–169.
45. William James to Charles William Eliot, December 2, 1875, in *Correspondence,* 4:528. For more on James's balance, see Daniel W. Bjork, *The Compromised Scientist: William James in the Development of American Psychology* (New York: Columbia University Press, 1983). See also Gerald E. Myers, "Pragmatism and Introspective Psychology," in *The Cambridge Companion to William James,* ed. Ruth Anna Putnam (Cambridge: Cambridge University Press, 1997), 11–24.
46. For "humbugs," see William James to James Jackson Putnam, May 26, 1877, in *Correspondence,* 4:564. On James's long engagement with Spencer, see Robert J.

Richards, *Darwin and the Emergence of Evolutionary Theories of Mind and Behavior* (Chicago: University of Chicago Press, 1987), 424–430. See also McGranahan, *Darwinism and Pragmatism,* 58–60.

47. For James's recollection, see William James, "Herbert Spencer," *Atlantic Monthly,* July 1904, 99–108, here 104; for "immense credit," see 103. For "variability," see William James, "Notes for Natural History 2: Physiological Psychology (1876–1877)," in *Manuscript Lectures,* 126–129, here 127. For "crowded," see [William James], "Herbert Spencer's Data of Ethics," *The Nation,* September 11, 1879, 178–179, here 178. For "crimes," see William James to Charles Sanders Peirce, March 16 [1890], in *Correspondence,* 7:8.
48. On Youmans's life, see Charles M. Haar, "E. L. Youmans: A Chapter in the Diffusion of Science in America," *Journal of the History of Ideas* 9, no. 2 (April 1948): 193–213; and William E. Leverette Jr., "E. L. Youmans' Crusade for Scientific Autonomy and Respectability," *American Quarterly* 17, no. 1 (April 1965): 12–32. On the American visit, see Barry Werth, *Banquet at Delmonico's: Great Minds, the Gilded Age, and the Triumph of Evolution in America* (New York: Random House, 2009). For Youmans's own account, see Edward Livingston Youmans, *Herbert Spencer on the Americans and the Americans on Herbert Spencer* (New York: D. Appleton and Co., 1883).
49. For "no other philosopher," see James, "Herbert Spencer," 104. For "worthless," see William James to Thomas Wren Ward, December 30, 1876, in *Correspondence,* 4:552, emphasis in the original. For "Uncle Spencer," see William James to Charles Sanders Peirce, March 16 [1890], in *Correspondence,* 7:8.
50. William James, "Remarks on Spencer's Definition of Mind as Correspondence," *Journal of Speculative Philosophy* 12, no. 1 (January 1878): 1–18, here 17. For a critical account of the mirror, see Richard Rorty, *Philosophy and the Mirror of Nature* (Princeton: Princeton University Press, 1979).
51. For "unsystematic," see William James to James Ward, November 1, 1892, in *Correspondence,* 7:329, emphasis in the original. For "premonitory," see William James, *The Principles of Psychology,* vol. 1 (New York: Henry Holt and Co., 1890), 255. For his self-description, see William James to Alice Howe Gibbens James, May 4, 1890, in *Correspondence,* 7:38, emphasis in the original.
52. For "whole system," see William James, "Spencer on Persistence of Force," bMS Am 1092.9 (4489), A.MS. and MS., [n.p., n.d.], 5s.(18p.), in the William James Papers (MS Am 1092.9–1092.12), Houghton Library, Harvard University (hereafter cited as James Papers), emphasis in the original. For "Spencer's real motive," see William James, "Spencer's First Principles," bMS Am 1092.9 (4492), A.MS. and MS., [n.p., n.d.], 16s.(49p.), James Papers.

53. For James's put-downs, see James, "Herbert Spencer," 100. For "consecrated," see William James, "Spencer's First Principles," bMS Am 1092.9 (4492), A.MS. and MS., [n.p., n.d.], 16s.(49p.), James Papers. On systems *as a genre,* see Clifford Siskin, *System: The Shaping of Modern Knowledge* (Cambridge, MA: MIT Press, 2016). See also Nasser Zakariya, *A Final Story: Science, Myth, and Beginnings* (Chicago: University of Chicago Press, 2017).
54. On this boundary-work, see Ruth Barton, *The X Club: Power and Authority in Victorian Science* (Chicago: University of Chicago Press, 2018). See also Bernard V. Lightman and Michael S. Reidy, eds., *The Age of Scientific Naturalism: Tyndall and His Contemporaries* (London: Pickering and Chatto, 2014).
55. [William James], review of *Lectures on the Elements of Comparative Anatomy: On the Classification of Animals, and on the Vertebrate Skull, North American Review,* January 1865, 290–298, here 291–292.
56. For "presumptuous," see [William James], review of Darwin's *The Variation of Animals and Plants under Domestication, North American Review,* July 1868, 362–368, here 367–368. For "instinctive guesses," see [William James], review of Darwin's *The Variation of Animals and Plants under Domestication, The Atlantic,* July 1868, 122–124, here 122, 124.
57. William James, "Notes for Philosophy 4: Psychology (1878–1879)," in *Manuscript Lectures,* 129–146, here 136, 140–141.
58. William James, "Great Men, Great Thoughts and the Environment," *Atlantic Monthly,* October 1880, 441–459, here 441, 443, emphasis in the original. The other two pieces appeared earlier. See William James, "The Sentiment of Rationality," *Mind* 4, no. 15 (July 1879): 317–346; and William James, "Brute and Human Intellect," *Journal of Speculative Philosophy* 12, no. 3 (July 1878): 236–276.
59. For "geniuses as data," see James, "Great Men," 445. For "draughtsman," see James, *Principles,* 2:625. On the role of free will in James's Darwinism, see Richards, *Darwin and Emergence,* 409–450.
60. For "brot in," see William James, "*Will,*" bMS Am 1092.9 (4418), A.MS. and MS., [n.p., n.d.], 12s.(23p.), James Papers. For "Positivists," see Wright, "Evolution of Self-Consciousness," 289. For "random images" and the remaining lines, see James, "Great Men," 456.
61. For "associations," see James, *Principles,* 1:586, emphasis in the original; "lists of instances" is at 1:346–347, "revery" is at 2:325. For "pluralistic evolution," see William James, "Introduction [for a systematic work on philosophy]," bMS Am 1092.9 (4439), A.MS., [n.p., ca. 1904], 12s.(13p.), James Papers. For his comments on method, see James, *Principles,* 1:191–194. Philip Wiener called spontaneity the "Ariadne's thread of James's philosophy of evolution." See Wiener, *Evolution and the Founders of Pragmatism,* 101.

62. William James to Hugo Münsterberg, July 8, 1891, in *Correspondence,* 7:180, emphasis in the original. For "walking" and "conceiving," see James, "Great Men," 456–457.
63. James, "Great Men," 457. For another example of James's affective thinking, see William James, "The Place of Affectional Facts in a World of Pure Experience," *Journal of Philosophy, Psychology and Scientific Methods* 2, no. 11 (May 1905): 281–287. On the importance of affect for James, see Alexander Livingston, "Excited Subjects: William James and the Politics of Radical Empiricism," *Theory & Event* 15, no. 4 (December 2012). On the affective turn, see Otniel E. Dror, "The Affect of Experiment: The Turn to Emotions in Anglo-American Physiology, 1900–1940," *Isis* 90, no. 2 (June 1999): 205–237.
64. For the back-and-forth between James and Holt, see James to Holt, November 22 [1878]; Holt to James, November 23, 1878; and James to Holt, November 25 [1878]; all in *Correspondence,* 5:24–28.
65. James, "Brute and Human Intellect," 236, 256, 260.
66. Ibid., 274–276.
67. William James, "Are We Automata?," *Mind* 4, no. 13 (January 1879): 1–22, here 9, 13. James was responding to T. H. Huxley, "On the Hypothesis That Animals Are Automata, and Its History," ed. John Morley, *Fortnightly Review, May 1865–June 1934* 16, no. 95 (November 1874): 555–580. For "all the while," see James, *Principles,* 1:284; for "theatre," see 288, emphasis in the original.
68. For "quick invention," see [William James], review of Jevons's *Principles of Science, Atlantic Monthly,* April 1875, 500–501, here 501. For "precisely," see C. S. Peirce, review of *Logical Machines, American Journal of Psychology* 1, no. 1 (November 1887): 165–170, here 165. Peirce was responding to W. Stanley Jevons, "On the Mechanical Performance of Logical Inference," *Philosophical Transactions of the Royal Society of London* 160 (January 1870): 497–518. On the importance of instruments to Jevons, see Harro Maas, "An Instrument Can Make a Science: Jevons's Balancing Acts in Economics," *History of Political Economy* 33 (December 2001): 277–302.
69. For "method of methods," see Charles S. Peirce, "Introductory Lecture on the Study of Logic," *Johns Hopkins University Circular,* November 1882, 11–12. For "generalization" and "laboratory," see Charles Sanders Peirce to Daniel Coit Gilman, January 13, 1878, quoted in Brent, *Peirce,* 120–121.

6. Animal Intelligence

1. [William James], "Herbert Spencer's Data of Ethics," *The Nation,* September 11, 1879, 178–179, here 178. On psychology's transformation in this period, see Gary Hatfield, "Psychology: Old and New," in *The Cambridge History of*

Philosophy, 1870–1945, ed. Thomas Baldwin (Cambridge: Cambridge University Press, 2003), 93–106. See also David E. Leary, "Wundt and After: Psychology's Shifting Relations with the Natural Sciences, Social Sciences, and Philosophy," *Journal of the History of the Behavioral Sciences* 15, no. 3 (July 1979): 231–241.

2. My discussion of comparative psychology is indebted to Robert J. Richards, *Darwin and the Emergence of Evolutionary Theories of Mind and Behavior* (Chicago: University of Chicago Press, 1987), esp. 331–407; and Gregory Radick, *The Simian Tongue: The Long Debate about Animal Language* (Chicago: University of Chicago Press, 2007), esp. 50–83.
3. To take one example of a topic that was not reducible to evolutionary concerns, intelligence testing was a major focus of interest in the decades around 1900 for some of the same reasons that evolutionary studies gained prominence. See John Carson, *The Measure of Merit: Talents, Intelligence, and Inequality in the French and American Republics, 1750–1940* (Princeton: Princeton University Press, 2007), 167–172. See also John Carson, "Minding Matter / Mattering Mind: Knowledge and the Subject in Nineteenth-Century Psychology," *Studies in History and Philosophy of Science Part C* 30, no. 3 (September 1999): 345–376.
4. On the prehistory of behaviorism, see John M. O'Donnell, *The Origins of Behaviorism: American Psychology, 1870–1920* (New York: NYU Press, 1985). The field of "theory of mind" coalesced in the 1980s in response to David Premack and Guy Woodruff, "Does the Chimpanzee Have a Theory of Mind?," *Behavioral and Brain Sciences* 1, no. 4 (December 1978): 515–526. For a reflection, see Josep Call and Michael Tomasello, "Does the Chimpanzee Have a Theory of Mind? 30 Years Later," *Trends in Cognitive Sciences* 12, no. 5 (May 2008): 187–192.
5. On the (long) history of anthropomorphism, see the essays collected in Lorraine Daston and Gregg Mitman, eds., *Thinking with Animals: New Perspectives on Anthropomorphism* (New York: Columbia University Press, 2005).
6. On evidence and experience, see Joan W. Scott, "The Evidence of Experience," *Critical Inquiry* 17, no. 4 (July 1991): 773–797. On the philosophical issues with imagining other minds, see Thomas Nagel, "What Is It Like to Be a Bat?," *Philosophical Review* 83, no. 4 (October 1974): 435–450.
7. John Stuart Mill, *An Examination of Sir William Hamilton's Philosophy and of the Principal Philosophical Questions Discussed in His Writings* (London: Longman, Green, Longman, Roberts and Green, 1865), 208–209. For a defense of Mill, see A. Hyslop and F. C. Jackson, "The Analogical Inference to Other Minds," *American Philosophical Quarterly* 9, no. 2 (April 1972): 168–176.
8. On the messy distinctions between subject and object, see the essays in Jill G. Morawski, ed., *The Rise of Experimentation in American Psychology* (New Haven:

Yale University Press, 1988). See also Jill G. Morawski, "Self-Regard and Other-Regard: Reflexive Practices in American Psychology, 1890–1940," *Science in Context* 5, no. 2 (Autumn 1992): 281–308.

9. For a general characterization of these reports on behavior, including the amateur status of many who made them, see Philip Howard Gray, "The Early Animal Behaviorists: Prolegomenon to Ethology," *Isis* 59, no. 4 (December 1968): 372–383. A representative sample of some of the period's influential authors and their later impact is discussed in Timothy D. Johnston, "Three Pioneers of Comparative Psychology in America, 1843–1890: Lewis H. Morgan, John Bascom, and Joseph LeConte," *History of Psychology* 6, no. 1 (February 2003): 14–51.
10. Charles Darwin, *On the Origin of Species by Means of Natural Selection, or the Preservation of Favoured Races in the Struggle for Life* (London: J. Murray, 1859), 184. The original story appeared in Samuel Hearne, *A Journey from Prince of Wales's Fort in Hudson's Bay* (London: A. Strahan and T. Cadell, 1795), 370. On the persistence of such methods in otherwise experimental approaches, see Anne C. Rose, "Animal Tales: Observations of the Emotions in American Experimental Psychology, 1890–1940," *Journal of the History of the Behavioral Sciences* 48, no. 4 (Autumn 2012): 301–317.
11. The review was [Richard Owen], "Darwin on the Origin of Species," *Edinburgh Review* 111 (April 1860): 487–532, here 517–518. For Darwin's debts to Lyell, see Darwin to Charles Lyell, 25 [November 1859], in *The Correspondence of Charles Darwin,* ed. Frederick Burkhardt et al., vol. 7: *1858–1859* (Cambridge: Cambridge University Press, 1992), 400–401, here 400. For "foolish" and "abominable," see Darwin to Andrew Murray, April 28 [1860], in *The Correspondence of Charles Darwin,* ed. Frederick Burkhardt et al., vol. 8: *1860* (Cambridge: Cambridge University Press, 1993), 175–179, here 176.
12. On the field's development in the wake of Darwin's publications in the 1870s, see Robert A. Boakes, *From Darwin to Behaviourism: Psychology and the Minds of Animals* (Cambridge: Cambridge University Press, 1984).
13. On Romanes and Morgan, see Richards, *Darwin and Emergence,* 334–406. See also J. David Pleins, *In Praise of Darwin: Georges Romanes and the Evolution of a Darwinian Believer* (New York: Bloomsbury Academic, 2014). On Romanes's life, see Ethel Duncan Romanes, *Life and Letters of George John Romanes* (London: Longmans, Green and Co., 1895).
14. George J. Romanes, "Intellect in Brutes," *Nature* 20 (4 June 1879): 122–125, here 125. His early essays in *Nature*—of which there were dozens—covered topics as varied as "Rainbows," "Hypnotism," and "Singing Mice." Much of the evidence Romanes related in these articles was compiled in Romanes, *Animal Intelligence* (London: K. Paul, Trench, 1882).

15. George John Romanes, *Mental Evolution in Animals* (London: Kegan Paul, Trench, 1883), 18, 60. The second volume picked up where the first left off (with nonhuman animals) before shifting to an emphasis on language. See Romanes, *Mental Evolution in Man: Origin of Human Faculty* (London: Kegan Paul, Trench, 1888). On the relationship between the two, see Radick, *The Simian Tongue,* 70–73.
16. Romanes, *Mental Evolution in Animals,* 59, 108, 109. Here Romanes was hinting at a principle that Morgan and the American James Mark Baldwin would jointly enunciate just a decade later. Called "organic selection" at the time (but now known, after one of its authors, as "the Baldwin effect"), the idea was that certain features acquired in the course of an individual's life might be "inherited" if changes in the creature's behavior are such that any subsequent *genetic* alterations that enable this sort of change will be selected for, based on its fitness advantages. See J. Mark Baldwin, "A New Factor in Evolution," *American Naturalist* 30, no. 354 (June 1896): 441–451. For a clarifying discussion of the idea, see John Watkins, "A Note on Baldwin Effect," *British Journal for the Philosophy of Science* 50, no. 3 (September 1999): 417–423.
17. Romanes, *Mental Evolution in Animals,* 341, 328, 336.
18. William Kingdon Clifford, *Lectures and Essays* (London: Macmillan Co., 1879), 72–73. For Morgan's summary, see C. Lloyd Morgan, "Instinct," *Nature* 29 (February 14, 1884): 370–374, here 370. See also Alan Costall, "Lloyd Morgan, and the Rise and Fall of 'Animal Psychology,'" *Society & Animals* 6, no. 1 (January 1998): 13–29.
19. Morgan, "Instinct," 370, 371. For "mind-story," see Carl Murchison, ed., *A History of Psychology in Autobiography,* vol. 2 (Worcester, MA: Clark University Press, 1930), 249. The canonical interpretation of the links between Romanes and Morgan is one of antagonism. See, for example, Richards, *Darwin and the Emergence,* 377–381. Some have argued, instead, for a kind of tacit agreement between the two. See Alan Costall, "How Lloyd Morgan's Canon Backfired," *Journal of the History of the Behavioral Sciences* 29, no. 2 (February 2006): 113–122. Radick notes there is some truth to both. See Radick, *The Simian Tongue,* 397n109 and 399–400n138.
20. For the back-and-forth, see George J. Romanes, "Mr. Lloyd Morgan on Instinct," *Nature* 29 (February 21, 1884): 379–381, here 380; C. Lloyd Morgan, "Instinct," *Nature* 29 (February 28, 1884): 405; and Romanes, "Instinct," *Nature* 29 (March 20, 1884): 477. For Morgan's later piece, see Morgan, "On the Study of Animal Intelligence," *Mind* 11, no. 42 (April 1886): 174–185, here 181.
21. On the methodological nature of the dispute, see Simon Fitzpatrick and Grant Goodrich, "Building a Science of Animal Minds: Lloyd Morgan, Experimentation,

and Morgan's Canon," *Journal of the History of Biology* 50, no. 3 (Summer 2017): 525–569.

22. C. Lloyd Morgan, *The Springs of Conduct: An Essay on Evolution* (London: Kegan Paul, Trench and Co., 1885), 135, 151.
23. C. Lloyd Morgan, "The Limits of Animal Intelligence," *Nature* 46 (September 1, 1892): 417. For Morgan's account of "the new psychology" in the United States, see Morgan, "Psychology in America," *Nature* 43 (April 2, 1891): 506–509. On his debts to James, see, for example, Gregory Radick, "Morgan's Canon, Garner's Phonograph, and the Evolutionary Origins of Language and Reason," *British Journal for the History of Science* 33, no. 1 (March 2000): 3–23, here 22.
24. C. Lloyd Morgan, *An Introduction to Comparative Psychology* (London: Walter Scott, 1894), 53, emphasis in the original. Morgan announced his "canon" in 1892 at the International Congress of Experimental Psychology in London, where he circulated a version of his remarks in pamphlet form. See Edward T. Dixon, "The Limits of Animal Intelligence," *Nature* 46, no. 1191 (August 25, 1892): 392–393. For a review of various versions, see Edward Newbury, "Current Interpretation and Significance of Lloyd Morgan's Canon," *Psychological Bulletin* 51, no. 1 (January 1954): 70–74.
25. Morgan, *Introduction,* 241. On this "anti-anthropomorphic" reading, see R. H. Waters, "Morgan's Canon and Anthropomorphism," *Psychological Review* 46, no. 6 (November 1939): 534–540.
26. For the "anecdotal stage," see Morgan, *Introduction,* 291–293. On the use of anecdotes in science, see Robert W. Mitchell, Nicholas S. Thompson, and H. Lyn Miles, eds., *Anthropomorphism, Anecdotes, and Animals* (Albany: SUNY Press, 1997). For "the majority of men," see Romanes, *Mental Evolution in Man,* 59.
27. William James, "On Some Omissions of Introspective Psychology," *Mind* 9, no. 33 (January 1884): 1–26, here 2, 20. James incorporated the essay in two places in the text of his *Principles of Psychology.* The bulk of the description of mind found its way into the book's most famous chapter, "The Stream of Thought"; the methodological points (including part of the quotation above) were cut out and inserted into the chapter "The Methods and Snares of Psychology."
28. In his *Principles,* James referred to the failure to keep these aspects of investigation straight as "The Psychologist's Fallacy." See William James, *The Principles of Psychology,* vol. 1 (New York: Henry Holt and Co., 1890), 196. For more on the fallacy in the context of James's work, see Edward Reed, "The Psychologist's Fallacy as a Persistent Framework in William James's Psychological Theorizing," *History of the Human Sciences* 8, no. 1 (February 1995): 61–72.
29. On Thorndike's life, see Geraldine M. Jonçich, *The Sane Positivist: A Biography of Edward L. Thorndike* (Middletown, CT: Wesleyan University Press, 1968); the

explanation of the title is on 4, the diary entry is on 122. For the colleague's recollection, see Murchison, *History of Psychology*, 2:366, and for Thorndike's recollection, see 3:266.

30. For "only book," see Murchison, *History of Psychology*, 3:263. On Thorndike's deflationary role, see Boakes, *From Darwin to Behaviourism*, esp. 68–78. It is easy to take Thorndike at his (grumpy) word, without noticing how his approach underwrote some of the slipperiness he decried.
31. The dissertation was published as Edward L. Thorndike, "Animal Intelligence: An Experimental Study of the Associative Processes in Animals," *Psychological Review: Monograph Supplements* 2, no. 4 (June 1898). The references to Romanes's claims appear on 40–42; his imagination of the cat's response to the situation is on 45. Thorndike republished the dissertation a decade later in Thorndike, *Animal Intelligence: Experimental Studies* (New York: Macmillan Co., 1911), 20–155.
32. For Thorndike's description, see Thorndike, "Animal Intelligence," 6; a representative set of curves appears on 18–26. According to Thorndike, "Cats (or rather kittens), dogs and chicks were the subjects of the experiments. All were apparently in excellent health, save an occasional chick" (7). For pictures and analysis, see John C. Burnham, "Thorndike's Puzzle Boxes," *Journal of the History of the Behavioral Sciences* 8, no. 2 (Spring 1972): 159–167. See also M. E. Bitterman, "Thorndike and the Problem of Animal Intelligence," *American Psychologist* 24, no. 4 (April 1969): 444–453.
33. See Thorndike, "Animal Intelligence," 94. He expanded his claims about (imagined) humans in a research summary printed around the same time. See Edward Thorndike, "Some Experiments on Animal Intelligence," *Science* 7, no. 181 (17 June 1898): 818–824, esp. 822–823.
34. For the cats' motivations, see Thorndike, "Animal Intelligence," 95. For homologous processes, see, for example, 83–84. On the importance of homologies (rather than analogies), see Stephen Jay Gould, "Evolution and the Triumph of Homology, or Why History Matters," *American Scientist* 74, no. 1 (January 1986): 60–69.
35. The "personal equation" was a well-known "problem" in a range of scientific fields—though what exactly the problem was changed from field to field and from period to period. Its main point of reference in the nineteenth century was astronomy. See Simon Schaffer, "Astronomers Mark Time: Discipline and the Personal Equation," *Science in Context* 2, no. 1 (March 1988): 115–145. While "comparative" approaches can be contrasted with "experimental" ones, comparative psychology brought the two together. On the two as distinct ways of knowing, see John V. Pickstone, "Museological Science? The Place of the

Analytical / Comparative in Nineteenth-Century Science, Technology and Medicine," *History of Science* 32, no. 2 (June 1994): 111–138. On their interrelations, see Bruno J. Strasser, "Laboratories, Museums, and the Comparative Perspective: Alan A. Boyden's Quest for Objectivity in Serological Taxonomy, 1924–1962," *Historical Studies in the Natural Sciences* 40, no. 2 (May 2010): 149–182.

36. Thorndike, "Animal Intelligence," 2. Although Thorndike's detachment seems to fit the account of self-abnegation offered by Daston and Galison in their history of "mechanical objectivity," leaving out particular details in favor of abstraction seems more like the earlier ideal of "truth-to-nature" or the subsequent celebration of "trained judgment." See Lorraine Daston and Peter Galison, *Objectivity* (New York: Zone Books, 2007), 234–246.
37. The preceding passages appear (in the order quoted) in Thorndike, "Animal Intelligence," 14, 16, 45, 29, 86–87, 109. These methodological precepts were, within the decade, deemed to be the work's most significant aspects. See S. J. Holmes, review of *Animal Intelligence: Experimental Studies, Psychological Bulletin* 9, no. 8 (August 1912): 318–320.
38. Wesley Mills, "The Nature of Animal Intelligence and the Methods of Investigating It," *Psychological Review* 6, no. 3 (May 1899): 262–274, here 267–268. Mills had written his own overview of the field, which directly opposed Thorndike's. See Wesley Mills, *The Nature and Development of Animal Intelligence* (New York: Macmillan Co., 1898). For Thorndike's view of their differences, see Edward L. Thorndike, review of *Animal Intelligence, Science* 8, no. 198 (1898): 520. See also Thorndike, "A Reply to 'The Nature of Animal Intelligence and the Methods of Investigating It,'" *Psychological Review* 6, no. 4 (July 1899): 412–420.
39. C. Lloyd Morgan, review of Thorndike's *Animal Intelligence, Nature* 58, no. 1498 (July 14, 1898): 249–250, here 249. Thorndike's account of his experiments for a general audience were similarly criticized. See Edward Thorndike, "Do Animals Reason?," *Popular Science Monthly,* August 1899, 480–490. The correspondence appears in the ensuing October issue.
40. For the kitten, see Thorndike, "Animal Intelligence," 2; for tennis, see 2, 84, 90, 101, and 108. Morgan noted the shared image in Morgan, review of Thorndike, 250.
41. For evidence of the durable impact of Thorndike's approach, see Kennon A. Lattal, "A Century of Effect: Legacies of E. L. Thorndike's Animal Intelligence Monograph," *Journal of the Experimental Analysis of Behavior* 70, no. 3 (November 1998): 325–336. Lattal makes it clear that Thorndike's practice of abstraction shaped much of the work that came after him.
42. C. O. Whitman, "Animal Behavior," in *Biological Lectures Delivered at the Marine Biological Laboratory of Woods Hole,* vol. 6 (Boston: Ginn and Co., 1899),

285–338, here 302. Thorndike's lectures "Instinct" and "The Associative Process in Animals" were delivered in 1899 and printed in the same series. See *Biological Lectures Delivered at the Marine Biological Laboratory of Woods Hole,* vol. 7 (Boston: Ginn and Co., 1900), 57–91. On Whitman's impact, see Richard W. Burkhardt Jr., *Patterns of Behavior: Konrad Lorenz, Niko Tinbergen, and the Founding of Ethology* (Chicago: University of Chicago Press, 2005), 17–68, esp. 17–33. See also Philip J. Pauly, "From Adventism to Biology: The Development of Charles Otis Whitman," *Perspectives in Biology and Medicine* 37, no. 3 (Spring 1994): 395–408.

43. Linus W. Kline, "Methods in Animal Psychology," *American Journal of Psychology* 10, no. 2 (January 1899): 256–279, here 257, 276. For an example of Whitman's skepticism, see C. O. Whitman, "Myths in Animal Psychology," *Monist* 9, no. 4 (July 1899): 524–537. Whitman and Kline were not alone in using unicellular organisms for psychological study in this period. See Judy Johns Schloegel and Henning Schmidgen, "General Physiology, Experimental Psychology, and Evolutionism: Unicellular Organisms as Objects of Psychophysiological Research, 1877–1918," *Isis* 93, no. 4 (December 2002): 614–645.

44. See, e.g., Kline, "Methods in Animal Psychology," 274. It is possible that two competing ideals of the laboratory notebook are at work in Kline's publication. On the one hand, his choice of adjectives would seem to violate the impulse, which he shared with Thorndike, to separate observation from interpretation. On the other hand, including words like "sumptuous" may preserve the effect that the notes themselves are exact recordings of what went through his mind. Given later claims about our inability to avoid such anthropomorphic editorializing in our observations, there is at least a plausible defense of his choice of words in a "warts-and-all" approach to publishing.

45. Linus W. Kline, "Suggestions toward a Laboratory Course in Comparative Psychology," *American Journal of Psychology* 10, no. 3 (April 1899): 399–430, here 399.

46. The accounts of the rat's utility (as well as of the priority for developing the maze apparatus in the next paragraph) are drawn from first-person reminiscences collected in Walter R. Miles, "On the History of Research with Rats and Mazes: A Collection of Notes," *Journal of General Psychology* 4 (January 1930): 324–336; for "cheap" and "gnawer," see 326. On the adoption of the (albino) rat in psychology, see Cheryl A. Logan, "'[A]re Norway Rats . . . Things?': Diversity versus Generality in the Use of Albino Rats in Experiments on Development and Sexuality," *Journal of the History of Biology* 34, no. 2 (Summer 2001): 287–314; Frederick J. Wertz, "Of Rats and Psychologists: A Study of the History and Meaning of Science," *Theory & Psychology* 4, no. 2 (May 1994): 165–197; and

Bonnie Tocher Clause, "The Wistar Rat as a Right Choice: Establishing Mammalian Standards and the Ideal of a Standardized Mammal," *Journal of the History of Biology* 26, no. 2 (Summer 1993): 329–349.

47. For "home-finding," see Miles, "On the History of Research," 331; for the discussion of Thorndike's influence, see 333. For more on these developments, see Cheryl A. Logan, "The Altered Rationale for the Choice of a Standard Animal in Experimental Psychology: Henry H. Donaldson, Adolf Meyer, and 'the' Albino Rat," *History of Psychology* 2, no. 1 (February 1999): 3–24.
48. Willard S. Small, "An Experimental Study of the Mental Processes of the Rat," *American Journal of Psychology* 11, no. 2 (January 1900): 133–165, here 133. For more on the consequences of adapting the white rat in the line of research, see Cheryl A. Logan, "The Legacy of Adolf Meyer's Comparative Approach: Worcester Rats and the Strange Birth of the Animal Model," *Integrative Physiological & Behavioral Science* 40, no. 4 (October 2005): 169–181.
49. Willard S. Small, "Experimental Study of the Mental Processes of the Rat. II," *American Journal of Psychology* 12, no. 2 (January 1901): 206–239, here 206, 209, 218; the image of the maze appears on 207. For more pictures of apparatus from the period, see Boakes, *From Darwin to Behaviourism,* 145.
50. Small, "Experimental Study II," 230, 225–226, 224. On links between maze studies and the wider world in this and later periods, see Rebecca M. Lemov, *World as Laboratory: Experiments with Mice, Mazes, and Men* (New York: Hill and Wang, 2005), esp. 24–45.
51. Small, "Experimental Study II," 227–228, emphasis in the original. The reference is to David Hume, *An Enquiry concerning Human Understanding,* ed. Peter Millican (Oxford: Oxford University Press, 2007), 78.
52. Small, "Experimental Study II," 228, 210. Thorndike, "Animal Intelligence," 6, 44, 90, 94.
53. Thorndike, "Animal Intelligence," 109, 63. For the follow-up, see Edward L. Thorndike, "The Mental Life of the Monkeys," *Psychological Review: Monograph Supplements* 3, no. 5 (May 1901), 12, 14. On Thorndike's impact on early primate studies, see Radick, *The Simian Tongue,* 202–214.
54. Thorndike, "Mental Life of the Monkeys," 56–57.
55. On Yerkes, see Murchison, *History of Psychology,* 2:381–407. On his laboratories, see Donald A. Dewsbury, *Monkey Farm: A History of the Yerkes Laboratories of Primate Biology, Orange Park, Florida, 1930–1965* (Lewisburg, PA: Bucknell University Press, 2006). See also Marion Thomas, "Yerkes, Hamilton and the Experimental Study of the Ape Mind: From Evolutionary Psychiatry to Eugenic Politics," *Studies in History and Philosophy of Science Part C* 37, no. 2 (June 2006):

273–294. On his role in intelligence testing, see Carson, *The Measure of Merit,* esp. 201–219. Yerkes was also crucial in publicizing Ivan Pavlov's work in the United States. See Robert M. Yerkes and Sergius Morgulis, "The Method of Pavlov in Animal Psychology," *Psychological Bulletin* 6, no. 8 (August 1909): 257–273. On Pavlov's impact on American psychologists, see Daniel P. Todes, *Ivan Pavlov: A Russian Life in Science* (Oxford: Oxford University Press, 2014), esp. 452–458, 566–570.

56. Robert Mearns Yerkes, *Chimpanzees: A Laboratory Colony* (New Haven: Yale University Press, 1943), 3. For "house of mirrors," see Donna J. Haraway, *Primate Visions: Gender, Race, and Nature in the World of Modern Science* (New York: Routledge, 1989), 64. Haraway quotes the Yerkes passage on 62.

57. For the field's name, see Robert M. Yerkes, "Recent Progress and Present Tendencies in Comparative Psychology," *Journal of Abnormal Psychology* 2, no. 6 (February–March 1908): 271–279, here 271n1. For "artificial," see Robert M. Yerkes, "Objective Nomenclature, Comparative Psychology, and Animal Behavior," *Journal of Comparative Neurology and Psychology* 16, no. 5 (September 1906): 380–389, here 381; for "business," see 388. For a general statement of Yerkes's views on the matter, see Yerkes, "Animal Psychology and Criteria of the Psychic," *Journal of Philosophy, Psychology and Scientific Methods* 2, no. 6 (Spring 1905): 141–149.

58. C. Lloyd Morgan, *Animal Behaviour* (London: Edward Arnold, 1900), 147. On Yerkes's usage of this turn of phrase (and an analysis thereof), see Haraway, *Primate Visions,* 61. On the long history of vivisection and its opponents, see Anita Guerrini, *Experimenting with Humans and Animals: From Galen to Animal Rights* (Baltimore: Johns Hopkins University Press, 2003). On the nineteenth century, see Richard D. French, *Antivivisection and Medical Science in Victorian Society* (Princeton: Princeton University Press, 1975); and Coral Lansbury, *The Old Brown Dog: Women, Workers, and Vivisection in Edwardian England* (Madison: University of Wisconsin Press, 1985).

59. For a textbook definition of the "ideal" method in this period, see Margaret Floy Washburn, *The Animal Mind: A Text-Book of Comparative Psychology* (New York: Macmillan, 1908), 12–13.

60. L. W. Sackett, review of *Animal Intelligence, American Journal of Psychology* 23, no. 1 (January 1912): 142–144, here 143. On Mills, see Thorndike, review of *Animal Intelligence,* 520.

61. The letter to the editor was printed as a response to Thorndike, "Do Animals Reason?"

62. On the impact of *Dr. Moreau,* see Laura Otis, "Monkey in the Mirror: The Science of Professor Higgins and Doctor Moreau," *Twentieth Century Literature*

55, no. 4 (December 2009): 485–509. On the gendered dimensions of the dispute over vivisection, see Carla Bittel, "Science, Suffrage, and Experimentation: Mary Putnam Jacobi and the Controversy over Vivisection in Late Nineteenth-Century America," *Bulletin of the History of Medicine* 79, no. 4 (Winter 2005): 664–694. On the specter of human experimentation, see Susan E. Lederer, *Subjected to Science: Human Experimentation in America before the Second World War* (Baltimore: Johns Hopkins University Press, 1995).

63. For "child-mind," see Thorndike, "Animal Intelligence," 106.

7. Laboratory School

1. Significantly, this expansion in American education was not confined to the classroom. During the same period, a pedagogical approach called "nature study" emphasized direct work with objects and immersion in natural environments as the best means to teach children science. See Sally Gregory Kohlstedt, *Teaching Children Science: Hands-On Nature Study in North America, 1890–1930* (Chicago: University of Chicago Press, 2010). Though "nature study" appealed to ideas (including some from psychology) about the virtues of teaching science outdoors and students' "intuitive" understanding of natural phenomena, the development was largely parallel to the theories of children as naturally scientific thinkers, explored below.
2. The best survey of child study is Sally Shuttleworth, *The Mind of the Child: Child Development in Literature, Science, and Medicine, 1840–1900* (Oxford: Oxford University Press, 2010). For more on the United States, see Hamilton Cravens, *Before Head Start: The Iowa Station and America's Children* (Chapel Hill: University of North Carolina Press, 1993); and Alice Boardman Smuts, *Science in the Service of Children, 1893–1935* (New Haven: Yale University Press, 2006). See also Felix Rietmann et al., "Knowledge of Childhood: Materiality, Text, and the History of Science—An Interdisciplinary Round Table Discussion," *British Journal for the History of Science* 50, no. 1 (March 2017): 111–141.
3. Ernest H. Lindley, "A Study of Puzzles with Special Reference to the Psychology of Mental Adaptation," *American Journal of Psychology* 8, no. 4 (July 1897): 431–493.
4. Ibid., 483, 433, 472, 479. On the continued power of such links, see Alison Gopnik, "Scientific Thinking in Young Children: Theoretical Advances, Empirical Research, and Policy Implications," *Science* 337, no. 6102 (September 2012): 1623–1627.
5. Lindley, "Study of Puzzles," 472, 474. Lindley's spelling of "scientest" was not a typo: the term's meaning (and spelling) had not yet been cemented.

6. The most famous example of this genre is William James, *Talks to Teachers on Psychology, and to Students on Some of Life's Ideals* (New York: Henry Holt and Co., 1899). See also C. Lloyd Morgan, *Psychology for Teachers* (London: Edward Arnold, 1894); and Edward L. Thorndike, *Educational Psychology* (New York: Lemcke and Buechner, 1903).
7. On the emergence of "applied psychology," see Jeremy T. Blatter, "The Psychotechnics of Everyday Life: Hugo Munsterberg and the Politics of Applied Psychology, 1887–1917" (PhD diss., Harvard University, 2014). For a prehistory of "applied science," see Eric Schatzberg, "From Art to Applied Science," *Isis* 103, no. 3 (September 2012): 555–563. On the opposition of "pure" and "applied" science, see Paul Lucier, "The Origins of Pure and Applied Science in Gilded Age America," *Isis* 103, no. 3 (September 2012): 527–536. See also Robert Bud, "Framed in the Public Sphere: Tools for the Conceptual History of 'Applied Science': A Review Paper," *History of Science* 51, no. 4 (December 2013): 413–433.
8. The quotation is from Henry Maudsley, *The Physiology and Pathology of the Mind* (London: Macmillan, 1867), 269. On Maudsley's role in child study, see Shuttleworth, *Mind of the Child,* 181–185. On his work more generally, see Lorraine J. Daston, "British Responses to Psycho-Physiology, 1860–1900," *Isis* 69, no. 2 (June 1978): 192–208, esp. 199–200.
9. Charles Darwin, "A Biographical Sketch of an Infant," *Mind* 2, no. 7 (July 1877): 285–294. On the piece's role in child study, see Sally Shuttleworth, "Tickling Babies: Gender, Authority and Baby Science," in *Science in the Nineteenth-Century Periodical: Reading the Magazine of Nature* (Cambridge: Cambridge University Press, 2004), 199–215. On the role of these notes in Darwin's work, see Paul White, "Darwin's Emotions: The Scientific Self and the Sentiment of Objectivity," *Isis* 100, no. 4 (December 2009): 811–826, esp. 813–817.
10. Literature played a key role in child study's development. See Sally Shuttleworth, "Inventing a Discipline: Autobiography and the Science of Child Study in the 1890s," *Comparative Critical Studies* 2, no. 2 (June 2005): 143–163, esp. 144–145.
11. Darwin, "Biographical Sketch," 289. For Bain, see [Alexander Bain], "On Toys," *Westminster Review* 37, no. 1 (January 1842): 97–121, here 97, 105.
12. This is still one of the ways developmental psychologists discuss play today. The major recent statement of this view is Alison Gopnik, "The Scientist as Child," *Philosophy of Science* 63, no. 4 (December 1996): 485–514. For an overview, see Olivia N. Saracho and Bernard Spodek, "Children's Play and Early Childhood Education: Insights from History and Theory," *Journal of Education* 177, no. 3 (October 1995): 129–148.

13. Karl Groos, *The Play of Man,* trans. Elizabeth L. Baldwin (New York: D. Appleton and Co., 1901), 151; for Groos's favorable citations, see 46–47, 55–56, 281–283. Baldwin, as the editor of a book his wife translated, inserted several references. See, for example, ibid., 64–65, 329–332.
14. G. Stanley Hall, *Adolescence: Its Psychology and Its Relations to Physiology, Anthropology, Sociology, Sex, Crime, Religion and Education,* vol. 1 (New York: D. Appleton and Co., 1904), 202. On Hall's aversion to Groos (rather than their shared interests), see Frank J. Sulloway, *Freud, Biologist of the Mind: Beyond the Psychoanalytic Legend* (Cambridge, MA: Harvard University Press, 1992), 250–251.
15. On Hall's life, see Dorothy Ross, *G. Stanley Hall: The Psychologist as Prophet* (Chicago: University of Chicago Press, 1972). See also David E. Leary, "G. Stanley Hall, a Man of Many Words: The Role of Reading, Speaking, and Writing in His Psychological Work," *History of Psychology* 9, no. 3 (August 2006): 198–223; and David E. Leary, "Between Peirce (1878) and James (1898): G. Stanley Hall, the Origins of Pragmatism, and the History of Psychology," *Journal of the History of the Behavioral Sciences* 45, no. 1 (Winter 2009): 5–20. Hall's autobiography is also an important source. See G. Stanley Hall, *The Life and Confessions of a Psychologist* (New York: D. Appleton and Co., 1923), which expanded on Hall, "Confessions of a Psychologist," *Pedagogical Seminary* 8, no. 1 (March 1901): 92–143.
16. It should be noted that although Hall spent time with Wundt, his German studies focused more on physiology. See Ross, *Stanley Hall,* 81–99. On "genetic psychology," see Sheldon H. White, "G. Stanley Hall: From Philosophy to Developmental Psychology," in *Evolving Perspectives on the History of Psychology,* ed. Wade E. Pickren and Donald A. Dewsbury (Washington, DC: American Psychological Association, 2002), 279–302. For a contemporaneous account, see Robert M. Yerkes, "Behaviorism and Genetic Psychology," *Journal of Philosophy, Psychology and Scientific Methods* 14, no. 6 (March 1917): 154–160.
17. [G. Stanley Hall], "College Instruction in Philosophy," *The Nation,* September 21, 1876, 180; [Hall], "The Philosophy of the Future," *The Nation,* 27, November 7, 1878, 283–284; Hall, "Philosophy in the United States," *Mind* 4, no. 13 (January 1879): 89–105, here 95. The piece occasioned an anonymous reply from William James. See [William James], "The Teaching of Philosophy in Our Colleges," *The Nation,* September 21, 1876, 178–179.
18. Hall, "Philosophy in the United States," 101–103.
19. Hall, "Confessions of a Psychologist," 113–114 and 138–139.
20. For a list of Hall's publications, see Ross, *Stanley Hall,* 439–450.

21. On Hall's journals, see Jacy L. Young and Christopher D. Green, "An Exploratory Digital Analysis of the Early Years of G. Stanley Hall's *American Journal of Psychology* and *Pedagogical Seminary*," *History of Psychology* 16, no. 4 (November 2013): 249–268. On Clark, see Michael M. Sokal, "G. Stanley Hall and the Institutional Character of Psychology at Clark, 1889–1920," *Journal of the History of the Behavioral Sciences* 26, no. 2 (April 1990): 114–124.
22. Hall, *Adolescence,* 1:vii–ix. On the book's production and reception, see Hamilton Cravens, "The Historical Context of G. Stanley Hall's *Adolescence* (1904)," *History of Psychology* 9, no. 3 (August 2006): 172–185. On its impact more broadly, see Jeanne Brooks-Gunn and Anna Duncan Johnson, "G. Stanley Hall's Contribution to Science, Practice and Policy: The Child Study, Parent Education, and Child Welfare Movements," *History of Psychology* 9, no. 3 (August 2006): 247–258.
23. Hall, *Adolescence,* 2:153, 229–230.
24. On "moral kinds" in general, see Laura Stark, "Out of Their Depths: 'Moral Kinds' and the Interpretation of Evidence in Foucault's Modern Episteme," *History and Theory* 55, no. 4 (December 2016): 131–147.
25. In his survey, Bruce Kuklick mentions Hall only in passing—as "more promoter than thinker." See Kuklick, *The Rise of American Philosophy, Cambridge, Massachusetts, 1860–1930* (New Haven: Yale University Press, 1977), 124. On Hall as both inspiration and foil, see Louis Menand, *The Metaphysical Club: A Story of Ideas in America* (New York: Farrar, Straus and Giroux, 2001), 267–272, 281–284. See also Jeffrey Sklansky, *The Soul's Economy: Market Society and Selfhood in American Thought, 1820–1920* (Chapel Hill: University of North Carolina Press, 2002), 137–170, esp. 162–170.
26. On Dewey's life, see Robert B. Westbrook, *John Dewey and American Democracy* (Ithaca, NY: Cornell University Press, 1991); and Alan Ryan, *John Dewey and the High Tide of American Liberalism* (New York: W. W. Norton, 1995). For Holmes's remark, see *Holmes–Pollock Letters: The Correspondence of Mr. Justice Holmes and Sir Frederick Pollock, 1874–1932,* vol. 2, ed. Mark Dewolf Howe (Cambridge, MA: Harvard University Press, 1941), 287. Westbrook, *John Dewey,* cites it on 341; Ryan, *John Dewey,* alludes to it on 20, 106, and, obliquely, 239.
27. John Dewey, "From Absolutism to Experimentalism," in *Contemporary American Philosophy: Personal Statements,* ed. George P. Adams and Wm. Pepperell Montague (New York: Russell and Russell, 1930), 13–27, here 19. On these shifts, see Neil Coughlan, *Young John Dewey: An Essay in American Intellectual History* (Chicago: University of Chicago Press, 1975). Hegel's line is from Georg Wilhelm Friedrich Hegel, *The Science of Logic,* trans. George di Giovanni

(Cambridge: Cambridge University Press, 2010), 735. On British Hegelianism, see Peter Robbins, *The British Hegelians, 1875–1925* (New York: Garland, 1982). On British idealism more generally, see W. J. Mander, *British Idealism: A History* (Oxford: Oxford University Press, 2011). On American counterparts, see William H. Goetzmann, ed., *The American Hegelians: An Intellectual Episode in the History of Western America* (New York: Knopf, 1973). See also James A. Good, "A 'World-Historical Idea': The St. Louis Hegelians and the Civil War," *Journal of American Studies* 34, no. 3 (December 2000): 447–464. On Dewey's debts, see Trevor Pearce, "The Dialectical Biologist, circa 1890: John Dewey and the Oxford Hegelians," *Journal of the History of Philosophy* 52, no. 4 (October 2014): 747–777. On James's links, see Burleigh Taylor Wilkins, "James, Dewey, and Hegelian Idealism," *Journal of the History of Ideas* 17, no. 3 (June 1956): 332–346. The most famous example of James's skepticism is William James, "On Some Hegelisms," *Mind* 7, no. 26 (April 1882): 186–208.

28. For "too much" and "unstable," see Dewey, "From Absolutism to Experimentalism," 22. It is worth noting that Dewey, Peirce, James, *and* Hall all published some of their first articles in what amounted to the official organ of the St. Louis Hegelians: William Torrey Harris's *Journal of Speculative Philosophy (JSP)*. See C. S. Peirce, "Questions concerning Certain Faculties Claimed for Man," *JSP* 2, no. 2 (January 1868): 103–114; William James, "Remarks on Spencer's Definition of Mind as Correspondence," *JSP* 12, no. 1 (January 1878): 1–18; G. Stanley Hall, "Notes on Hegel and His Critics," *JSP* 12, no. 1 (January 1878): 93–103; and John Dewey, "The Metaphysical Assumptions of Materialism," *JSP* 16, no. 2 (April 1882): 208–213. On the journal and its editor, see Kurt F. Leidecker, *Yankee Teacher: The Life of William Torrey Harris* (New York: Philosophical Library, 1946).

29. For Dewey's recollection, see John Dewey to Henri Robet, May 2, 1911, published in Gérard Deledalle, ed., *John Dewey's Unpublished Items* (Paris: Presses Universitaires de France, 1967), 36–37. For "concrete relations" and "individual consciousness," see John Dewey, "The Psychological Standpoint," *Mind* 11, no. 41 (January 1886): 1–19, here 19. See also the piece's sequel: Dewey, "Psychology as Philosophic Method," *Mind* 11, no. 42 (April 1886): 153–173. For the transitional lines, see Dewey, "The Present Position of Logical Theory," *Monist* 2, no. 1 (October 1891): 1–17, here 10, 17, 2. For another account, see Richard Rorty, "Dewey between Hegel and Darwin," in *Modernist Impulses in the Human Sciences, 1870–1930,* ed. Dorothy Ross (Baltimore: Johns Hopkins University Press, 1994), 54–68.

30. William James, *The Principles of Psychology,* vol. 1 (New York: Henry Holt and Co., 1890), 192. On the reaction-time experiments, see David Kent Robinson,

"Reaction-Time Experiments in Wundt's Institute and Beyond," in *Wilhelm Wundt in History: The Making of a Scientific Psychology,* ed. R. W. Rieber and David Kent Robinson (Dordrecht: Kluwer, 2001), 161–204. See also Henning Schmidgen, "Time and Noise: The Stable Surroundings of Reaction Experiments, 1860–1890," *Studies in History and Philosophy of Science Part C* 34, no. 2 (June 2003): 237–275.

31. For the definition, see E. B. Titchener, "Simple Reactions," *Mind* 4, no. 13 (January 1895): 74–81, here 74. The clearest statement of Titchener's structuralism is Titchener, "The Postulates of a Structural Psychology," *Philosophical Review* 7, no. 5 (September 1898): 449–465. For an overview, see Christian Beenfeldt, *The Philosophical Background and Scientific Legacy of E. B. Titchener's Psychology: Understanding Introspectionism* (New York: Springer, 2013), esp. 27–44. See also Christopher D. Green, "Scientific Objectivity and E. B. Titchener's Experimental Psychology," *Isis* 101, no. 4 (December 2010): 697–721. Baldwin announced his theory in J. Mark Baldwin, "Types of Reaction," *Psychological Review* 2, no. 3 (May 1895): 259–273, here 271. For Titchener's response, see Titchener, "The Type-Theory of the Simple Reaction," *Mind* 4, no. 16 (October 1895): 506–514. For "machinery," see Baldwin, "The 'Type-Theory' of Reaction," *Mind* 5, no. 17 (January 1896): 81–90, here 81. On Baldwin's career, see Robert J. Richards, *Darwin and the Emergence of Evolutionary Theories of Mind and Behavior* (Chicago: University of Chicago Press, 1987), 451–503, esp. 480–495.

32. James Rowland Angell and Addison W. Moore, "Reaction-Time: A Study in Attention and Habit," *Psychological Review* 3, no. 3 (May 1896): 245–258, here 246, 250, 252–254. The most influential "functionalist" manifesto was Angell, "The Province of Functional Psychology," *Psychological Review* 14, no. 2 (March 1907): 61–91. See also Angell, "The Relations of Structural and Functional Psychology to Philosophy," *Philosophical Review* 12, no. 3 (May 1903): 243–271. For an overview, see Christopher D. Green, "Darwinian Theory, Functionalism, and the First American Psychological Revolution," *American Psychologist* 64, no. 2 (February 2009): 75–83.

33. On "schools," see David L. Krantz, "The Baldwin-Titchener Controversy: A Case Study in the Functioning and Malfunctioning of Schools," in *Schools of Psychology: A Symposium of Papers,* ed. David L. Krantz (New York: Appleton-Century-Crofts, 1969), 1–19. For a reappraisal, see Christopher D. Green, Ingo Feinerer, and Jeremy T. Burman, "Beyond the Schools of Psychology 1: A Digital Analysis of *Psychological Review,* 1894–1903," *Journal of the History of the Behavioral Sciences* 49, no. 2 (Spring 2013): 167–189. See also Green, Feinerer, and Burman, "Beyond the Schools of Psychology 2: A Digital Analysis of

Psychological Review, 1904–1923," *Journal of the History of the Behavioral Sciences* 50, no. 3 (Summer 2014): 249–279.

34. Angell and Moore, "Reaction-Time," 252. On Chicago's central role in functionalism, see Lawrence Richard Carleton, "The Rise of Chicago Functionalism," *Erkenntnis* 18, no. 1 (July 1982): 3–23. On Dewey's role, see Andrew Backe, "John Dewey and Early Chicago Functionalism," *History of Psychology* 4, no. 4 (November 2001): 323–340.
35. John Dewey, "The Reflex Arc Concept in Psychology," *Psychological Review* 3, no. 4 (July 1896): 357–370, here 360, 370. On the role of this paper in Dewey's career, see Allen K. Smith, "Dewey's Transition Piece: The 'Reflex Arc' Paper," *Tulane Studies in Philosophy* 22 (1973): 122–141. See also Eric Bredo, "Evolution, Psychology, and John Dewey's Critique of the Reflex Arc Concept," *Elementary School Journal* 98, no. 5 (May 1998): 447–466. On its place in functionalism, see Alfred C. Raphelson, "The Pre-Chicago Association of the Early Functionalists," *Journal of the History of the Behavioral Sciences* 9, no. 2 (Spring 1973): 115–122.
36. For Angell's compliment, see Angell to William James, November 13, 1898, #18 in William James Papers (MS Am 1092.9), Houghton Library, Harvard University (hereafter cited as James Papers). For "disappointment," see William James to Thomas Davidson, January 12, 1887, #860 in James Papers. See also James to G. Stanley Hall, January 30, 1887, in *The Correspondence of William James,* vol. 6, ed. Ignas K. Skrupskelis and Elizabeth M. Berkeley (Charlottesville: University Press of Virginia, 1992–), 197–198. For Dewey's conversion efforts, see letters #128, #130, #133, and for James's skepticism, see #887, in James Papers.
37. William James, "The Chicago School," *Psychological Bulletin* 1, no. 1 (January 1904): 1–5, here 1. On James's radical empiricism in this context, see Wayne Viney, "The Radical Empiricism of William James and Philosophy of History," *History of Psychology* 4, no. 3 (August 2001): 211–227. On his long-term engagement with spiritualism, see Paul Jerome Croce, "Between Spiritualism and Science: William James on Religion and Human Nature," *Journal for the History of Modern Theology* 4, no. 2 (January 2010): 197–220. On his boundary-work, see Francesca Bordogna, "Inner Division and Uncertain Contours: William James and the Politics of the Modern Self," *British Journal for the History of Science* 40, no. 4 (December 2007): 505–536.
38. James, "Chicago School," 4. James and Dewey both reflect the impact of Chauncey Wright. See Edward H. Madden, "Chauncey Wright's Functionalism," *Journal of the History of the Behavioral Sciences* 10, no. 3 (July 1974): 281–290.
39. Dewey, "Reflex Arc," 357. On James's doubting of introspection and, later, consciousness, see William James, "On Some Omissions of Introspective

Psychology," *Mind* 9, no. 33 (January 1884): 1–26; and James, "Does 'Consciousness' Exist?," *Journal of Philosophy, Psychology and Scientific Methods* 1, no. 18 (September 1904): 477–491.

40. For Dewey's impressions of Ford, see John Dewey to William James, June 3, 1891, published in Ralph Barton Perry, *The Thought and Character of William James, as Revealed in Unpublished Correspondence and Notes, Together with His Published Writings,* vol. 2 (Boston: Little, Brown, and Co., 1935), 517–518. On Ford's life, see Corydon L. Ford, *The Child of Democracy: Being the Adventures of the Embryo State* (Ann Arbor, MI: J. V. Sheehan and Co., 1894). See also David H. Burton, ed., *Progressive Masks: Letters of Oliver Wendell Holmes, Jr., and Franklin Ford* (Newark: University of Delaware Press, 1982).
41. For "inquiry as a *business,*" see Dewey to James, June 3, 1891, in Perry, *Thought and Character,* 517–518. For "scoundrel," see Corliss Lamont, ed., *Dialogue on John Dewey* (New York: Horizon Press, 1959), 30. Neil Coughlan calls Ford a "quack." See Coughlan, *Young John Dewey,* 92–108, here 96. For "overenthusiastic," see Willinda H. Savage, "John Dewey and 'Thought News' at the University of Michigan," in *Studies in the History of Higher Education in Michigan,* ed. Claude Eggertsen (Ann Arbor: University of Michigan Press, 1950), 12–17, here 16. See also Savage, "The Evolution of John Dewey's Philosophy of Experimentalism as Developed at the University of Michigan" (EdD diss., University of Michigan, 1950). On the place of *Thought News* in Dewey's career, see Westbrook, *John Dewey,* 51–58; and Ryan, *John Dewey,* 107–110.
42. While most biographers—including Westbrook, Coughlan, and to a certain extent Ryan—see this episode as embarrassing for Dewey, it is also a sign of his gradual political awakening. See Lewis S. Feuer, "John Dewey and the Back to the People Movement in American Thought," *Journal of the History of Ideas* 20, no. 4 (October 1959): 545–568; and Steven C. Rockefeller, *John Dewey: Religious Faith and Democratic Humanism* (New York: Columbia University Press, 1991). See also Andrew Feffer, *The Chicago Pragmatists and American Progressivism* (Ithaca, NY: Cornell University Press, 1993), 82–86.
43. On the rise of Chicago, see William Cronon, *Nature's Metropolis: Chicago and the Great West* (New York: W. W. Norton, 1992). On the university's founding, see Thomas Wakefield Goodspeed, *A History of the University of Chicago: The First Quarter-Century* (Chicago: University of Chicago Press, 1916). On Harper, see Richard J. Storr, *Harper's University, the Beginnings: A History of the University of Chicago* (Chicago: University of Chicago Press, 1966). On graduate education, see Robert E. Kohler, "The Ph.D. Machine: Building on the Collegiate Base," *Isis* 81, no. 4 (December 1990): 638–662. On Chicago, Clark, and Johns Hopkins, see W. Carson Ryan, *Studies in Early Graduate Education: The Johns Hopkins,*

Clark University, the University of Chicago (New York: Carnegie Foundation, 1939). On "Harper's Raid," see Hall, *Life and Confessions*, 295–297.

44. John Dewey, "The Chaos in Moral Training," *Popular Science Monthly*, August 1894, 433–443, here 443. For his recollection, see John Dewey, "Biography of John Dewey," in *The Philosophy of John Dewey*, ed. Paul Arthur Schilpp and Jane M. Dewey (Evanston, IL: Northwestern University, 1939), 3–45, here 27. On the move, see George Dykhuizen, "John Dewey and the University of Michigan," *Journal of the History of Ideas* 23, no. 4 (October 1962): 513–544; Dykhuizen, "John Dewey: The Chicago Years," *Journal of the History of Philosophy* 2, no. 2 (October 1964): 227–253; and Dykhuizen, "John Dewey in Chicago: Some Biographical Notes," *Journal of the History of Philosophy* 3, no. 2 (October 1965): 217–233.
45. Citations to the correspondence of the Deweys in this section are drawn from digital reproductions from the Intelex "Past Masters" series. See http://www.nlx.com/collections/132. John Dewey to Alice Chipman Dewey, November 1, 1894, #218. On Parker, see Jack Kenagy Campbell, *Colonel Francis W. Parker, the Children's Crusader* (New York: Teachers College Press, 1967); and Michael B. Katz, "The 'New Departure' in Quincy, 1873–1881: The Nature of Nineteenth-Century Educational Reform," *New England Quarterly* 40, no. 1 (March 1967): 3–30. On his school and its place in the movement, see Marie Kirchner Stone, *The Progressive Legacy: Chicago's Francis W. Parker School, 1901–2001* (New York: P. Lang, 2001). The canonical account of "progressive education" is Lawrence A. Cremin, *The Transformation of the School: Progressivism in American Education, 1876–1957* (New York: Vintage Books, 1964). On American education in general, see Cremin, *American Education: The Metropolitan Experience, 1876–1980* (New York: Harper and Row, 1988). For a more recent account, see David B. Tyack, *Tinkering toward Utopia: A Century of Public School Reform* (Cambridge, MA: Harvard University Press, 1995).
46. For "complete experimental school," see John Dewey to Alice Chipman Dewey, November 22, 1894. For "starting point," see John Dewey to Alice Chipman Dewey, November 1, 1894, 218. On psychologists' move into pedagogy and the challenges they encountered, see Ellen Condliffe Lagemann, *An Elusive Science: The Troubling History of Education Research* (Chicago: University of Chicago Press, 2000).
47. On psychology's impact in this regard (in the twentieth century), see Rebecca M. Lemov, *World as Laboratory: Experiments with Mice, Mazes, and Men* (New York: Hill and Wang, 2005). On the broader artistic and literary meanings attached to experimentalism in these years, see Mark S. Micale, ed., *The Mind of Modernism: Medicine, Psychology, and the Cultural Arts in Europe and America, 1880–1940* (Stanford: Stanford University Press, 2004).

48. For Dewey's original use of the phrase, see John Dewey to Clara I. Mitchell, December 24, 1895, #275. On the school's early years, see Katherine Camp Mayhew and Anna Camp Edwards, *The Dewey School: The Laboratory School of the University of Chicago, 1896–1903* (New York: D. Appleton-Century Co., 1936); Dewey's crediting of Young appears on 7. On their relationship, see Ellen Condliffe Lagemann, "Experimenting with Education: John Dewey and Ella Flagg Young at the University of Chicago," *American Journal of Education* 104, no. 3 (May 1996): 171–185. On Young's life, see L. Dean Webb and Martha McCarthy, "Ella Flagg Young: Pioneer of Democratic School Administration," *Educational Administration Quarterly* 34, no. 2 (April 1998): 223–242.
49. For "cooking," see John Dewey, *The School and Society* (Chicago: University of Chicago Press, 1899), 57. For "acts," see Dewey, "The Results of Child-Study Applied to Education," in *Transactions of the Illinois Society for Child-Study* 1 (Chicago: Werner Co., 1894): 18–19.
50. For "two sides," see Dewey, "A Pedagogical Experiment," *Kindergarten Magazine* 8 (1896): 739–741. For "reducible" and "law," see John Dewey, *My Pedagogic Creed* (New York: E. L. Kellogg and Co., 1897), 13.
51. For "image," see John Dewey, "Imagination and Expression," *Kindergarten Magazine* 8 (1896): 61–69. For "practical exhibition," see Dewey, "The Need for a Laboratory School," in *The Early Works of John Dewey*, vol. 5 (Carbondale: Southern Illinois University Press, 1972–1985), 433–435. For "trial-and-error," see Mayhew and Edwards, *The Dewey School*, 366–367.
52. For the promotional material, see John Dewey, "The University School," *The University Record* 1 (November 1896): 417–419. For Dewey's budget, see John Dewey to Thomas W. Goodspeed, November 23, 1896, #555, http://www.nlx.com/collections/132. On the laboratory's power, see Bruno Latour, "Give Me a Laboratory and I Will Raise the World," in *Science Observed: Perspectives on the Social Study of Science*, ed. Michael Mulkay and Karin D. Knorr-Cetina (London: Sage, 1983), 141–170. See also John Harley Warner, "The Fall and Rise of Professional Mystery: Epistemology, Authority, and the Emergence of Laboratory Medicine in Nineteenth-Century America," in *The Laboratory Revolution in Medicine*, ed. Andrew Cunningham and Perry Williams (Cambridge: Cambridge University Press, 1992), 110–141.
53. Jane Addams, *Twenty Years at Hull-House* (New York: Macmillan, 1910), 308, 125, 126. For "desire," see Addams, "A New Impulse to an Old Gospel," *The Forum*, November 1892, 345–358, here 345. On Addams's life, see Louise W. Knight, *Citizen: Jane Addams and the Struggle for Democracy* (Chicago: University of Chicago Press, 2005). On her affinities with Dewey, see Charlene

Haddock Seigfried, *Pragmatism and Feminism: Reweaving the Social Fabric* (Chicago: University of Chicago Press, 1996). On Addams's impact, see Mary Jo Deegan, *Jane Addams and the Men of the Chicago School, 1892–1918* (New Brunswick, NJ: Transaction Books, 1988).

54. George H. Mead, "The Working Hypothesis in Social Reform," *American Journal of Sociology* 5, no. 3 (November 1899): 367–371, here 367, 369–370. On Mead's life, see Hans Joas, *G. H. Mead: A Contemporary Re-Examination of His Thought* (Cambridge, MA: MIT Press, 1985). On his subsequent "reconstruction" by students and followers, see Daniel R. Huebner, *Becoming Mead: The Social Process of Academic Knowledge* (Chicago: University of Chicago Press, 2014). See also the essays in Hans Joas and Daniel R. Huebner, eds., *The Timeliness of George Herbert Mead* (Chicago: University of Chicago Press, 2016).
55. For "practical," see George H. Mead, "The Social Settlement: Its Basis and Function," *University of Chicago Record* 12 (1908): 108–110, here 110. For "reflective consciousness," see Mead, "Working Hypothesis," 371. This loop was a major object of Mead's work, as embodied in Mead, *Mind, Self, and Society: The Definitive Edition,* ed. Hans Joas, Daniel R. Huebner, and Charles W. Morris (Chicago: University of Chicago Press, 2015 [1934]).
56. The relationship between "subjective" and "objective" in Addams's work has been obscured by a decision to publish the two essays together, as though they represented different approaches. See Jane Addams, "The Subjective Necessity for Social Settlements" and "The Objective Value of a Social Settlement," in *Philanthropy and Social Progress: Seven Essays,* ed. Henry C. Adams (New York: Thomas Y. Crowell and Co., 1893), 1–26, 27–56.
57. John Dewey, "Psychology and Philosophic Method," *University of California Chronicle* 2 (1899): 159–179, here 173, 177.
58. For "technical," see John Dewey, "Science as Subject-Matter and as Method," *Science* 31, no. 787 (January 1910): 121–127, here 127. For "plain," see Dewey, "Thought and Its Subject-Matter: The General Problem of Logical Theory," in *Studies in Logical Theory,* ed. John Dewey (Chicago: University of Chicago Press, 1903), 1–22, here 10.
59. John Dewey, *How We Think* (Boston: D. C. Heath and Co, 1910), iii. On Dewey's departure, see Westbrook, *John Dewey and American Democracy,* 111–118.
60. Dewey, *How We Think,* 72.
61. For "free play," see Dewey, *School and Society,* 65. For "other selves," see George H. Mead, "What Social Objects Must Psychology Presuppose?," *Journal of Philosophy, Psychology and Scientific Methods* 7, no. 7 (March 1910): 174–180, here 179.
62. For "social conduct," see George H. Mead, "Social Consciousness and the Consciousness of Meaning," *Psychological Bulletin* 7, no. 12 (December 1910):

397–405, here 403. In the same year, Mead extrapolated lessons for teaching. See Mead, "The Psychology of Social Consciousness Implied in Instruction," *Science* 31, no. 801 (May 1910): 688–693. These publications followed on an article that is now treated as a founding document in social psychology: Mead, "Social Psychology as Counterpart to Physiological Psychology," *Psychological Bulletin* 6, no. 12 (December 1909): 401–408. Mead famously synthesized his early work in his account of the "social self." See Mead, "The Social Self," *Journal of Philosophy, Psychology and Scientific Methods* 10, no. 14 (July 1913): 374–380.

8. A Method Only

1. John L. Rudolph, "Epistemology for the Masses: The Origins of 'The Scientific Method' in American Schools," *History of Education Quarterly* 45, no. 3 (Fall 2005): 341–376, here 369. See also Rudolph, *How We Teach Science: What's Changed, and Why It Matters* (Cambridge, MA: Harvard University Press, 2019). I see my analysis as a prehistory of sorts, an explanation of the changes in method's meaning that enabled the shift documented by Ruldolph.
2. Charles Sanders Peirce, "The Fixation of Belief," *Popular Science Monthly*, November 1877, 1–15, here 2. John Dewey, "Darwin's Influence upon Philosophy," *Popular Science Monthly*, July 1909, 90–98, here 93. Dewey reprinted this article the following year under an altered title ("The Influence of Darwinism on Philosophy"), in a book with a third version of the title. See John Dewey, *The Influence of Darwin on Philosophy and Other Essays in Contemporary Thought* (New York: Henry Holt and Co., 1910), 1–19. Significantly, the lecture was actually delivered under a *fourth* title: "Charles Darwin and His Influence on Science." This original title shows how Dewey saw philosophy and science converging in these years—and the role evolution played in that convergence.
3. C. Lloyd Morgan, "Mental Factors in Evolution," in *Darwin and Modern Science*, ed. A. C. Seward (Cambridge: Cambridge University Press, 1909), 424–445, here 443, 445; for "out-Darwined," see, in the same volume, C. Bougle, "Darwin and Sociology," 466–476, here 474. For "future historians," see Raphael Meldola, "Evolution: Old and New," *Nature* 80, no. 2069 (June 24, 1909): 481–485, here 481. On the 1909 proceedings, see Marsha L. Richmond, "The 1909 Darwin Celebration: Reexamining Evolution in the Light of Mendel, Mutation, and Meiosis," *Isis* 97, no. 3 (September 2006): 447–484. On these commemorations in general, see Janet Browne, "Presidential Address: Commemorating Darwin," *British Journal for the History of Science* 38, no. 3 (September 2005): 251–274. On the "eclipse," see Peter J. Bowler, *The Eclipse of Darwinism: Anti-Darwinian Evolution Theories in the Decades around 1900* (Baltimore: Johns Hopkins

University Press, 1983). For a reassessment, see Peter J. Bowler, "Revisiting the Eclipse of Darwinism," *Journal of the History of Biology* 38, no. 1 (April 2005): 19–32. See also Mark A. Largent, "The So-Called Eclipse of Darwinism," in *Descended from Darwin: Insights into the History of Evolutionary Studies, 1900–1970*, ed. Joe Cain and Michael Ruse (Philadelphia: American Philosophical Society, 2009), 3–21.

4. James Mark Baldwin, "The Influence of Darwin on Theory of Knowledge and Philosophy," *Psychological Review* 16, no. 3 (May 1909): 207–218, here 207. For Hall, see G. Stanley Hall, "Evolution and Psychology," in *Fifty Years of Darwinism: Modern Aspects of Evolution* (New York: Henry Holt and Co., 1909), 251–267, here 266–267, 264.

5. On "the revolt against formalism," see Morton Gabriel White, *Social Thought in America: The Revolt against Formalism* (New York: Viking Press, 1949). For a reassessment, see James T. Kloppenberg, "Morton White's Social Thought in America," *Reviews in American History* 15, no. 3 (September 1987): 507–519. On the same issues today, see John L. Rudolph, "Inquiry, Instrumentalism, and the Public Understanding of Science," *Science Education* 89, no. 5 (September 2005): 803–821. For a broader consideration of instrumentalism, see Cathryn Carson, "Science as Instrumental Reason: Heidegger, Habermas, Heisenberg," *Continental Philosophy Review* 42, no. 4 (March 2010): 483–509.

6. For "existence," see William James, "Philosophical Conceptions and Practical Results," *University of California Chronicle* 4 (September 1898): 287–310, here 290, 291. For "hypothesis," see William James, "The Will to Believe," *New World* 5, no. 18 (June 1896): 327–347, here 328–329. James published the latter essay, together with a few other pieces, as *The Will to Believe, and Other Essays in Popular Philosophy* (New York: Longmans, Green, and Co., 1897), 1–31.

7. William James, *Pragmatism: A New Name for Some Old Ways of Thinking* (New York: Longmans, Green, 1907), v, 6, 11–12. Peirce introduced his new term in his "What Pragmatism Is," *Monist* 15, no. 2 (April 1905): 161–181, here 165–166. On his Kantianism, see James Feibleman, "Peirce's Use of Kant," *Journal of Philosophy* 42, no. 14 (July 1945): 365–377. See also John Kaag, "Continuity and Inheritance: Kant's 'Critique of Judgment' and the Work of C. S. Peirce," *Transactions of the Charles S. Peirce Society* 41, no. 3 (July 2005): 515–540.

8. James, *Pragmatism*, 20, 33. In his account of the period's *via media*, James T. Kloppenberg does not highlight this connection between James and Mill, in part because he identifies Mill so strongly with associationism. See Kloppenberg,

Uncertain Victory: Social Democracy and Progressivism in European and American Thought, 1870–1920 (Oxford: Oxford University Press, 1986), 51–55.

9. James, *Pragmatism,* 58, 79. On meliorism, see Scott R. Stroud, "William James on Meliorism, Moral Ideals, and Business Ethics," *Transactions of the Charles S. Peirce Society* 45, no. 3 (July 2009): 378–401. On pragmatism as *transition,* see Colin Koopman, *Pragmatism as Transition: Historicity and Hope in James, Dewey, and Rorty* (New York: Columbia University Press, 2009), esp. 11–49.
10. Arthur O. Lovejoy, "The Thirteen Pragmatisms. I," *Journal of Philosophy, Psychology and Scientific Methods* 5, no. 1 (January 1908): 5–12, here 12. See also Lovejoy, "The Thirteen Pragmatisms. II," *Journal of Philosophy, Psychology and Scientific Methods* 5, no. 2 (January 1908): 29–39. Lovejoy did not put an end to debates over "who owns pragmatism," which are ongoing today. See Bruce Kuklick, "Who Owns Pragmatism?," *Modern Intellectual History* 14, no. 2 (August 2017): 565–583. For "dominant attitude," see John Dewey, "What Does Pragmatism Mean by Practical?," *Journal of Philosophy, Psychology and Scientific Methods* 5, no. 4 (February 1908): 85–99, here 86.
11. Dewey, "What Does Pragmatism Mean by Practical?" 99. On these financial metaphors, see James Livingston, *Pragmatism and the Political Economy of Cultural Revolution, 1850–1940* (Chapel Hill: University of North Carolina Press, 1994). See also Livingston, "The Politics of Pragmatism," *Social Text,* no. 49 (December 1996): 149–172. On critical reviews, see Livingston, "War and the Intellectuals: Bourne, Dewey, and the Fate of Pragmatism," *Journal of the Gilded Age and Progressive Era* 2, no. 4 (October 2003): 431–450. The most famous were Randolph Bourne, "Twilight of Idols," *Seven Arts* 11 (October 1917): 688–702; and [Bertrand Russell], "Pragmatism (A Review)," *Edinburgh Review* 209, no. 428 (April 1909): 363–388.
12. This way of reading pragmatism is suggested by David A. Hollinger, "The Problem of Pragmatism in American History," *Journal of American History* 67, no. 1 (June 1980): 88–107.
13. John B. Watson, "Psychology as the Behaviorist Views It," *Psychological Review* 20, no. 2 (March 1913): 158–177, here 158. William James, "A Plea for Psychology as a 'Natural Science,'" *Philosophical Review* 1, no. 2 (March 1892): 146–153, here 147, 148. On James's divisions in an earlier period, see Alexander Klein, "Divide et Impera! William James's Pragmatist Tradition in the Philosophy of Science," *Philosophical Topics* 36, no. 1 (June 2010): 129–166.
14. James, "A Plea for Psychology," 150–151.
15. Watson, "Psychology as the Behaviorist Views It," 170. For more on Watson's reductionism, see Michael Pettit, "The Problem of Raccoon Intelligence in

Behaviourist America," *British Journal for the History of Science* 43, no. 3 (September 2010): 391–421.

16. Watson, "Psychology as the Behaviorist Views It," 159. On Watson's impact, see Franz Samelson, "Struggle for Scientific Authority: The Reception of Watson's Behaviorism, 1913–1920," *Journal of the History of the Behavioral Sciences* 17, no. 3 (July 1981): 399–425. For Skinner's balance, see B. F. Skinner, *About Behaviorism* (New York: A. A. Knopf, 1974), 16–17. On the distinction, see Willard Day, "On the Difference between Radical and Methodological Behaviorism," *Behaviorism* 11, no. 1 (Spring 1983): 89–102. See also Alexandra Rutherford, *Beyond the Box: B. F. Skinner's Technology of Behavior from Laboratory to Life, 1950s–1970s* (Toronto: University of Toronto Press, 2009).
17. On the disturbing aspects of Watson's approach, see Cathy Gere, *Pain, Pleasure, and the Greater Good* (Chicago: University of Chicago Press, 2017), 169–170.
18. For sales curves, see Carl Murchison, ed., *A History of Psychology in Autobiography,* vol. 3 (Worcester, MA: Clark University Press, 1930), 271–281, here 280. On Watson's career change, see Rebecca M. Lemov, *World as Laboratory: Experiments with Mice, Mazes, and Men* (New York: Hill and Wang, 2005), 24–45. On "methodological imperative," see Kurt Danziger, "The Methodological Imperative in Psychology," *Philosophy of the Social Sciences* 15, no. 1 (March 1985): 1–13.
19. On "scientific democrats," see Andrew Jewett, *Science, Democracy, and the American University: From the Civil War to the Cold War* (Cambridge: Cambridge University Press, 2012). On the eugenic context, see Daniel J. Kevles, *In the Name of Eugenics: Genetics and the Uses of Human Heredity* (Cambridge, MA: Harvard University Press, 1995).
20. Richard Hofstadter, *Social Darwinism in American Thought, 1860–1915* (Philadelphia: University of Pennsylvania Press, 1944). On the book's myth-building role, see Thomas C. Leonard, "Origins of the Myth of Social Darwinism: The Ambiguous Legacy of Richard Hofstadter's Social Darwinism in American Thought," *Journal of Economic Behavior and Organization* 71, no. 1 (July 2009): 37–51. See also Leonard, *Illiberal Reformers: Race, Eugenics, and American Economics in the Progressive Era* (Princeton: Princeton University Press, 2016). On Du Bois, see Adolph L. Reed, *W. E. B. Du Bois and American Political Thought: Fabianism and the Color Line* (Oxford: Oxford University Press, 1997).
21. Abraham Flexner, *Medical Education in the United States and Canada: A Report to the Carnegie Foundation for the Advancement of Teaching* (New York: Carnegie Foundation for the Advancement of Teaching, 1910), 68n2. The quoted passage is from John Dewey, "Science as Subject-Matter and as Method," *Science* 31, no. 787 (January 1910): 121–127, here 122. For his later view, see Abraham

Flexner, *Medical Education: A Comparative Study* (New York: Macmillan, 1925), 6–7. On the Report's history, see Kenneth M. Ludmerer, "Commentary: Understanding the Flexner Report," *Academic Medicine: Journal of the Association of American Medical Colleges* 85, no. 2 (February 2010): 193–196. On Flexner's background in (progressive) education, see Thomas Neville Bonner, *Iconoclast: Abraham Flexner and a Life in Learning* (Baltimore: Johns Hopkins University Press, 2002). On Dewey's impact, see Ludmerer, *Time to Heal: American Medical Education from the Turn of the Century* (Oxford: Oxford University Press, 2005), esp. 64–68. On the rise of scientific medicine, see John Harley Warner, "Ideals of Science and Their Discontents in Late Nineteenth-Century American Medicine," *Isis* 82, no. 3 (September 1991): 454–478. On the Report's role, see Howard S. Berliner, *A System of Scientific Medicine: Philanthropic Foundations in the Flexner Era* (New York: Tavistock, 1985), esp. 101–127. On curricular standardization, see Andrew H. Beck, "The Flexner Report and the Standardization of American Medical Education," *Journal of the American Medical Association* 291, no. 17 (May 2004): 2139–2140.

22. Frederick Winslow Taylor, *The Principles of Scientific Management* (New York: Harper and Bros., 1911), 25. The classic account of scientific management is Samuel Haber, *Efficiency and Uplift: Scientific Management in the Progressive Era, 1890–1920* (Chicago: University of Chicago Press, 1964). See also Maarten Derksen, "Turning Men into Machines? Scientific Management, Industrial Psychology, and the 'Human Factor,'" *Journal of the History of the Behavioral Sciences* 50, no. 2 (March 2014): 148–165. On Taylor's methodological ambitions, see Robert Kanigel, *The One Best Way: Frederick Winslow Taylor and the Enigma of Efficiency* (New York: Viking, 1997).

23. Walter Lippmann, *A Preface to Politics* (New York: Mitchell Kennerley, 1913), 23, 69, 286; Lippmann, *Drift and Mastery: An Attempt to Diagnose the Current Unrest* (New York: Henry Holt and Co., 1914), 276, 145. On the development of Lippmann's views, see David A. Hollinger, "Science and Anarchy: Walter Lippmann's *Drift and Mastery*," *American Quarterly* 29, no. 5 (Winter 1977): 463–475. See also John Patrick Diggins, "From Pragmatism to Natural Law: Walter Lippmann's Quest for the Foundations of Legitimacy," *Political Theory* 19, no. 4 (November 1991): 519–538.

24. Walter Lippmann, *Public Opinion* (Harcourt, Brace, 1922), 370. For Dewey's views, see John Dewey, *The Public and Its Problems* (New York: Henry Holt, 1927); and Dewey, "The Pragmatic Acquiescence," *New Republic* (January 5, 1927): 186–189. Earlier their views had been more alike. See Jewett, *Science, Democracy, and the American University,* 142–145. On Lippmann's disillusionment, see Edward A. Purcell, *The Crisis of Democratic Theory: Scientific Naturalism and*

the Problem of Value (Lexington: University Press of Kentucky, 1973), 152–154.

25. One example of this approach to science was the rise of logical empiricism. On its convergence with pragmatism, see Alan W. Richardson, "Engineering Philosophy of Science: American Pragmatism and Logical Empiricism in the 1930s," *Philosophy of Science* 69, no. S3 (September 2002): S36–S47. Although I agree with Richardson, Dewey's later work is distinct enough that one can describe the relationship as being simultaneously a continuation and a break. Cybernetics entailed a similar view of science, insofar as it blurred the boundaries between rationality and mechanism. See Peter Galison, "The Ontology of the Enemy: Norbert Wiener and the Cybernetic Vision," *Critical Inquiry* 21, no. 1 (October 1994): 228–266. Donna Haraway recognized the same blurriness a decade earlier, albeit with a different ethical twist. See Donna Haraway, "A Manifesto for Cyborgs: Science, Technology, and Socialist Feminism in the 1980s," *Socialist Review* 80 (1985): 65–108.

ACKNOWLEDGMENTS

Method gets you only so far. Learning its limits, and learning how to write about them, took the advice of mentors and colleagues, the support of friends and family, and assistance from the odd stranger. Anyone who undertakes a project of this scope knows it is both a dream and a nightmare to sit down and record one's debts, trying to capture gratitude in words. What follows is a small and imperfect attempt to trace the web of people and institutions that made it possible for me to try and fail to do so. Of course, any flaws in the final product are mine alone, not theirs.

Though I find it hard to believe, the paper that started this project was submitted a decade ago. The recipient was Dan Rodgers, and from that moment forward he has shaped my thinking more than anyone else. Janet Browne and Sheila Jasanoff introduced me to the history of science, while Angela Creager and Michael Gordin showed me how to read, write, and act in that field. Later, Ted Porter helped me see what was missing. Princeton, and especially Program Seminar, taught me that asking questions was often the best way to add something to a conversation. I still believe it.

This project was remade at Yale, where colleagues and students—especially in the Program in the History of Science and Medicine—helped me write with an eye to what mattered. My biggest debts are to John Warner for showing me the ropes and to Joanna Radin for making it possible. My new home in Ann Arbor has been the perfect place to bring this project to completion. Colleagues and students in the History Department, the Program in Anthropology and History, and the Science, Technology, and

Society Program at the University of Michigan have supported me in the final push. Thanks especially to my chairs Kathleen Canning, Geoff Eley, and Jay Cook for helping make this dream a reality, and to all the new friends who have made it a reality worth waking up to—even in the dead of winter!

I was invited to present versions of my argument at the University of California–Berkeley, the University of Cambridge, the University of Chicago, Columbia University, the University of Georgia, Harvard University, the University of Leeds, the Max Planck Institute for the History of Science, the University of Michigan, the University of Pennsylvania, Princeton University, Stanford University, the University of Wisconsin–Madison, and Yale University. Audiences there and at many conferences pushed and pulled this book into its present form. Chapters 2 and 5 build on ideas first discussed in "The Age of Methods: William Whewell, Charles Peirce, and Scientific Kinds," *Isis* 107, no. 4 (December 2016): 722–737; Chapter 3 develops concepts treated in "On the Origins of Theories: Charles Darwin's Vocabulary of Method," *American Historical Review* 122, no. 4 (October 2017): 1079–1104; Chapters 4 and 6 expand on "Hypothesis Bound: Trial and Error in the Nineteenth Century," *Isis* 106, no. 3 (September 2015): 635–645; and Chapter 7 touches on ideas first presented in *Aeon* on January 9, 2017. I thank the University of Chicago Press, Oxford University Press, and *Aeon,* as well as their editors, referees, and audiences, for shaping my thinking.

Networking is a bad word for a good thing: meeting others, in person and online, who share a passion for talking shop. For their feedback over the years, I thank Babak Ashrafi, Michael Barany, Joshua Bauchner, Jeremy Blatter, Francesca Bordogna, Dan Bouk, Daniel Braun, Ben Breen, Alex Csiszar, Jamie Cohen-Cole, Paul Croce, Anthony Cross, Helen Curry, Will Deringer, Catherine Evans, Christian Flow, Gregory Ferguson-Cradler, John Guillory, Evan Hepler-Smith, David Hollinger, Sarah Igo, Andrew Jewett, Adrian Johns, Matt Jones, James Kloppenberg, Susan Lindee, Elizabeth Lunbeck, John Levi Martin, Jill Morawski, Mary Morgan, Wangui Muigai, Carla Nappi, Lynn Nyhart, Carola Ossmer, Trevor Pearce, Greg Radick, Marco Ramos, Jennifer Ratner-Rosenhagen, Robert Richards, Felix Rietmann, Michael Rossi, David Russell, Claudio Saunt, Padraic Scanlan, David Sepkoski, Chris Shannon, Alistair Sponsel, Matt Stanley, Laura Stark, Katie Tabb, John Tresch, Caitlin Tully, Lee Vinsel, Nasser Zakariya, and Dora Zhang. Lukas Rieppel and Stephanie Dick changed my mind the most, often the latest at night. Sam Lewallen was there from the start.

My work has depended on enormous institutional support. For fellowships and grants, I am grateful to the Department of History, Program in History of Science, Program in American Studies, Center for Human Values, and Graduate School at Princeton University; the Program on Science, Technology, and Society at Harvard University; the Department of History, Program in the History of Science and Medicine, Section of the History of Medicine, and Whitney Humanities Center at Yale University; and the History Department, Weinberg Institute for Cognitive Science, and Eisen-berg Institute for Historical Studies at the University of Michigan. Two months in Department II of the Max Planck Institute for the History of Science launched this project into book form. I will always be grateful to Lorraine Daston for the opportunity, and for her comments and encouragement over the years. What follows is based, in part, upon research supported by a National Science Foundation Graduate Research Fellowship under Grant No. DGE-0646086.

I wrote a first draft of the book while Distinguished Junior External Fellow at the Stanford Humanities Center, where I was nurtured by an idyllic setting, copious cuisine, and inspiring fellows. For their generosity and wit during my year in the Bay Area, I want to thank Angèle Christin, Phil Hammack, Brielle Johnck, Philippa Levine, Benjamin Morgan, Andrei Pesic, Steve Schmidt, Kyla Schuller, and, most of all, Caroline Winterer and Colin Webster. Final revisions were supported by a Summer Faculty Fellowship at the Institute for the Humanities here at Michigan, where I benefited from the perceptive comments of Phil Christman, Enrique García, Annette Joseph-Gabriel, Shelley Manis, Christine Modey, David Morse, and especially Peggy McCracken and Antoine Traisnel.

Books are bound for libraries—and cannot be written without them. This one was formed (often in subtle ways) by librarians and archivists at the British Library, the University of Cambridge, the University of California–Berkeley, the University of California–Los Angeles, the University of Chicago, Columbia University, the Library of Congress, the University of Michigan, Stanford University, and Yale University. In 2007, I interned for the Darwin Correspondence Project at the Cambridge University Library, where I learned how to think *with,* and not just *about,* historical figures. I owe that skill to Shelley Innes, Jim Secord, Paul White, and especially Alison Pearn.

Since she acquired this project for Harvard University Press, Janice Audet has read more drafts (and responded to more emails) than she ever should have had to. The finished product is not her responsibility, but it bears her fingerprints throughout. The same is true of my wonderful research assistants, for whom no task was too tall: Sam Franz, Lynsey Randolph, Sarah Thomas, and above all Julia Menzel. Audra Wolfe got me off on the right foot; Adam Johnson was there for me at the end. The unearned generosity of Karen Darling, Laura Davulis, and Matt McAdam made a huge difference. In addition to all the friends and colleagues who read chapters, a few went above and beyond, reading the whole manuscript and insisting on the potential I had not yet realized in it. In addition to two anonymous readers for Harvard University Press, I depended upon John Carson, Michael Cowles, Anna Freidin, Joel Isaac, Jamie Kreiner, and Paul Johnson. You know where to find me when it's time to return the favor!

Family is what it is all for. Mine has put up with the scientific method as long as I have, with better humor. They have welcomed me during conferences and archival trips (in one case, for two summers!), reminding me what really matters. I wish Marge could see what kept me off the beach. Vicki and Grisha started coaching me at our very first dinner—they are the in-laws of my dreams. Jane and Mark have taught me more about the human side of science than any book ever could. Parenting is an experiment, and I am grateful every day that Michael and Katie took a chance on me. And then there is Anna, who read and talked about so much of this thing that it is as much hers as mine. She knows better than anyone how much more there is to life than method. That she will not let me forget it is a wonderful fact that I could never have presumed to expect.

INDEX